Fire Hazard and Fire Resistance of Wooden Structures

Sivenkov Andrey Borisovich •
Berlin Alexander Alexandrovich •
Mukhamedgaliev Bakhtiyor Abdukadirovich •
Almenbayev Mirzhan Maratovich •
Makishev Zhandos Kuandykovich •
Berik Zhambulovich Rakhmetulin

Fire Hazard and Fire Resistance of Wooden Structures

Springer

Sivenkov Andrey Borisovich
Academy of the State Fire Service of the Ministry of the Russian Federation for Civil Defense, Emergencies and Elimination of Consequences of Natural Disasters
Moscow, Russia

Berlin Alexander Alexandrovich
N.N. Semenov Federal Research Center for Chemical Physics of the Russian Academy of Sciences
Moscow, Russia

Mukhamedgaliev Bakhtiyor Abdukadirovich
Tashkent University of Architecture and Civil Engineering
Tashkent, Uzbekistan

Almenbayev Mirzhan Maratovich
Malik Gabdullin Academy of Civil Protection of the Ministry of Emergency Situations of the Republic of Kazakhstan
Kokshetau, Kazakhstan

Makishev Zhandos Kuandykovich
Malik Gabdullin Academy of Civil Protection of the Ministry of Emergency Situations of the Republic of Kazakhstan
Kokshetau, Kazakhstan

Berik Zhambulovich Rakhmetulin
Malik Gabdullin Academy of Civil Protection of the Ministry of Emergency Situations of the Republic of Kazakhstan
Kokshetau, Kazakhstan

ISBN 978-3-031-24073-7 ISBN 978-3-031-24074-4 (eBook)
https://doi.org/10.1007/978-3-031-24074-4

This Springer imprint is published by the registered company Springer Nature Switzerland AG
The registered company address is: Gewerbestrasse 11, 6330 Cham, Switzerland

The monograph presents the results of many years of research in the study of the effect of the duration of operation of wooden structures on their fire danger and fire resistance. A new hazard, not previously considered in the norms and estimates of fire resistance, was found—the reduction of fire resistance of wooden structures for a long period of operation. The influence of thermal influence on the change of mechanical, physicochemical and thermophysical properties of wooden structures of different duration of operation is shown. Formed coke layers are characterized by a higher oxidation capacity and high values of the rate of heat loss, it contributes to a significant thermal damage of wooden structures. Characteristic features of the combustion process of long-term aging wood, caused by abnormal heat release and a significant degree of thermal damage (charring), have been established. Significant influence of time factor of operation is conditioned by increase of charring rate of wooden structures (on the average 1.5–2 times). Occurring changes in physical, chemical and mechanical properties of wood of a long service life under the influence of high temperatures or fire can lead to a decrease in the fire-resistance limit of wooden structures on the loss of carrying capacity by more than twice. The paper shows the possibility of prognostic assessment of the behavior of wooden structures in fire conditions using the methods of thermal analysis. The peculiarities of the development and application of various fire protection means for wooden constructions of long service life are considered. The monograph can be useful to researchers, teachers, postgraduates and listeners of educational and scientific institutions of fire-technical, construction and wood-technical profile, as well as to designers, builders, experts in the field of fire safety and fire department workers.

Preface

During the millennial history of mankind, wood has truly become not only a traditional building material but also an inseparable epochal part of the whole building industry. It is difficult to overestimate the importance of wood in human life, given the scale of its use in the field of construction over a long period to the present day. This natural composite material has gained exceptional popularity in such vital areas of science and technology as medicine, the chemical industry, energy, and woodworking. Special interest in the last century appeared in the development of the sphere of wood science, which is the basis for many related industrially important fields all over the world.

Taking into account the importance of wood as a material for building structures, the authors of the book have attempted to form the main scientific results of research on the fire hazard and fire resistance of wooden structures, taking into account the duration of operation, as well as the use of various types and methods of fire protection. The monograph is prepared on the basis of the authors' studies of peculiarities of the wood aging process and its effect on fire hazard properties and fire resistance limits of wooden structures.

In their studies, the authors focus on the huge fund of old buildings and structures, made in whole or in part with the use of wooden structures. According to preliminary data, the number of such objects for various countries of the world can be in the range of 30–70% of the total number of buildings with different useful life of 30–50 years and more, and for the monuments of wooden architecture more than 100 or even 300 years.

The main prerequisites for the consideration of the hypothesis of increasing fire danger and reducing fire resistance of wooden structures are based on earlier studies of physical and chemical processes of wood aging, as well as the specifics of fires at objects involving wooden structures, subjected to long-term environmental influences and mechanical stress.

The work presents for the first time an in-depth study of the issues related to the emergence of a new threat to human safety, contained in a particular state of the material, wood-based structures under conditions of long-term aging. The influence of this process on the fire hazard and fire resistance of wooden structures is shown, as well as the peculiarities of the development of effective fire protection means for these structures.

The authors pay special attention to the importance of the obtained results for the safety of people with permanent or temporary stay in the buildings and constructions of various terms of exploitation and also to the necessity of inclusion of the obtained results into the normative and technical documents in the sphere of fire safety.

The book was written by a team of authors known for their work in the field of low flammability polymeric materials and the development of innovative chemical technologies in the field of fire protection of various materials and structures.

This monograph consists of 6 chapters and a conclusion.

Chapter 1 of the monograph presents general information about the structure and properties of wood.

Chapter 2 shows the relationship between the chemical composition of wood and its heat of combustion. The features of pyrolysis and thermal oxidative decomposition of wood, its fire hazard, fire danger and fire resistance of wooden structures, and the main ways and types of fire protection for wood, materials, and structures based on it are presented.

Chapter 3 is devoted to the influence of the aging process on the properties and structure of long-term aging wood; a critical analysis of the works in the field of aging of wood-based materials and structures is presented; the issues of the influence of the natural and artificial aging process of wood on its physico-chemical and fire hazard properties are considered.

Chapter 4 presents the results of investigating the influence of long service life on the main characteristics of fire hazard and the charring process of long-term aging wood. The fundamental differences in the structure and thermal-physical characteristics of charring layers of long-lived wood have been established.

Chapter 5 contains the results of research on the effect of long service life on the fire resistance of wooden structures.

Chapter 6 considers the development of effective fire protection to reduce fire hazards and increase the fire resistance of wooden structures for a long service life. The conclusion shows the prospects for further development of the direction of wood combustion science, taking into account the process of long-term aging.

The authors express their gratitude to the scientists and specialists who took part in carrying out research on the subject of the monograph. The authors hope that the monograph for its relevance and significance will be met with interest by a wide range of readers. The monograph can be useful for students, graduate students, and scientific and engineering employees of construction, chemical, and forestry technical institutions; specialists in the field of fire safety and fire protection of wooden structures; and chemists engaged in the development of new effective means to reduce the fire hazard of wood, as well as polymeric materials of carbonized type, taking into account the duration of their aging.

The monograph presents the results of studies within the framework of a new scientific direction in the field of fire safety of construction objects, connected with the study of the effect of operating time on the peculiarities of the behavior of various materials and structures in conditions of fire, as well as the development of effective solutions to implement their fire protection.

We hope that the monograph offered to the readers will encourage further development of the considered direction of science in the field of fire safety of wooden structures.

Moscow, Russia	Sivenkov Andrey Borisovich
Moscow, Russia	Berlin Alexander Alexandrovich
Tashkent, Uzbekistan	Mukhamedgaliev Bakhtiyor Abdukadirovich
Kokshetau, Kazakhstan	Almenbayev Mirzhan Maratovich
Kokshetau, Kazakhstan	Makishev Zhandos Kuandykovich
Kokshetau, Kazakhstan	Berik Zhambulovich Rakhmetulin

Introduction

Wood is a unique natural polymeric composite. It has long been widely used in residential and civil construction and serves as a source of thermal energy and a variety of chemicals. It is especially attractive because of its use as an enormous renewable, environmentally friendly resource.

The modern practice of wooden architecture is characterized by the development of new progressive technologies of industrial production and application of new construction materials of solid wood with bearing and fencing functions. This opens up broad prospects for the construction of high-span frame structures, medium and multi-storey buildings of different purposes.

Wooden structures remain one of the most used and demanded in the construction of buildings and structures for residential, public, industrial, storage, livestock, and agricultural purposes. The most dynamic development of the construction industry with the use of building structures made of wood is observed in Russia, a number of European countries, Australia, Japan, Canada, and the USA. In many respects, it is explained by aesthetic decorativeness, relatively high mechanical strength, and resistance of these structures to the influence of the external environment.

The results of numerous studies of natural wood testify to the influence of various physical and chemical factors on its performance properties. Questions related to the study of the influence of the structure (volume weight (density), moisture, and chemical composition) on the mechanical and other important physical properties of wood are a traditional subject of research in the field of wood science. The most significant results of scientific research in this area are associated with the names of such scientists as Varfolomeev Yu. A., Galkin V.P., Kovalchuk L.M., Nikitin V.M., Nikitin N.I., Ugolev B.N., Perelygin L.M., Pischik I.I., Poluboyarinov O.I., Rasev A.I., Sanayev V.G., Sergovsky P.S., Shubin G.S., Siau J.F., Hodgins J.W., Meyer J. A., and Dinwoodie J.M.

The above factors also affect the ability of wood to ignite and develop fire. Theoretical and experimental studies of the regularities of the formation and combustion processes of organic natural, artificial, and synthetic polymeric materials began to develop actively in the 60s of the last century. Much attention was also paid to the development of effective ways to reduce their fire hazard. Significant contribution to the development of this direction was made by such scientists as Abduragimov I.M., Aseeva R.M., Baratov A.N., Bakhman N.N., Berlin A.A., Biryukov V.G., Isaeva L.K., Kodolov V.I., Leonovich A.A., Molchadsky I.S.,

Pokrovskaya E.N., Serkov B.B., Taubkin S.I., Khalturinsky N.A., Khrulev V.M., Di Blasi, Parker W.J., Kashiwagi T., Tran H.C., Hilado C.J., Tewarson A., De Ris J, and Roberts A.M.. The results obtained are of scientific and practical importance for the application of polymeric materials in the construction industry and other areas of engineering.

Unique examples of the durability of wooden structures include buildings and structures such as Grinstead Church (Grinstead, Essex, UK, 1045), Lazarus Resurrection Church (Kizhi, Arkhangelsk region, 1391), St. Nicholas Wonderworker Church in Kolodny (Transcarpathia, Ukraine, 1470), Dormition Church (Nelazkoye village, Vologda region, 1694), Michael Archangel Cathedral (Uralsk, Republic of Kazakhstan, 1750), and many others.

At the same time, it is necessary to testify to the existence of huge housing and public buildings in the world, mostly pre-war buildings (100–150 years of service life), as well as residential buildings of the first half of the 50s. Given the cultural and historical significance of such sites, the cultural authorities, the World Organization of UNESCO, and the construction supervisory authorities are constantly monitoring the technical condition of these buildings and structures. In addition, in recent decades the change in the quality characteristics of wood as a result of long-term operation of wooden structures of buildings and structures has attracted a lot of attention of the scientific community. These are the works of such scientists as Varfolomeev A.Y., Varfolomeev Y.A., Gusev B.P., Keskulla T.E., Kisternaya M. V., Kovalchuk L.M., Okamoto K., Pischik I.I., Streltsov D.Y., Terentyev V.Y., Fefilov V.V., Fukada E, Shapovalova L.G., and others. These works show a significant impact of long-term natural aging on physical, chemical, mechanical, and other properties of wood as a result of transformation of its structure and chemical transformations occurring in wood material.

Despite this, the question of the effect of prolonged natural aging on fire properties of wood remains poorly studied at present. Analysis of fires that have occurred testifies to the unusual behavior of long-lived objects with wooden structures under conditions of high temperatures. The most characteristic features of such fires are abnormally high temperature, rapid fire development, and a high degree of thermal damage to structures. Examples of wooden structures with a useful life of 100 years or more lost to destructive fires include the Church of St. Catherine (Ostrava, Czech Republic, 1543), the Church of the Epiphany (Voskresensky Monastery of New Jerusalem (Moscow region, 1673)), the Church of the Savior on Piles (Ipatiev Monastery, Kostroma, 1713), the Church of the Protection of the Blessed Virgin (Uzden district, Minsk region, Republic of Belarus, 1898), the house of fabrics "Kyzyl-Tan" (Almaty, Republic of Kazakhstan, 1912), and dozens of other historical sites.

Few works are devoted to the study of non-standard behavior in conditions of fire of wooden constructions of a long line of operation and their fire hazard. In these works, the results were obtained, testifying to the possible increase in the fire hazard of wooden structures with the increase of their service life. At the same time, one of the main requirements that are imposed on the enclosing and bearing wooden structures is to ensure the normative indicators of fire resistance. The decrease in the load-carrying capacity of the elements and units of wooden structures, expressed by the fire resistance limit, is due to the change in the strength and geometric characteristics of the sections of structures under fire conditions. However, how the features of the charring process, which determines the loss of carrying capacity and fire resistance of wooden structures, change as a result of their long-term operation remains unstudied. Lack of scientific research in this direction determines the relevance of this monograph.

Contents

1 General Information About the Structure and Properties of Wood

1.1 Features of Macro- and Microstructure of Different Wood Species

Wood is a natural polymeric composite whose renewable raw material source is wood plants. They are amazing living organisms with enormous biosynthetic capabilities. Consuming water and carbon dioxide, microelements, and simple inorganic nutrients supplying the plant with only six elements, namely, carbon, hydrogen, oxygen, nitrogen, sulfur, and phosphorus, they are capable of synthesizing all complex organic substances necessary for the construction of plant tissue components, growth, and reproduction of trees.

The first ancient forests emerged in the Paleozoic Era of Earth evolution more than 300 million years ago and were evergreen conifers. Deciduous trees came later, in the Jurassic Period of the Mesozoic Era 165 million years ago. They are thought to have evolved from conifers, before evolving into plants with seasonal emergence and shedding of leaves. Their distribution became more significant in the Cenozoic Era 75 million years ago.

Russia is one of the most forested countries in the world. Almost a quarter of our planet's forests grow on its territory. With a relatively small number of major tree species, there is a huge intraspecific diversity. The major species include six conifers and 13 native deciduous species [1]. However, many subspecies, varieties, climatic populations, clones, spontaneous hybrids, and other biological forms of these major species have been identified and described [1].

The main forest-forming conifers in Russia are larch, pine, and spruce. Larch forests account for 2/5 of the country's forest area and one third of its timber reserves. At the same time, up to 14 different species of larch occur throughout Russia. The ranges of their distribution are isolated geographically. At the northern limit of the forest ecosystem (subarctic zone), there is an absolute dominance of Dahurian larch (Glaucous larch) and Cajander's larch. It is believed that in the process of evolution, these particular species of larch acquired such traits that

S. Andrey Borisovich et al., *Fire Hazard and Fire Resistance of Wooden Structures*,
https://doi.org/10.1007/978-3-031-24074-4_1

allowed the plants to adapt to the extreme permafrost conditions, as well as to the effects of natural fires [1].

Pine forests are the second most common coniferous species, occupying 1/6 of the area of all forests in the country, and spruce forests are the third (about 1/8 of the area, respectively). Among conifers, in addition to those mentioned, among the main species are cedar (as well as pine of the genus *Pinus*), fir (of the genus *Abies*), and yew (genus *Tachus*).

Although deciduous forests account for only 1/5 of the area of our country's forests, they are characterized by greater diversity than conifers. Of industrial importance for obtaining a variety of products are oak (genus *Quercus*), beech (genus *Fagus*), ash (genus *Fraxinus*), linden (genus *Tilia*), maple (genus *Acer*), birch (genus *Betula*), aspen and poplar (genus *Populus*), elm (genus *Ulmus*), alder (genus *Alnus*), walnut (genus *Junglas*), and others [1].

Studies of the diversity of morphological forms of woody species in forest natural populations from the perspective of the nature of the genetic structure of plants and their genetic conditioning began to actively develop since the second half of the last century. These studies are the scientific basis for the development of applied areas of forestry and are of great practical importance for solving the problems of selection of woody plants and for improving the productivity and quality of wood. Of particular interest here is the question of the interaction of a particular plant genotype with the conditions of the growing environment and the influence of genetic and environmental factors on the structure, chemical composition, and properties (distinctive features) of wood [2–5].

At present, a huge amount of information about the anatomical structure and microstructure of wood of different species and kinds has been accumulated. With all the commonality of the layer (zonal) picture of wood macrostructure, each species and kinds has its own characteristics. They are manifested, in particular, in the quantitative ratio of anatomical elements of macrostructure and their size, color, and properties [1].

It should be noted that some elements of wood structure are peculiar only to deciduous species or only to certain species of conifers. For example, resin canals are observed only in conifers. Only deciduous species are characterized by the presence of vessels visible on the cross section of wood in the form of round-shaped holes [1].

The microstructure of conifers consists of a system of fibers (tracheids) located along and across the trunk, core rays, and vertical and horizontal resin channels, which are placed in the intercellular space. Approximate composition of the elements is as follows:

Longitudinal tracheids—93%
Core rays and tracheids—6%
Resin canals—1%

The microstructure of hardwood is more diverse and complex compared to that of conifers. Distribution of structural elements by volume, in typical disseminated-vessel wood, is as follows [6]:

Vessels—55%
Libriform fibers and tracheids—26%
Wood parenchyma—1%
Core rays—18%

Libriform fibers got their name from the combination of the Latin words "libri"—bast and "form"—species. They are relatively short and have narrow cavities and thick walls. Parenchyma (elongated) cells are located in medullary rays [1], which partially perform the functions of storage of spare substances. In wood with ring-shaped microstructure, a significant volume is occupied by large vascular elements in the early wood zone (with a diameter of 200–400 μm) and smaller vessels in the late zone (30–40 μm). Usually, a small number of vascular tracheids are located near the vessels, which are a transition from between typical coniferous tracheids and vessels.

The basic elements of wood structure are plant cells, which form tissues of different structure and functional purpose.

The structural framework of wood is formed by cell membranes (walls), which perform mainly mechanical functions. Cellulose is the main substance that gives stability and strength to the cell wall. An additional strengthening component is lignin.

Of particular interest are studies not only of the physical microstructure of wood but also of the chemical composition and structure of its individual elements during the growth of wood plants and their common and distinctive features for different species [2–6].

Changes in the composition and ratio of structural elements in wood of different species and kinds affect many properties of wood. In particular, such important properties that affect the patterns of wood combustion process are density, moisture content, thermophysical, thermochemical, diffusion, and some others [6–9].

1.2 Chemical Composition of Wood

Among the organic chemical components of wood, there are usually four main components: cellulose, hemicelluloses, lignin, and extractives. Inorganic compounds are present in the amount, as a rule, not exceeding 1%.

The chemical composition of wood, even of the same botanical species in terms of the content of components, may differ significantly depending on the soil and climatic conditions of woody plants [1, 8].

The content and ratio of the main chemical components is different in separate parts of plants (leaves, needles, branches, bark, bows, in different zones of annual rings, etc.) and depends on the variety and species of trees [1, 8]. The cellulose content is considered to be almost constant in coniferous and deciduous wood, averaging 42 ± 2% per absolute dry weight of the sample [4, 9]. Analysis of more recent published data and those obtained in this study leads, however, to the

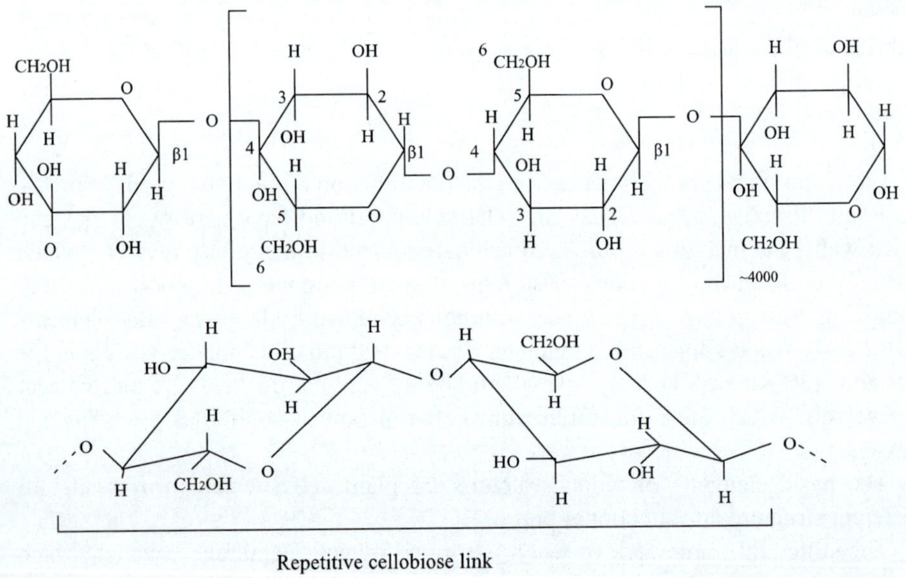

Fig. 1.1 Structure and conformation of the cellulose molecule

conclusion that the cellulose content in coniferous wood can vary from 32.6% to 55% and in deciduous wood from 33.7% to 49% [1, 8, 10].

Cellulose is a linear rigid-chain polymer composed of 1,4-β-D-glucopyranose units with the general formula $[C_6H_7O_3(OH)_3]_n$ (Fig. 1.1).

Lignins are three-dimensional polymers that are products of the polymerization and condensation reactions of coumarilic, coniferous, and synaptic alcohols.

Due to the participation of functional groups of these alcohols in polymerization and condensation reactions, the following main fragments of the spatial structure of lignins are formed (Fig. 1.2):

Coniferous lignin has a predominantly guaiacyl type of three-dimensional polymer structure with phenylpropane links containing one phenolic hydroxyl group and one methoxyl group in the aromatic core. The source of such links is coniferyl alcohol. The presence of syringyl type links (most significant in early wood lignin) and n-coumar phenylpropane links was also found. Their sources are synaptic and n-coumaric alcohols, respectively. The phenol-containing aromatic structures (coniferyl alcohol 4, synapse alcohol 5, and n-coumar alcohol 6) are as follows.

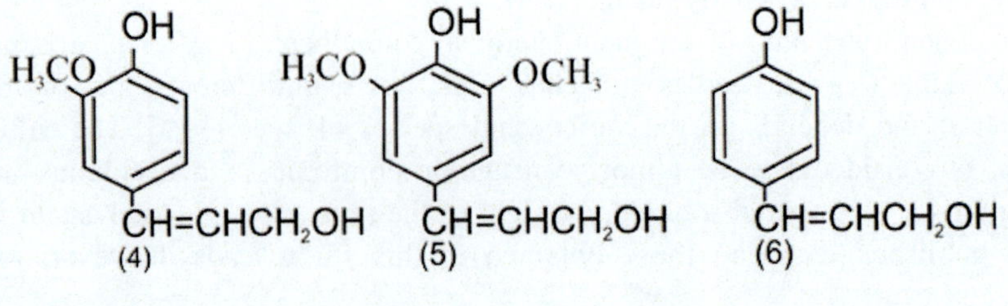

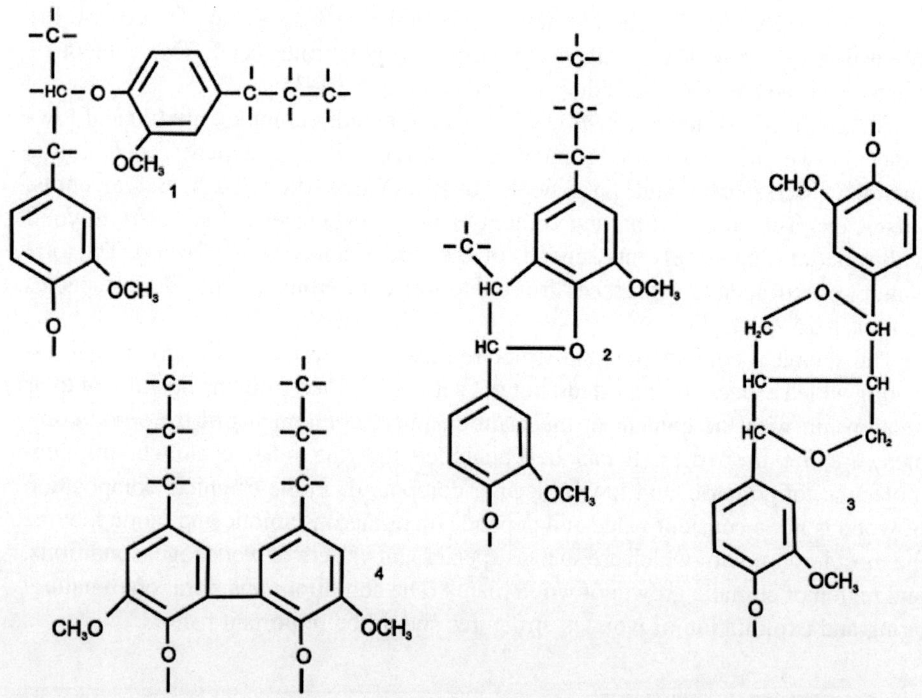

Fig. 1.2 Four main types of chemical bonds and fragments of the coniferous lignin structure: 1—arylglycerol-β-aryl ether; 2—phenylcumarone; 3—pinoresinol; 4—diphenyl

Hardwood lignin is a mixed guaiacyl-syringyl type, being essentially a product of copolymerization and condensation of coniferous and synaptic alcohols in different proportions depending on the wood species. Because of the presence of two methoxyl groups in the aromatic core, lignin of deciduous wood has a less dense spatial reticular structure compared to lignin of conifers. Most coniferous woods have a fairly high lignin content (30 ± 4%) compared to deciduous woods (less than 22–24%) [1].

Hemicelluloses are essentially oligosaccharides with a degree of polymerization of $n = 30–200$. This group of oligosaccharides includes both pentosans with 5 carbon atoms in the main cyclic link with the structural formula $(C_5H_{10}O_5)_n$ and hexosans $(C_6H_{10}O_5)_n$. Among hemicelluloses of deciduous wood, pentosans predominate, while hexosans predominate in coniferous wood.

Hemicelluloses are mainly copolymers of different carbohydrates. In conifers of wood species, for example, a family of galactoglucomannans consisting of 1,4-β-D-glucopyranose and 1,4-β-D-mannopyranose units as well as a family of arabinogalactans with a branched structure including conventional (1,4)-β-D-carbon units as well as lateral (1,3)- and (1,6)-β-D-galactopyranose residues were identified.

Among hemicelluloses in wood of all deciduous species, xylans, in particular, glucuronoxylans, predominate. The main molecular chains of these oligosaccharides consist of 1,4-β-D-xylopyranose links with acetyl substituents by hydroxyl groups in

2 (or 3) positions of the carbohydrate cycle and rare side groups formed by the interaction of xyloglucans with 4-O-methyl-α-D-glucuronic acid. The content of xylans in some species of deciduous wood can reach 30–35%.

The group of extractive constituents of wood includes complex mixtures of low-molecular-weight organic substances of different classes: terpenes, polyatomic phenols, esters, resin and fatty acids, unsaturated hydrocarbons and alcohols, waxes, etc. The composition and content of these substances extracted from wood with water or organic solvents depends on the species and variety of wood. The total content of extractable substances from wood of coniferous and deciduous species can reach 16–20% [1].

The chemical composition and structure of wood determine its many properties, among which especially important, but not actually studied from the position of their relationship with the content of the main chemical components of the wood complex, are fire properties. It can be concluded that the relative content of high-molecular, oligomeric, and low-molecular compounds in the chemical composition of wood is not a constant value and depends on numerous abiotic and biotic factors, the most important of which are such as species and variety of wood, age, conditions, and region of climatic growth of wood plants. The conditions and duration of natural aging and exploitation of wooden structures should be important factors.

1.3 Physical and Chemical Properties of Wood

1.3.1 Wood Physical Properties

Wood density is the most important physical characteristic of wood, closely related to its anatomical structure. The structural features of wood directly affect not only its density but also its ability to interact with water molecules. This in turn affects the thermal and mechanical properties of wood. The density of wood and its humidity have a great influence on the flammability and combustibility of this material. It is known, however, that the density of wood and its moisture content are, to a certain extent, interrelated characteristics [11].

Moistening of absolutely dry wood leads not only to an increase in the weight of the sample but also to an increase in its volume due to swelling of the material. It has been noticed that when wood stays in water or in the air of high humidity, the increase in the volume of the sample occurs only up to a certain limit. This limit corresponds to complete saturation of wood cell walls with water in the form of so-called bound water. Subsequent increase of water content in wood due to filling of pores, capillaries, and other empty space in the structure of wood, the so-called free water, leads only to increase of total sample mass, but without increasing the volume of wet wood.

Analysis of the obtained average values of absolutely dry wood density, basic wood density, and wood density at normalized moisture content for the most common wood species growing in Russia indicates a significant influence of moisture on the density of wood material [1, 7, 11]. Of great importance is not only the species and kinds of wood but also the climatic region of its origin. Wood even of the same species and kinds from different climatic regions differs in the content of chemical components and macrostructure, an indicator of which is density. So, the pine growing in the Ural region (Russia) has a density value $\rho_{12} = 505$ kg/m^3, and the density of the same species from Eastern Siberia is actually 12% lower (465 kg/m^3).

Among Russian species, the wood with a very low density has Siberian fir from Eastern Siberia ($\rho_{12} = 346$ kg/m^3) and with the highest density iron birch from Primorsky Krai ($\rho_{12} = 970$ kg/m^3). Wood density of the same species can vary depending on the ratio of the size of layers of early and late wood in the annual layer, heartwood, and sapwood, which in turn depends on the conditions, soil, and climatic factors of tree growth [1, 6].

According to the normalized density, different types of wood are divided into groups: species with low density ($\rho_{12} < 540$ kg/m^3), medium density ($550 < \rho_{12} < 740$ kg/m^3), and high density ($\rho_{12} > 750$ kg/m^3). The lowest and highest densities of wood are characterized by deciduous foreign species balsa ($\rho_{12} = 120$ kg/m^3) and backout ($\rho_{12} = 1300$ kg/m^3), respectively [1].

The true density of wood matter refers to the density of cell shells. Measured by water, the average for all species is 1.53 g/cm^3, and measured by helium and in non-polar liquids (benzene, toluene, mineral oil) is 1.46 and 1.44 g/cm^3, respectively [1].

Based on data on the density of the woody substance of sequoia (redwood) and white cedar before and after removal of extractive components of wood (on average for these species 1.505 and 1.530 g/cm^3, respectively), it was suggested that the density value is influenced by the chemical composition of wood. The influence of extractive substances on the density of wood matter was also revealed in other species [7].

The statement that the chemical components of the cell walls of woody plants differ little in their true density from each other does not seem entirely valid. Indeed, even cellulose of cell shells, which forms the framework of wood and has an amorphous-crystalline structure, is characterized by the presence of ordered crystallites and amorphous parts of macromolecules with different densities.

True crystallite density of known polymorphic modifications of native cellulose 1α and 1β reaches 1.55 and 1.59 g/cm^3. Disordered cellulose macromolecules, usually localized in the space between microfibrils, have a markedly lower density [12]. The true density of lignin is approximately 1.25–1.36 g/cm^3 [6]. Unfortunately, no data on the true density of hemicelluloses have been found. However, if the densities of crystalline modifications of monosaccharides (mannitol, xylose, glucose and arabinose) are 1.489; 1.53; 1.544; 1.585 g/cm^3 or disaccharides (sucrose and maltose) have values of 1.588; 1.540 g/cm^3 [13], then one would expect that hemicelluloses, which are oligomeric carbohydrates with a less ordered amorphous structure, should have a true density of approximately 1.32–1.41 g/cm^3.

The volume weight (density) of wood is closely related to other properties of the wood material, such as moisture content, porosity, sound conductivity, thermal conductivity, heat capacity, electrical conductivity, etc.

1.3.2 Thermal Physical Properties of Wood

Thermal physical properties largely determine the fire hazard characteristics when using wood and wood-based materials. The thermophysical properties of wood depend on its species, volume weight, moisture content, and material temperature.

Experimental and theoretical studies of polymeric materials, including wood, usually include such thermophysical characteristics as heat capacity, thermal conductivity, thermal conductivity, and thermal inertia (thermal activity). The last two indicators are complex, taking into account the density of the material.

On average, the specific heat capacity for all wood species at 0 °C is taken to be $c_{Po} = 1.55$ kJ/kg °C [1]. According to other sources, $c_{Po} = 1.20$ [14] and 1.36 kJ/kg °C [15]. When the saturation limit of cell walls (30% moisture content) is reached at zero temperature, specific heat capacity of wood increases to 2.2 kJ/kg °C and at 100% moisture content to 3.0 kJ/kg °C [1]. Naturally, increasing temperature causes an increase in the specific heat capacity of wood, since when wood is heated, certain oscillatory movements of structural elements of macromolecules are unfrozen. Specific heat capacity of dry wood changes with temperature according to the linear law, reaching almost 2.0 kJ/kg °C at 100 °C. According to [16], the temperature dependence of specific heat capacity can be represented by the relation

$$c_{Po}(T) = c_{pr}(T/T_r), \tag{1.1}$$

where c_{pr} is the specific heat capacity of dry wood at temperature $T_g = 293$ K.

The authors of [15] estimated the specific heat capacity and thermal conductivity of wood samples cut from the trunk in transverse and longitudinal direction relative to the fiber orientation from the ignition experiments (Table 1.1).

The influence of wood anisotropy on the obtained characteristics is noticeable. In the longitudinal direction relative to the orientation of the fibers in the wood, the thermophysical properties of the material are significantly higher than in the transverse direction. The observed effect is due to different heat transfer conditions when the external heat flux is brought to the surface of the sample perpendicular or parallel to the wood fiber axis and different response of the material to temperature changes.

Table 1.1 Values of heat transfer coefficient and specific heat capacity of wood with different direction of heat flow

Type of wood	Direction heat flow	Density, ρ_w, kg/m^3	k_w, W/m·grad	C_{pw}, J/kg·grad
Sequoia	Across the fibers	354	0.19	3200
	Along the fibers	328	0.85	7400
Douglas fir	Across the fibers	502	0.23	2200
	Along the fibers	455	0.80	4000
Red oak	Across the fibers	753	0.44	3100
	Along the fibers	678	0.86	3200
Maple	Across the fibers	741	0.35	2500
	Along the fibers	742	–	7100

Table 1.2 Values for thermal diffusivity and thermal activity of different wood species

Wood species	Direction heat flow	ρ_{12}, kg/m^3	a, m^2/s	$k\rho c$, kJ^2/K^2 m^4 s	References
Douglas fir	[a]	465	–	0.158	[16]
Spruce (Wansham), Vietnam	⊥	502	1.79×10^{-7}	0.25	[18]
	II	450	–	1.44	[18]
	⊥	400	–	0.148	[17]
Pine (Moscow region, Russia)	⊥	450	1.84×10^{-7}	0.070	[19]
Pine	⊥	460	–	0.156	[16]
White pine	⊥	380	6.6×10^{-7}	–	[20]
Southern pine	⊥	508	–	0.183	[21]
Pine (Thongkaribe), Vietnam	⊥	430	–	0.173	[17]
Sequoia	⊥	430	–	0.141	[16]
	⊥	354	1.89×10^{-7}	0.22	[15]
	II	328	–	2.07	[15]
	[a]	312	–	0.073	[21]
Thuja	⊥	330	–	0.087	[16]
Red oak	[a]	660	–	0.360	[21]
	⊥	753	1.74×10^{-7}	1.01	[15]
	II	678	–	1.85	[15]
Maple	[a]	530	1.0×10^{-7}	–	[20]
	⊥	741	1.75×10^{-7}	0.67	[15]
	II	742	–	1.09	[15]
Lime	[a]	420	–	0.141	[16]
Eucalyptus (Bachdan), Vietnam	⊥	595	–	0.139	[17]
Eucalyptus	[a]	640	–	0.260	[16]
Eucalyptus	[a]	810	–	0.393	[16]
Acacia (Keolai), Vietnam	⊥	560	–	0.140	[17]

[a]The direction of the heat flux in relation to the wood fibrer orientation is not specified. The data seems to refer to the action of external heat flux in perpendicular direction to the fiber axis

There are very limited published data relating to direct measurements of the thermophysical properties of wood over a wide temperature range.

Table 1.2 shows the results of thermal diffusivity coefficient (a) and thermal activity coefficient ($k\rho c$) estimates obtained in [17] and in the works of other authors [15–21] for some samples of coniferous and deciduous wood. The influence of direction of external heat flux relative to wood fiber axis is shown.

At humidity levels below the saturation limit of cell walls, the thermal diffusivity coefficient in wood does not differ much from that in dry wood and varies in the range (1.6–2.1)·10^{-7} m^2/s for different species [7].

The influence of temperature is more significant, since both the thermal conductivity coefficient and the specific heat capacity of the material change with increasing temperature. However, it seems that due to adequate change in these values with temperature increase up to 250–290 °C, the thermal diffusivity coefficient of wood, according to different researchers, practically does not depend on temperature in this range, but is constant for each kind of wood [17].

In contrast to the thermal diffusivity coefficient, the $k\rho c$ is strongly temperature dependent. That is why the $k\rho c$ determined from ignition experiments is the effective value. It corresponds to the average temperature of the material between the ignition temperature and the ambient temperature. The $k\rho c$ value obtained this way is 1.6 to 2.6 times higher than the value at 293 K. The general rule of increasing the $k\rho c$ values of wood with increasing volume weight (density) of wood is observed. At the same time, the influence of moisture is noticeable.

1.4 Physical and Mechanical Properties of Wood

The notion of general fire safety when applied to building structures includes, in addition to the fire hazard category of the building materials as such, the concepts of fire resistance and fire hazard of the structures made of these materials.

The behavior of wooden building structures in fire (resistance, fire resistance) is largely determined by the level and variability of the physical and mechanical properties of wood when heated under load and the rate of charring. Wood is a viscoelastic anisotropic polymer composite. Depending on the direction of the applied mechanical force, the values of the parameters that determine the occurring stresses and strains in the material vary.

Important parameters that characterize the quality of a building material are the strength properties corresponding to the deformation at the breaking point of the material; the elastic moduli in bending, stretching, and shearing; and the Poisson's ratio. The latter is the ratio of the deformation of the material in transverse and longitudinal direction to the applied load.

As a structural building material, wood works mainly in compression and bending. Tensile stresses occur simultaneously in compression and bending as well as shear across the fibers, to which wood has a poor resistance. Therefore, it is not practical to use wood in tensile applications.

In the present work, the most important mechanical properties of Russian coniferous wood samples (pine and spruce), deciduous wood with a scattered-vessel structure (birch), and ring-vessel structure (oak) from the central part of the country were evaluated. In addition, the mechanical properties and influence of moisture content of tropical wood samples from Vietnam were determined [22].

In the tropical climate of Vietnam, high humidity and high temperature affect not only the physical and mechanical properties of wood of all species, but also the material's ability to resist the effects of various fungi and rotting. The eucalyptus bachdan wood is the most rot-resistant and very hard and is used in water engineering.

Under these growing conditions, the distinction between the early and late zones in the annual layers of the wood macrostructure becomes blurred. It is believed that increasing the thickness of the late wood layer in relation to the early wood layer has a favorable effect on the strength of the material.

Table 1.3 shows the mechanical properties of domestic species of pine and oak from different regions of Russia and Vietnam according to the results obtained by standard methods of GOST 16483 for samples of "clean" wood without defects from the sapwood part of the trunk. Tropical wood samples were examined according to Vietnamese standard TCVN 358–396-70, which is an analogue of the above Russian standard in terms of testing conditions.

Correlation equations established by V.N. Volynsky [23] were used to estimate the physical and mechanical properties of birch and spruce. Such equations were obtained by the mentioned author based on the analysis of basic data on mechanical properties of 167 varieties of wood of different species from the main climatic regions of growth on the territory of the former USSR. Coniferous and deciduous wood species were analyzed separately as two independent aggregates. Similarly, analysis of physical and mechanical properties of 112 wood species (47 coniferous and 65 deciduous) grown in the USA and Canada was made. In addition, data on the mechanical properties of tropical wood species were considered.

V.N. Volynsky found that wood density has the closest correlation with strength indices. The correlation of the relationship is expressed by an equation of a power form: $Y = a{\cdot}X^n$. The author [24] has shown that consideration of separate sets of wood species (coniferous and deciduous) gives a higher correlation coefficient exceeding $R^2 > 0.9$ than consideration of a single set of wood that differs only in material density.

Therefore, the following correlation equations were used in the present work to estimate the physical and mechanical properties of birch wood:

Compressive strength along the fibers:

$$\sigma_{\text{compression II}} = 0.445 \cdot {\rho_{\text{basic}}}^{0.737}, \text{MPa} \tag{1.2}$$

Static bending strength along the fibers:

$$\sigma_{\text{bend II}} = 0.325 \cdot {\rho_{\text{basic}}}^{0.879}, \text{MPa} \tag{1.3}$$

shear strength along the fibers:

$$\sigma_{\text{shear II}} = 0.02 \cdot {\rho_{\text{basic}}}^{1.082}, \text{MPa} \tag{1.4}$$

Recalculation of used samples density with moisture content $W(\%)$ to the basic one was carried out according to the formula:

$$\rho_{\text{basic}} = \rho_w\, 100(100 + K_\alpha \cdot W)/(100 + W) \cdot (100 + 30 \cdot K_\alpha), \text{kg/m}^3 \tag{1.5}$$

Table 1.3 Physical and mechanical properties of wood

Index	Pine Russia	Spruce Russia	Pine Vietnam	Spruce Vietnam	Eucalyptus Vietnam	Acacia Keo Tai Thiong Vietnam	Acacia Keo lai Vietnam
ρ_{12}, kg/m^3	470–520	455	449	417	618	543	583
$\sigma_{\text{compression along the fibers}}$, MPa	33–51	37.2	32.7	41.3	68.6	74.3	75.0
$\sigma_{\text{cross fiber compression}}$, MPa	–	–	5.6	4.2	9.7	8.4	8.1
$\sigma_{\text{fiber stretch}}$, MPa	78–115	91.1	78.0	37.2	140.7	140.1	112.0
$\sigma_{\text{static bending}}$, MPa	67.4–86	65.5	69.3	62.2	109.4	109.7	118.2
E_{bend}, GPa	6.7–8	9.5	7.1	4.7	12.3	11.7	12.1
$\sigma_{\text{shear along the fibers}}$, MPa	6.0-7.7	5.4-6.5	10.4	8.2	16.6	11.8	9.4
$H_{\text{facing}}/H_{\text{radial}}$, MPa	24–30	29.3/20.9	37.5	43.7	78.8	74.8	65.5

Here K_α coefficient for birch is taken equal to 0.6. Correspondingly for spruce 0.5, to evaluate the physical and mechanical properties of spruce wood, we used the following equations:

Compressive strength along the fibers:

$$\sigma_{\text{compression II}} = 0.309 \cdot \rho_{\text{базисная}}{}^{0.812}, \text{MPa} \tag{1.6}$$

Static bending strength along the fibers:

$$\sigma_{\text{bend II}} = 0.271 \cdot \rho_{\text{базисная}}{}^{0.930}, \text{MPa} \tag{1.7}$$

Shear strength along the fibers:

$$\sigma_{\text{shear II}} = 0.013 \cdot \rho_{\text{базисная}}{}^{1.022}, \text{MPa} \tag{1.8}$$

Lateral (radial) hardness was calculated by equation:

$$\rho_{\text{базисная}} = 267 \cdot H^{0.438}, \text{кN } (1 \text{ кN} = 10 \text{ MPa}) \tag{1.9}$$

The modulus of elasticity of wood along the fibers was determined by equation [24]:

$$E = 0.013 \cdot \rho_{12} + 3.59, \text{GPa} \tag{1.10}$$

And the end hardness according to equation [24]:

$$H_{\text{facing}} = 0.1583 \cdot \rho_{12} - 42.689, \text{MPa} \tag{1.11}$$

It should be noted that between bending strength, bending modulus of elasticity, and radial (lateral) hardness, a close correlation was found for softwood domestic wood [24] with $R^2 = 0.817$:

$$\sigma_{\text{bend II}} = 3.18 + 3.16 \cdot E_{\text{bend}} + 0.707, \text{N} \tag{1.12}$$

where σ of bending II is expressed in MPa, E_{bend} in GPa, and hardness in kN.

The effect of density on the mechanical properties of wood is due to changes in the thickness of the cell walls (shells) of the plant fibrous tissue that performs the mechanical functions in wood.

As can be seen from Table 1.3, in all cases coniferous wood is inferior to hardwood in its mechanical properties. The humidity and temperature of wood reduce the mechanical properties of wood. Thus, it was found that humidity of tropical wood affects the bending strength only in the ranges of cell wall saturation limit. To a greater extent, it affects the reduction of bending strength of hardwood, to a lesser extent—softwood.

Thus, when the moisture content increases from 12 to 21%, the bending strength of Thongkaribe wood decreases only by 21% and that of Bachdan and Keo Tai

Tiong by 32.4 and 32.8%, respectively. A similar trend is observed in the reduction of the static hardness of wood in the end direction. In hardwoods, when the moisture content increases to 21%, this index decreases by 1.35–1.4 times, while in softwoods, it decreases only by 1.1–1.2 times.

Reducing the strength of wood at room temperature under the influence of moisture is caused by a decrease in intermolecular interaction in the main components of wood, as well as changes in the concentration of stresses in the places of structural heterogeneities of the composite. When drying wood samples, reversible recovery of strength characteristics was observed [22].

Heating of dry wood affects the strength indicators in different ways, but especially noticeably on the bending strength along the fibers. Thus, when heating wood from 20 °C to 100 °C, the decrease of bending strength is about 0.7% with every 1°C increase of temperature. At 270–300 °C, wood strength is completely lost [20]. Therefore, it is recommended when calculating the bearing capacity of wooden structures in conditions of fire (fire-resistance limit) to take a layer of wood under a layer of coal, in which the temperature changes from 300 to 100 °C, 5–7 mm, and to exclude it from consideration of the section under load [25].

References

1. Ugolev, B. N. (2001). *Wood science with the bases of forest commodity science* (p. 340). Publishing House of Moscow State University of Forestry.
2. Antonova, G. F. (1999). *Cell growth of conifers* (p. 232). Nauka.
3. Antonova, G. F. (2000). *Comparative analysis of early and late lignification of Siberian larch. Materials of III International Symposium Structure, Properties, and Quality of Wood* (pp. 27–29). Institute of Forestry, KSC RAS.
4. Goodwin, T., & Mercer, E. (1986). *Introduction to plant biochemistry* (Vol. 2, p. 396). MIR.
5. Antonova, G. F., Stasova, V. V., Konovalov, N. T., Konovalova, N. N. (2002). Distribution of lignin in the structural elements of oak wood. In *Proceedings of the II International Conference on Plant Anatomy and Morphology, St. Petersburg, SPbGLTA*, pp. 331–332.
6. Siau, J. F. (1984). *Transport processes in wood* (p. 301). Springer.
7. Chudinov, B. S. (1984). *Water in wood* (p. 267). Nauka.
8. Zaytsev, D. A. (2004). *Proceedings of the IV International Symposium "Structure, properties and quality of wood – 2004"* (Vol. 2, p. 569). St. Petersburg State Forestry Academy.
9. Overend, R. P., Milne, T. A., & Mudge, L. K. (Eds.). (1985). *Fundamentals of biomass thermochemical conversion* (p. 1). Elsevier.
10. Serkov, B. B., Sivenkov, A. B., Thanh, B. D., & Aseeva, R. M. (2003). Heat emission during burning of wood. *Bulletin of Moscow State Forest University, Forest Bulletin, 5*, 74–79.
11. Poluboyarinov, O. I. (1976). *Density of wood* (p. 160). Forest Industry.
12. Alyoshina, L. A., Glazkova, S. V., Lugovskaya, L. A., et al. (2001). Modern concepts of cellulose structure in wood. *Chemistry of Vegetable Raw Materials, 1*, 5–36.
13. Perelman, V. I. (1955). *Short reference book for chemists* (p. 119). Scientific and Technical Publishing House of Chemical Literature.
14. USDA. (2010). *Wood handbook, Wood as an engineering materials, Chapter 4* (pp. 1–19). Forest Products Laboratory, FPL-GTR-190.
15. Spearpoint, M. J., & Quintiere, J. G. (2001). Prediction the piloted ignition of wood in the cone calorimeters using an integral model. *Fire Safety Journal, 36*, 391–415.

16. Janssens, M. A. (1991). Thermal model for piloted ignition of wood including variable thermophysical properties. In *Proceedings of the Third International Symposium on Fire Safety Science*, pp. 167–176.
17. Thanh, B. D. (2006). *Influence of chemical components on the fire hazardous properties of tropical wood in Vietnam*. Ph.D. thesis, Russian Ministry of Emergency Situations, p. 184.
18. Simms, D. L., & Law, M. (1967). The ignition of wet and dry wood by radiation. *Combustion and Flame, 11*, 377–388.
19. Sivenkov, A. B. (2002). *Reducing the fire hazard of materials based on cellulose*. Dissertation of Candidate of Technical Sciences, Academy of the Russian Ministry for Emergency Situations, p. 193.
20. Moghtadery, B., Novozhilov, V., Fletcher, D., & Kent, J. H. (1997). An integral model for the transient pyrolysis of solid materials. *Fire and Materials, 21*, 7–16.
21. Tran, H. C., & White, R. H. (1992). Burning rate of solid wood measured in a heat release rate calorimeter. *Fire and Materials, 16*, 197–206.
22. Aseeva, R. M., Serkov, B. B., Sivenkov, A. B., & Thanh, B. D. (2006). Research of physical and mechanical properties of some tropical wood species. In *Building materials, equipment, technology of the XXI century* (Vol. 6, pp. 42–43).
23. Volynsky, V. N. (2006). *The relationship and variability of physical and mechanical properties of wood* (2nd ed., p. 196). Arkhangelsk State Technical University Press.
24. Paul, E. E., & Kuhta, V. N. (2011). Dependence of mechanical properties of wood on its density. *Forestry and Hunting, 10*, 20–23.
25. Makishev, J. K. (2016). *Fire resistance of wooden structures of long service life: Dissertation. Candidate of Technical Sciences: 05.26.03* (p. 163). Academy of State Fire Service of EMERCOM of Russia.

Fire Hazard and Fire Resistance of Wooden Structures and Their Fire Protection

2

2.1 Effect of the Chemical Composition of Wood on the Lower Heat of Complete Combustion

A large number of works are devoted to the study of the chemical composition of wood [1]. This is not surprising, given the use of wood as a renewable raw material source for the production of many valuable chemical products, as well as a focused desire to create high-quality materials for various technical purposes. Particular attention in this aspect is paid to the problem of the relationship between the structure and chemical composition of softwood and hardwood from different geographical areas with the physical, mechanical, and other performance properties of wood materials [2].

The main components of wood are characterized by different heat of total combustion, so we can expect differences in the values of the heat of total combustion of wood of different species. According to [3], the typical average values of the higher heat of total combustion of hardwood are 19.734 ± 0.981 MJ/kg and that of softwood is 20.817 ± 1.479 MJ/kg.

The effect of the content of the main components of wood on the heat of complete combustion of wood was first considered in [4]. The author determined the content of lignin and extractive substances in the wood of four conifers (spruce, pine, thuja, and sequoia) and four hardwoods (maple, poplar, red oak, and linden). He showed that there is a close correlation between the higher heat of complete combustion of wood and the lignin content. The correlation coefficient $R^2 = 0.70$ increased to 0.97 if dry samples free of extractive substances were considered. In this case, the empirical correlation corresponded to the regression equation:

$$Q_{\mathrm{h}} = 17612.472 + 85.364\, X_{\mathrm{l}}, \mathrm{kJ/kg} \qquad (2.1)$$

where X_{l}—lignin content in the wood sample, %.

In this work, the Q_{h} values were determined using an adiabatic oxygen bomb calorimeter without taking into account the heat of water condensation as a product

S. Andrey Borisovich et al., *Fire Hazard and Fire Resistance of Wooden Structures*,
https://doi.org/10.1007/978-3-031-24074-4_2

of the combustion reaction. According to the author [4], the higher heat of complete combustion of different types of wood should be significantly influenced by the content of extractives, in particular, resins and terpenes in their composition. The Q_h values of resins from the wood of some coniferous species can reach 34.9–37.2 MJ/kg [4].

Since the heat of total combustion is a very important thermodynamic characteristic of substances and materials, its determination is of great practical interest. In addition to direct experimental methods of estimating the heat of total combustion of substances and materials by means of various kinds of calorimeters, engineering methods of calculation are also used [5, 6]. They are based on Hess's thermochemical law, according to which the total heat effect of a reaction depends only on the initial reacting substances and the final products, but does not depend on the intermediate reaction stages. Engineering methods of calculation can be divided into two groups. The first group should include the use of empirical equations linking the heat of complete combustion of organic substances to their elemental composition, primarily to the content of such elements as carbon, hydrogen, oxygen, sulfur, and nitrogen. The Dulong equation [6] is especially popular abroad, while in our country, the Mendeleev D.I. equation is usually used. The second group can include the calculation of the heat of total combustion of substances of known chemical structure using structural group contributions [5–7].

When studying the combustibility of materials, modeling the occurrence and development of fire, and solving many practical problems related to fire safety of facilities, the so-called lower heat of total combustion is used, which is absolutely less than Q_h (standard enthalpy of combustion) by the condensation heat of water formed during combustion. In this case, D.I. Mendeleev's equation, taking into account the moisture contained in the substance, has the form

$$Q_{low} = 339.4C + 1257H - 108.9(O + N - S) - 25.1(9H + W), \text{kJ/kg} \qquad (2.2)$$

where C, H, O, N, S, and W are the content of the corresponding elements and moisture, %.

Since lignin and extractive substances are the most high-energy components of wood, the total content of these components was taken into account when considering the effect of the chemical composition on the lower heat of complete combustion in the present work.

Experimental data on the chemical composition and values of the lower heat of complete combustion were obtained for 18 varieties of wood from different regions of Russia and the subequatorial region of Vietnam [8, 9]. Among them, in addition to samples of conifers, we studied six samples of deciduous wood with diffuse-vascular structure (beech, maple, birch, hornbeam, aspen, eucalyptus-bachdan) and six with ring-vascular structure (oak, ash, two species of acacia—keo lai and Keo tai tuon, chestnut, and elm Karagach).

Table 2.1 shows the results of the analysis of the chemical composition and the lower heat of total combustion of the indicated wood samples [8, 9].

Table 2.1 Lower calorific value and chemical composition of different wood species

Wood species	ρ, kg/m^3	W, %	Q_{low}, kJ/g	Holocellulose/ hemicellulose, %	Lignin, %	Extractable substances, %
Spruce	430	9.5	18.90	62.7/–	27.3	10
Pine	450	10.0	19.62	62.9/–	28.0	9.1
Larch	660	–	18.61	–/–	–	–
Cedar	400	10	18.84	61.4/–	30.8	7.8
Beech	600	9.2	18.26	74.1/–	21.0	4.9
Ash	740	–	18.40	–/–	–	–
Maple	610	8.0	18.04	71.9/–	23.2	4.9
Oak	570	7.0	18.66	68.7/–	23.6	7.7
Birch	540	6.5	18.08	73.3/–	20.6	6.1
Hornbeam	595	–	18.42	–/–	–	–
Aspen	480	6.8	18.14	74.6/–	21.5	3.9
Thongkaribe	430	4	18.62	65.5/10.5	27.0	7.93
Wansham	400	4	18.84	82.2/27.5	11.24	5.06
Bachdan	595	5	18.85	67.3/19.16	25.4	6.40
Keo tai tuon	420	4	18.11	70.0/21.03	24.75	5.09
Keo lai	560	4	18.53	69.1/20.10	25.16	5.56
Chestnut	490	4.0	17.86	–/–	–	–
Ilm	620	4.0	18.06	–/–	–	–

Tropical conifers are represented by Thongkaribe (*Pinus massoniana*) and vansham (*Picea koraiensis* nakai) specimens and deciduous species—bachdan (*Eucalyptus camaldulensis*), Keo tai tuon (*Acacia mangium*), and keo lai (*Acacia auriculiformis*). All of them are characterized by rapid tree growth (2–2.5 m/year). Because of the almost year-round growth, early and late wood in the annual layers is almost indistinguishable.

The lowest calorific value of total combustion of wood samples was determined by the standard method of GOST 147-74, as well as by elemental composition. To analyze the elemental composition of absolutely dry samples, we used an automatic device Carlo Erba 1106 (Italy). The content of the main chemical constituents of wood samples was determined by the known accepted methods [9, 10].

Figure 2.1 shows the graphical dependence of the lower heat of total combustion of the studied samples on the total content of lignin and extractive substances in the chemical composition of wood.

There is a linear relationship between the indicators under consideration, obeying the equation with a correlation coefficient of $R = 0.86$:

$$Q_{low} = 15.45 + 0.1\ (X_l + X_e), \mathrm{kJ/g} \tag{2.3}$$

where X_l and X_e are the content of Clason lignin and extractable substances in absolutely dry wood, %.

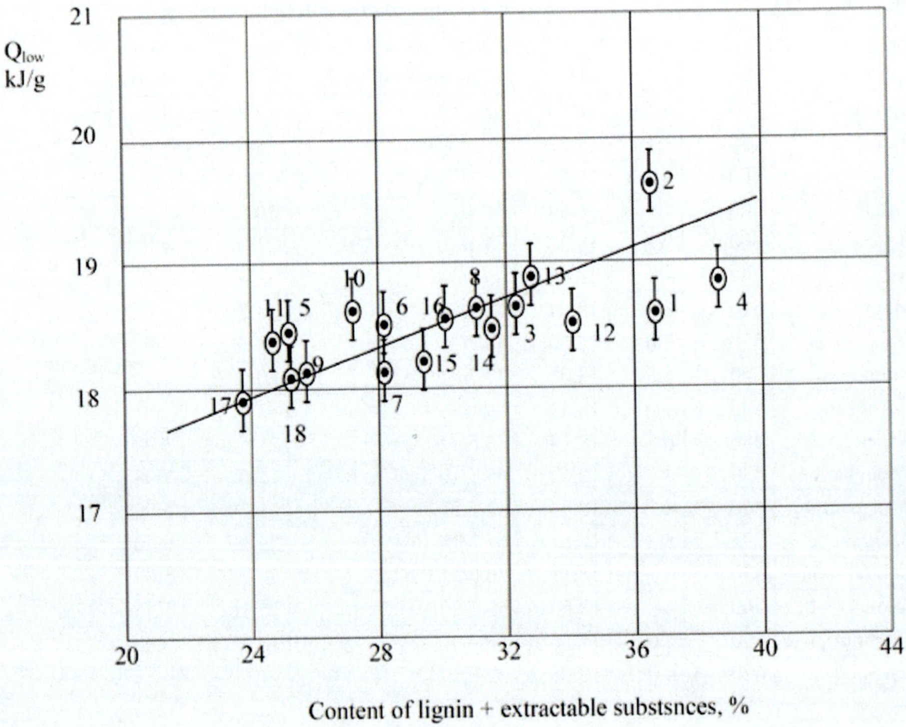

Fig. 2.1 Dependence of the lower heat of total combustion of wood samples on the total content of lignin and extractable substances: 1—spruce; 2—pine; 3—larch; 4—cedar; 5—beech; 6—ash; 7—maple; 8—oak; 9—birch; 10—hornbeam; 11—aspen; 12—pine (thongkaribe); 13—spruce (vansham); 14—eucalyptus (bachdan); 15—acacia (Keo tai tuon); 16—acacia (keo lai); 17—chestnut; 18—elm Karagach

The greatest scatter of Q_{low} values is characteristic of coniferous wood samples with high content of extractable substances. As can be seen from Fig. 2.1, samples with scattered-vessel structure occupy on the correlation line the middle position between conifers and hardwoods with ring-vessel structure.

The resulting equation makes it possible to estimate the contribution of various chemical components of wood to the total heat of combustion of wood material and to calculate the Q_{low} values not only of the mixture of extractable substances from individual wood species but also of hemicelluloses included in their composition. If the total content of lignin and extractives is zero, in this case the lowest heat of complete combustion for the remaining part of wood, i.e., the holocellulose consisting of cellulose and hemicellulose, should be equal to 15.45 kJ/g.

In the other limiting case, when the content $X_{\text{l}} + X_{\text{e}} = 100\%$, value $Q_{\text{low}}^{\text{l+e}} = 24.45$ kJ/g.

Then the corresponding equations for estimating the heat of total combustion of extractives, $Q_{\text{low}}^{\text{e}}$, and hemicellulose, $Q_{\text{low}}^{\text{hc}}$, for each type of wood can be represented as follows:

Table 2.2 Lower heat of total combustion of extractives and hemicelluloses from softwood and hardwood

Sample of wood	$Q_{low}{}^{e}$, kJ/g	Sample of wood	$Q_{low}{}^{e}$, kJ/g	$Q_{low}{}^{hc}$, kJ/g
Spruce	29.66	Spruce [10]	39.1	14.11
Pine	30.20	Pine [10]	30.97	14.14
Cedar	31.85	Larch [10]	28.39	14.69
Thuja [4]	31.53	Fir [10]	36.64	13.88
Southern pine [4]	33.07	Spruce (vansham)	33.78	12.40
Canadian spruce [4]	42.68	Pine (thongkaribe)	30.69	12.19
Sequoia [4]	31.80	Eucalyptus (bachdan)	31.56	13.87
Maple	32.76	Aspen tree	33.91	–
Maple [4]	31.75	Poplar [10]	33.75	–
Red Oak [4]	30.86	Acacia (Keo tai tuon)	32.93	14.0
White oak	30.17	Acacia (keo lai)	32.41	13.92
Birch	30.67	Beech	32.06	–
Lime [4]	29.59	Aspen [10]	34.72	14.13

$$Q_{low}{}^{l+e} = 24.45 = X_l\, Q_{low}{}^{l} + X_e\, Q_{low}{}^{e}, \text{kJ/g} \tag{2.4}$$

$$Q_{low}{}^{cell+hc} = 15.45 = X_{cell} Q_{low}{}^{cell} + X_{hc}\, Q_{low}{}^{hc}, \text{kJ/g} \tag{2.5}$$

where X_l and X_e—the relative proportion of lignin and extractives in their mixture; X_{cell} and X_{hc}—the relative proportion of cellulose and hemicellulose in cellulose; and $Q_{low}{}^{cell + hc}$—the lower heat of complete combustion of cellulose.

Thus, the relative energy contribution of the main wood components to the total heat release during complete combustion is proportional to their content in the wood.

The elemental composition of Clason lignin from different wood species is almost identical and includes 63.54% carbon, 5.54% hydrogen, and 30.92% oxygen [11]. The value $Q_{low}{}^{l}$, calculated by D.I. Mendeleev's equation, is 23.91 kJ/g, and $Q_h{}^{l} = 25.16$ kJ/g. The obtained result is in good agreement with the experimental value of the higher heat of total combustion of lignin $Q_h{}^{l} = 25.6$ kJ/g [11]. Calculation of the lower heat of total combustion of cellulose leads to the value $Q_{low}{}^{cell} = 16.078$ kJ/g.

Table 2.2 shows the results of estimating the values of the lower heat of total combustion of substances extracted from wood of different species, as well as hemicelluloses from tropical wood species.

In addition, Table 2.2 includes the values of $Q_{low}{}^{e}$ и $Q_{low}{}^{hc}$, which were calculated from the data on the chemical composition of hardwood and softwood presented in [4]. Due to the lack of information on the chemical composition of the substances extracted from wood, the contribution of each of the extractive components to the heat of complete combustion is rather uncertain. However, as can be seen from Table 2.2, in general, the values of the lower heat of total combustion of extractives are not the same for different wood samples. This may be caused not only by the presence of different substances in the chemical composition of extractives but also by their different ratios.

The values of the lowest heat of total combustion of substances extracted from coniferous wood vary from 28.39 to 42.68 kJ/g. In this case, by the example of spruce wood, it is possible to note the dependence of this value on the region of its growth. The heat of complete combustion of hemicelluloses in wood of different species differs little from each other and has a value of Q_{low}^{hc} of about 14 kJ/g, which is characteristic of pentosans. The exception is the samples of coniferous wood from the subequatorial region of Vietnam. The results suggest the presence in the chemical structure of hexosan-type hemicelluloses of fragments with increased oxygen content, such as galacturonic acid residues [8, 9].

2.2 Pyrolysis and Thermal Oxidative Decomposition of Wood

The decomposition of wood when heated in an inert environment or in the presence of air oxygen produces combustible volatile products and is one of the main stages of the combustion process of the material.

It should be expected that variations in the chemical composition and structural characteristics of wood depending on its species and variety will affect the direction of chemical reactions, macrokinetic features of wood decomposition and charring, respectively, on the quantitative yield (thickness), structure, and properties of the coke formed. Under equal conditions of thermal effects, it should affect the burn rate, flammability, and flame propagation over the surface of wood, formation of smoke, and toxic products of combustion, ultimately—on the main indicators of fire danger of wood materials. It is no coincidence that interest has recently increased in the comparative study of pyrolysis and thermal oxidative decomposition of wood of different species and its individual chemical constituents in a wide temperature range. The rate and degree of thermal decomposition of wood determine not only the rate of wood burning and charring but also the integrity and mechanical strength of wooden structures in fire conditions.

The detailed mechanism and kinetics of pyrolytic reactions and charring of wood have not yet been finally established. Early works in this field refer to the study of materials based almost exclusively on cellulose or its derivatives [10].

The main components of wood differ in their thermal stability and kinetic parameters of decomposition. Hemicelluloses decompose actively in the temperature range of 225–325 °C and cellulose—when heated in the range of 325–375 °C. Weight loss in lignin occurs gradually and is especially noticeable in the range of 250–500 °C [12].

Thermal analysis of wood samples in a nitrogen atmosphere was carried out with subsequent change of the medium to air at 500 or 750 °C. Air induction leads to almost complete loss of mass at a high rate.

Figure 2.2 shows as an example the effect of the environment on the degree of transformation and decomposition rate of spruce vansham wood (Vietnam) when

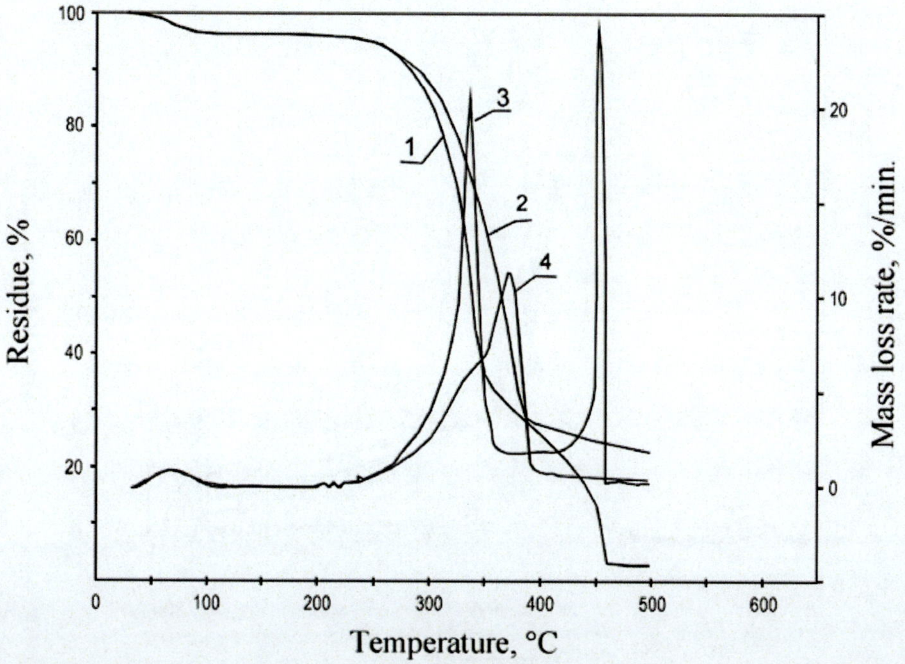

Fig. 2.2 TG and DTG decomposition curves of spruce wood vansham when heated in air (1, 3) and in an inert environment (2, 4) (heating rate 10 °C/min) [13]

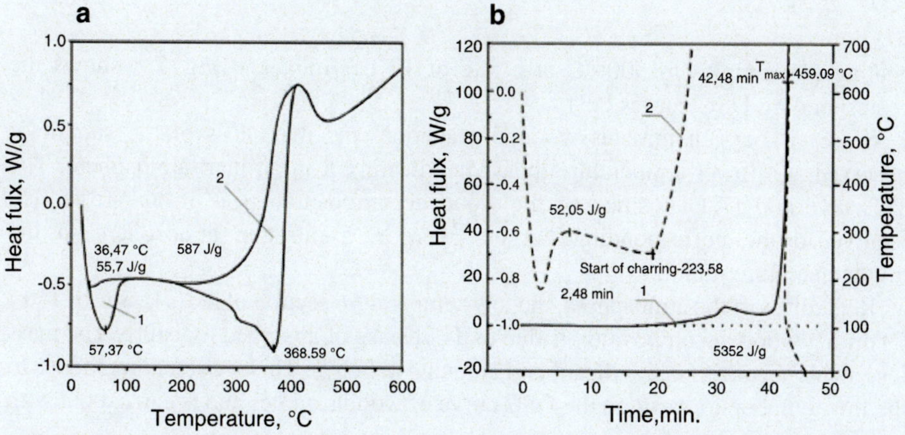

Fig. 2.3 DSC curves of decomposition of spruce vansham when heated at a rate of 10 deg./min in an inert atmosphere (**a**) and in air (**b**); (**a**) 1—endopics of moisture release; 2—endopics of pyrolysis; (**b**) 1—in scale 1:1; 2—in scale 10:1)

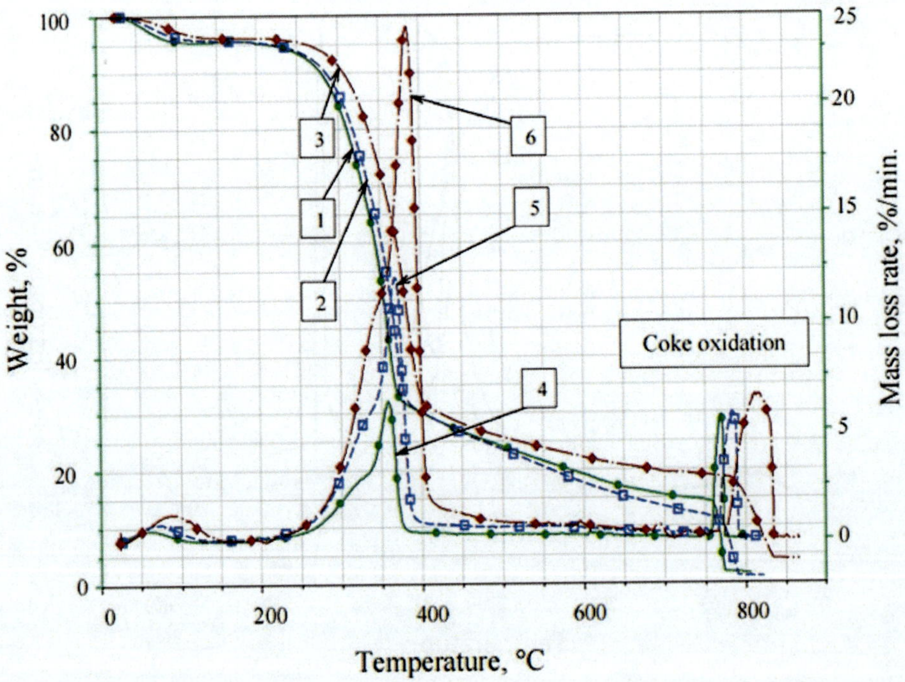

Fig. 2.4 TG (1,2,3) and DTG (4,5,6) curves of pine wood samples (Russia) with a change of atmosphere, nitrogen to 750 °C then air: 1,4—heating rate 5 °C/min; 2,5—heating rate 10 °C/min; 3,6—heating rate 20 °C/min

heating the samples to 500 °C at a rate of 10 deg/min, and Fig. 2.3 shows the corresponding DSC curves [13].

Since air oxygen initiates wood degradation, the mass loss of the sample is observed at a lower temperature than when heating it in an inert atmosphere. The TG curve and DTG position of the wood decomposition peak in air are shifted relative to the corresponding curves taken in a nitrogen atmosphere to the low-temperature region.

Regardless of the atmosphere, the low-temperature section of mass loss (30–150 °C) during wood decomposition is due to the release of absorbed moisture. The mass loss at this section is insignificant and amounts to 3–5%. This section corresponds to the low-temperature peak on the DSC curve of wood samples and is carried out with heat absorption (Fig. 2.3).

Gasification of wood in an inert medium is an endothermic process, unlike decomposition in air. As the rate of heating of the samples increases, there is a natural shift of TG and DTG curves towards high temperature. The results of the TG and DTG multiplot thermal decomposition curves for pine (Fig. 2.4), spruce (Fig. 2.5), birch (Fig. 2.6), and oak (Fig. 2.7) samples as well as the corresponding DSC curves (Figs. 2.8, 2.9 and 2.10) are presented in Tables 2.3, 2.4, 2.5 and 2.6 [12, 14].

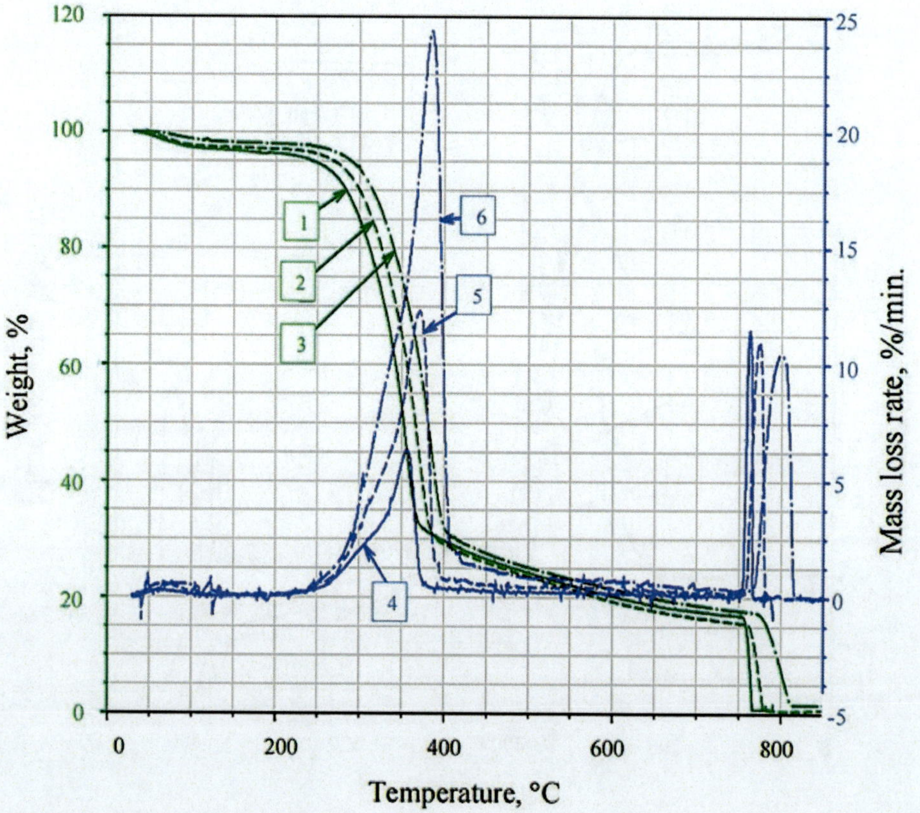

Fig. 2.5 TG (1,2,3) and DTG (4,5,6) curves of spruce wood samples (Russia): 1,4—heating rate 5 °C/min; 2,5—heating rate 10 °C/min; 3,6—heating rate 20 °C/min (atmosphere—nitrogen to 750 °C further air)

Table 2.7 shows the results of complex thermal analysis of tropical wood at a heating rate of 10 °C/min.

As can be seen from Table 2.7, higher extrapolated values of the decomposition onset temperature are characteristic of coniferous wood. However, if we take the temperature corresponding to the mass loss rate of 0.1%/min according to the DTG curve as the beginning of volatile product extraction, it turns out that it is almost the same for all wood species (210–211 °C) and does not depend on the type of environment (inert or oxidizing). At the same time, the initial release of combustible gases is recorded by the chromatograph, apparently as a result of evaporation of the low-boiling part of the extractives.

In general, the thermograms (TG and DTG curves) of different wood species obtained in a complex analysis with a change from an inert atmosphere at 500 °C to an oxidizing atmosphere have the same appearance, but differ in the values of the main thermal indicators:

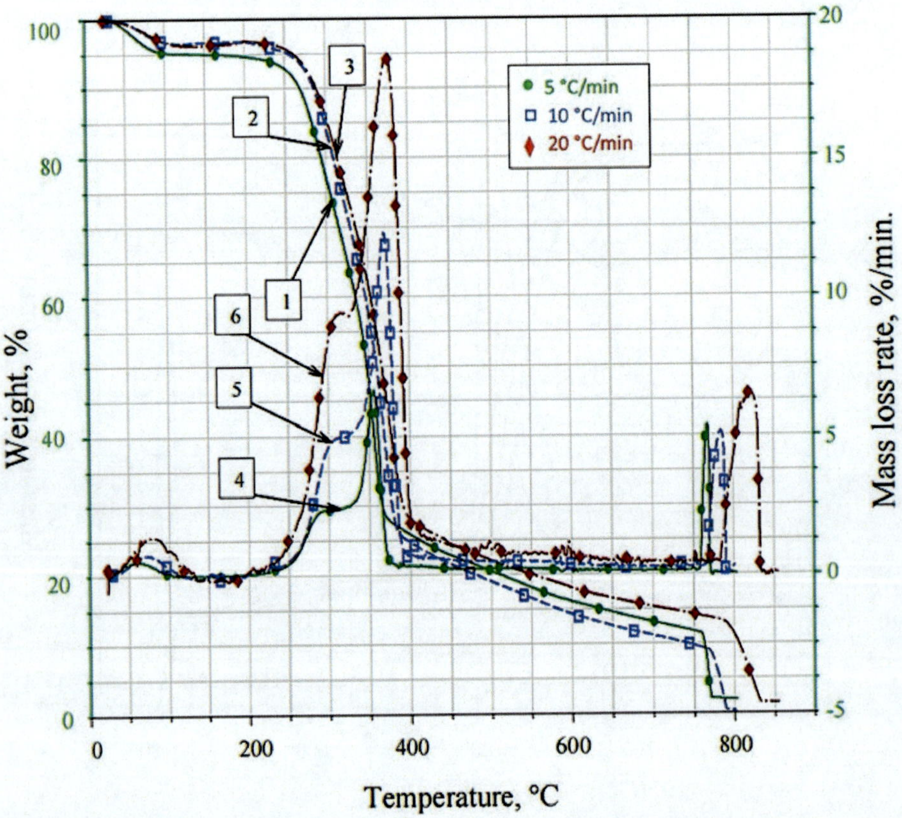

Fig. 2.6 TG (1, 2, 3) and DTG (4, 5, 6) curves of birch wood samples with a change of atmosphere, nitrogen to 750 °C then air: 1,4—heating rate 5 °C/min; 2,5—heating rate 10 °C/min; 3,6—heating rate 20 °C/min

- % mass loss due to moisture evaporation in the temperature range from 30 to 150 °C, WH_2O
- Maximum mass loss rate, V_{max}, %/min;
- Extrapolated temperature of the beginning of decomposition, $T_{beg.}$, °C
- Temperature corresponding to the maximum mass loss rate, T_{max}, °C
- Maximum rate of coke oxidation, V_{max}, %/min
- Yield of coke residue at 500 °C, CR (coking residue), %
- Maximum coke oxidation rate, DOC (degree of coke oxidation), %/min;
- Ash residue at 600 °C, AR (ash residue), %

Indeed, when heated in an inert environment, up to 400 °C wood decomposition proceeds with the absorption of heat. The pyrolysis heat of wood between 200 and 400 °C is 584 J/g. At the same time, the heat taken to release volatile products in the range of 200–275 °C is only 58.7 J/g. It can be assumed that at least the initial mass loss of dry wood (~2%) should be attributed to evaporation of part of the extractable substances.

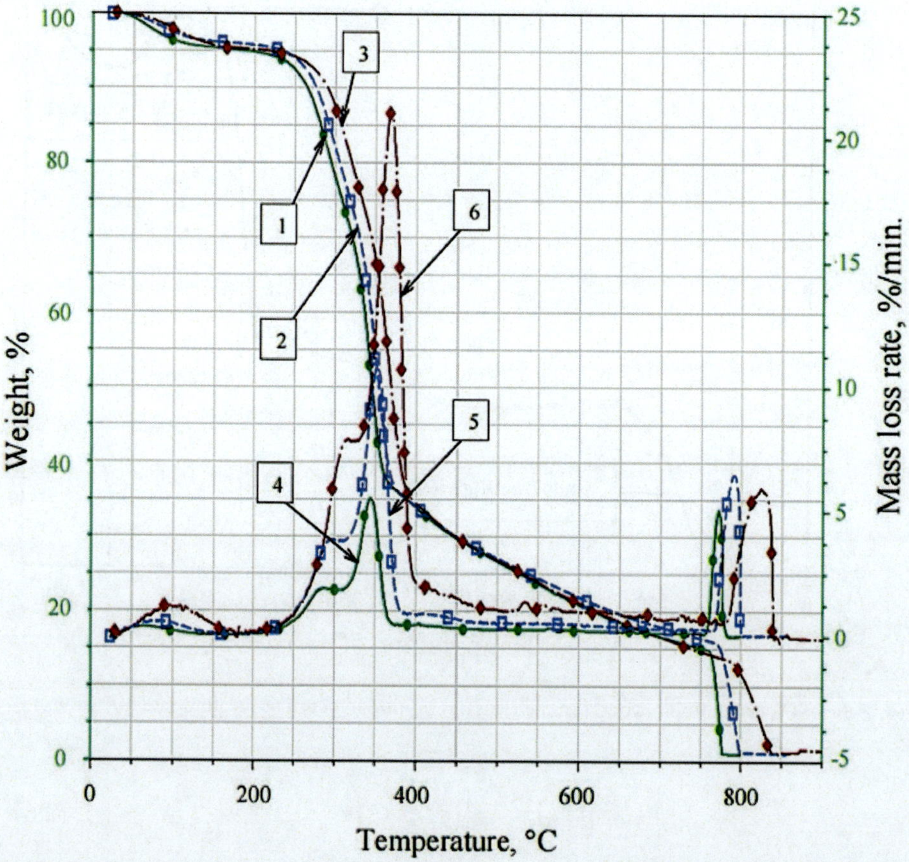

Fig. 2.7 TG (1, 2, 3) and DTG (4, 5, 6) curves of oak wood samples with a change of atmosphere, nitrogen to 750 °C then air: 1,4—heating rate 5 °C/min; 2,5—heating rate 10 °C/min; 3,6—heating rate 20 °C/min

The effect on the maximum weight loss rate and the temperature corresponding to it on the ratio between the content of low-molecular-weight oligomeric components (extractives and hemicellulose) and the high-molecular-weight part of the composite (cellulose and lignin) in coniferous and deciduous wood is noticeable. The higher the extractives content, the greater the maximum rate of wood weight loss in this low-temperature zone and the greater the decrease in the initial wood decomposition temperature. The position of the maximum of the first sometimes "hidden" peak (in the form of the shoulder on the DTG curve) also changes towards a lower temperature. The mentioned fact confirms the concept that substances extracted from wood are the least thermostable compared to other main components of wood [13]. The low-temperature peak is particularly evident on DTG pyrolysis curves of birch and oak wood samples (Figs. 2.6 and 2.7). Calculation of kinetic parameters of wood decomposition and its components based on the data of dynamic TG analysis is based on the application of the equation:

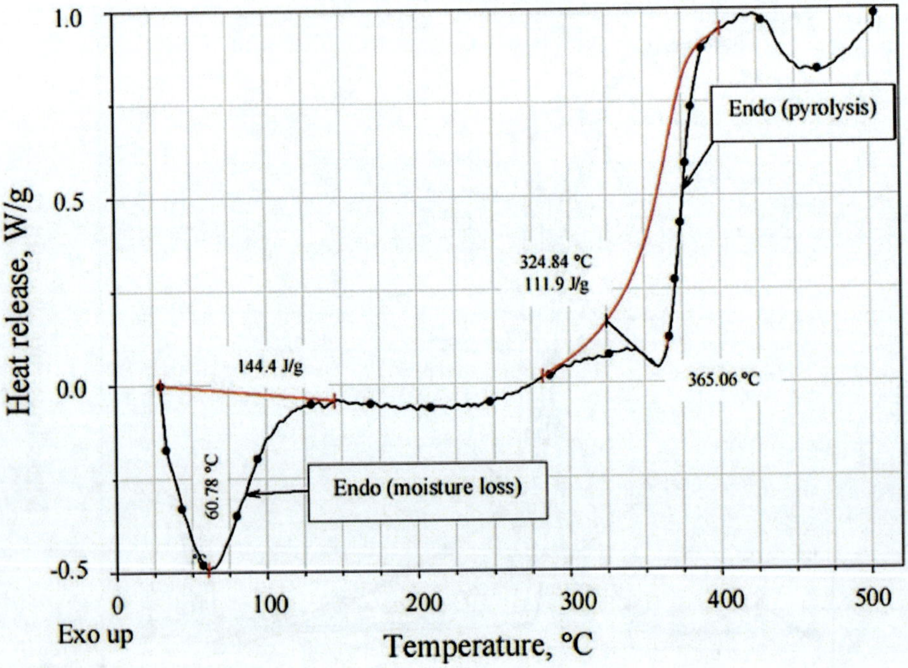

Fig. 2.8 DSC curve of pine wood sample (Russia) in nitrogen current (heating rate 10 °C/min)

$$d\alpha/dT = (k/\beta)f(\alpha), \tag{2.6}$$

where α is the fraction of reacted substance (degree of transformation, $\alpha = (m_0 - m_t)/(m_0 - m_k)$); β—heating rate; k—decomposition rate constant: $k = A \exp(-E/RT)$; A and E—pre-exponential multiplier and activation energy; and $f(\alpha)$—function that reflects individual law of decomposition rate depending on degree of substance transformation and reaction order. This function characterizes the physical mechanism of the process that controls the rate of pyrolytic reactions of the analyzed substance [13].

Table 2.8 shows the results of determining the pyrolysis kinetic parameters of wood samples (Russia) depending on the degree of transformation: for the low-temperature region, in which the extractives and hemicelluloses can be destructed and the main part of wood components—cellulose and lignin—decompose. The order of wood pyrolysis reaction was 1.

The change in activation energy of pyrolysis of wood samples (Russia) in the whole range of degrees of transformation is shown in Figs. 2.11 and 2.12 [14].

For comparison, Fig. 2.12 shows E_{eff} values during thermooxidative decomposition. A well-known approach [15] was used to establish the mechanism of pyrolysis and thermooxidative decomposition of wood.

Dividing the variables in the above kinetic equation by α and T leads to equation

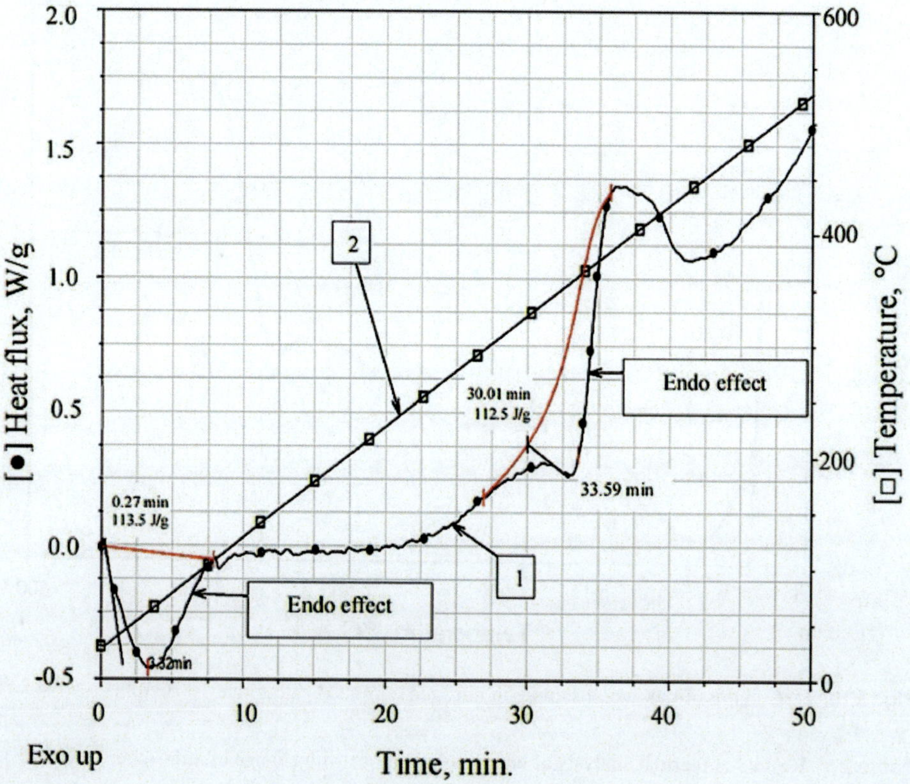

Fig. 2.9 DSC (1) curve of birch wood sample in nitrogen current; 2—temperature (heating rate 10 °C/min)

$$d\alpha/f(\alpha) = (A/\beta)\ \exp(-E/RT)\ dT. \tag{2.7}$$

The integral of the left-hand side of the equation is denoted as the function $g(\alpha)$. This integral has an exact solution for the currently established 18 kinetic functional dependences $f(\alpha)$. All of them may be subdivided into four groups according to the mechanism controlling the kinetics of heterogeneous process of substance thermal transformation [15, 16]: (1) diffusion (D), (2) reactions on the phase interface (B), (3) formation and growth of active centers nuclei (R), and (4) chemical reactions with power law of transformation degree change (P).

Integration of the right part of Eq. (2.7) presents significant difficulties. Various approximate solutions have been suggested [15].

In the present work, when calculating kinetic parameters of wood decomposition, Gorbachev's approximation [16] was used, in particular, equation:

$$g(\alpha) = \{\mathrm{ART}^2/\beta(E + 2RT)\}\ \exp(-E/RT). \tag{2.8}$$

For a realistic description of the dynamics of wood decomposition at different heating rates, it is important to identify the mechanism of the process, i.e., the

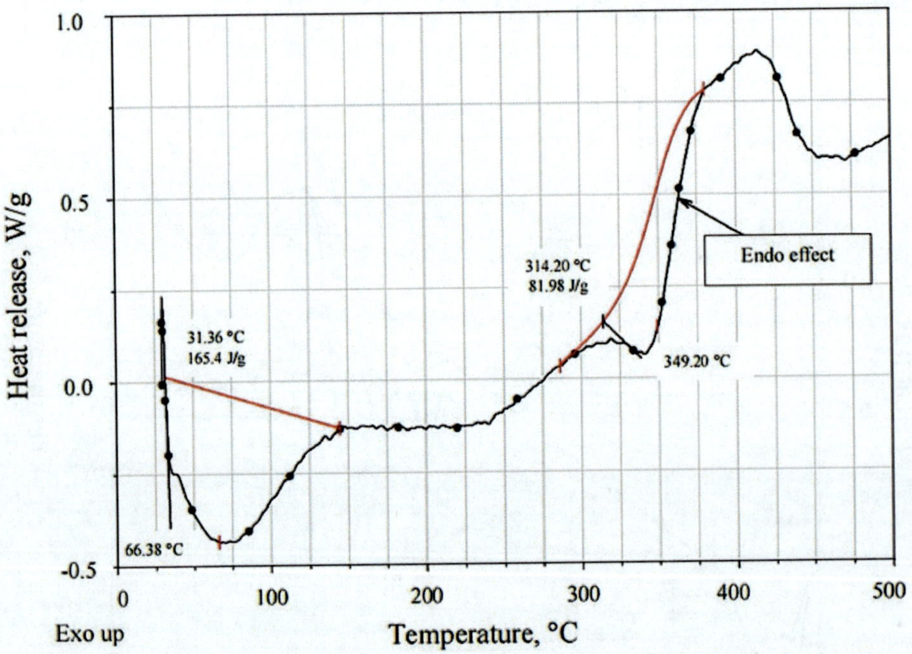

Fig. 2.10 DSC curve of oak wood sample in nitrogen current (heating rate 10 °C/min)

Table 2.3 Results of thermal analysis of pine wood (Russia) with change of atmosphere, nitrogen to 750 °C, and then air

Characteristic of TA curve	Heating rate, °C/min		
	5	10	20
Temperature interval of destruction, °C	30–158	30–160	30–169
Mass loss in the area, %	4.5	4.3	3.8
Temperature of maximum, °C	58	66	87
Peak amplitude, %/min	0.49	0.85	1.23
Temperature interval of destruction, °C	158–490	160–520	170–550
Mass loss at the site, %	70.9	73.3	72.3
Temperature of maximum, °C, T_{max1}/T_{max}	325/358	340/368	384
Amplitude of maximum, %/min, V_{max1}/V_{max}	3.1/6.36	6.8/12.0	23.5
Oxidation rate (%/min)/time (min)	5.6	5.8	6.6
Pyrolysis heat by DSC (ΔH, kJ/g)	–	111.9 (pyrolysis)	–

function $g(\alpha)$, and to determine the reaction order. To this end, we applied the method of DTG curves processing based on the analysis of reduced decomposition rate values depending on α. It was shown [17] that the reduced rate referred to the half-turn time of the substance $\alpha = 0.5$:

$$(T/T_{0.5})^2\ (d\alpha/dt)/(d\alpha/dt)_{0.5} \tag{2.9}$$

does not depend on the kinetic parameters and heating rate, but depends only on the mechanism of the decomposition reaction. Since at $\alpha = 0.5$ the value $(T/T_{0.5})^2$ is

Table 2.4 Results of thermal analysis of spruce wood (Russia) with a change of atmosphere, nitrogen to 750 °C, and then air

Characteristic of TA curve	Heating rate, °C/min		
	5	10	20
Temperature interval of destruction, °C	30–140	30–146	30–165
Mass loss in the area, %	3.1	2.8	1.8
Temperature of maximum, °C	62	66	87
Peak amplitude, %/min	0.29	0.49	0.57
Temperature range of destruction, °C	140–500	146–500	165–500
Mass loss on the plot, %	72.7	73.8	73.0
Temperature of maximum, °C, T_{max1}/T_{max}	315/356	321/371	–/385
Amplitude of maximum, %/min, V_{max1}/V_{max}	2.51/6.43	4.08/12.39	–/24.9
Velocity (%/min)/time (min) coke oxidation	11	11.3	10.6

Table 2.5 Results of thermal analysis of birch wood with change of atmosphere, nitrogen to 750 °C, and then air

Characteristic of TA curve	Heating rate, °C/min		
	5	10	20
Temperature interval of destruction, °C	30–140	30–155	30–160
Mass loss in the area, %	4.97	3.4	3.6
Temperature of maximum, °C	62	74	81
Peak amplitude, %/min	0.53	0.74	1.35
Temperature range of destruction, °C	140–490	155–520	160–550
Mass loss on the plot, %	74.0	78.6	76.3
Temperature of maximum, °C, T_{max1}/T_{max}	295/355	310/368	321/376
Amplitude of maximum, %/min, V_{max1}/V_{max}	2.4/6.6	5.0/12.3	8.23/18.6
Rate (%/min)/time (min) of coke oxidation	5.23	5.13	6.54
Pyrolysis heat by DSC (ΔH, kJ/g)	–	112.5	–

Table 2.6 Results of thermal analysis of oak wood with change of atmosphere, nitrogen to 750 °C, and then air

Characteristic of TA curve	Heating rate, °C/min		
	5	10	20
Temperature interval of destruction, °C	30–159	30–167	30–184
Mass loss in the area, %	4.55	3.9	4.71
Temperature of maximum, °C	66	85	102
Peak amplitude, %/min	0.36	0.56	1.15
Temperature range of destruction, °C	159–503	167–525	184–550
Mass loss on the plot, %	68.8	70.3	71.3
Temperature of maximum, °C, T_{max1}/T_{max}	298/343	301/353	320/367
Amplitude of maximum, %/min, V_{max1}/V_{max}	1.9/5.54	3.85/10.7	8.2/21.0
Oxidation rate (%/min)/time (min)	5.1	6.44	5.93
Pyrolysis heat by DSC (ΔH, J/g)	–	81.98	–

Table 2.7 Results of complex thermal analysis of tropical wood

Index	Wood species				
	Spruce vansham	Pine thongkaribe	Eucalyptus bachdan	Acacia keo lai	Acacia Keo tai tuon
Interval 30–150 °C, WH_2O, %	3.78	4.09	3.85	4.82	5.17
V_{max}, %/мин	0.81	0.79	0.81	1.02	0.99
$T_{\text{beg.}}$, °C	79	75	73	69	66
Interval 150–600 °C, $T_{\text{нр}}$, °C	275	277	246	257	248
$T_{\text{макс1}}/T_{\text{макс}}$, °C	333/375	322/369	294/362	297/371	300/359
$V_{\text{max1}}/V_{\text{max}}$, %/мин	4.8/11.23	4.0/10.5	3.0/9.59	3.27/12.4	3.05/9.67
CR, %	22.4	21.8	27.5	20.2	22.3
DOC, %/мин	11.8	11.1	11.1	11.7	11.04
AR, %	1.04	0.55	2.9	0.7	0.76

Table 2.8 Kinetic parameters of pyrolysis of wood samples (Russia)

Index	Wood species			
	Pine	Spruce	Birch	Oak
α = 20–30, %				
$E_{\text{eff.}}$, kJ/mol	159.4/163.5	155.6/162.0	182.9/221.6	218.8/262.3
$\lg k_0(k_0$, min$^{-1})$	13.09/13.29	12.83/13.25	15.48/18.62	19.25/22.82
α = 60–70 %				
$E_{\text{eff.}}$, kJ/mol	165.2/162.5	166.8/168.0	194.7/186.7	271.5/256.4
$\lg k_0(k_0$, min$^{-1})$	13.13/12.90	13.39/13.46	15.67/14.86	22.65/21.25

close to 1, the reduced rate can be determined from the DTG curve by the ratio of distances of the corresponding points from the baseline.

In [16], tabulated values of reduced velocities at $\alpha = 0.75$ and α as well as values of α corresponding to the maximum velocity are given, which have a diagnostic character for revealing the mechanism of the decomposition process of substances.

The above approach to the analysis of TG and the main peak of DTG curves in the thermal decomposition of tropical wood in the interval 200–500 °C allowed us to reveal the type of function $g(\alpha)$. The order of the reaction, n, was estimated from the shape factor of the DTG peak, S, [15]: $n = 1.26\ S^{1/2}$. The results obtained are summarized in Table 2.9.

With the exception of vansham coniferous wood, the formal mechanism of thermal decomposition of other wood samples in the temperature interval under study corresponds to the mechanism controlled by formation of active centers by the law of chance and their growth. Tabulated values of α_{max} and reduced rate at conversion degrees of 0.75 and 0.8 are practically close in this case to the theoretical values of 0.63, 1.00, and 0.93 according to [16]. The mechanism of thermal decomposition of vansham wood when comparing the experimental data with

A

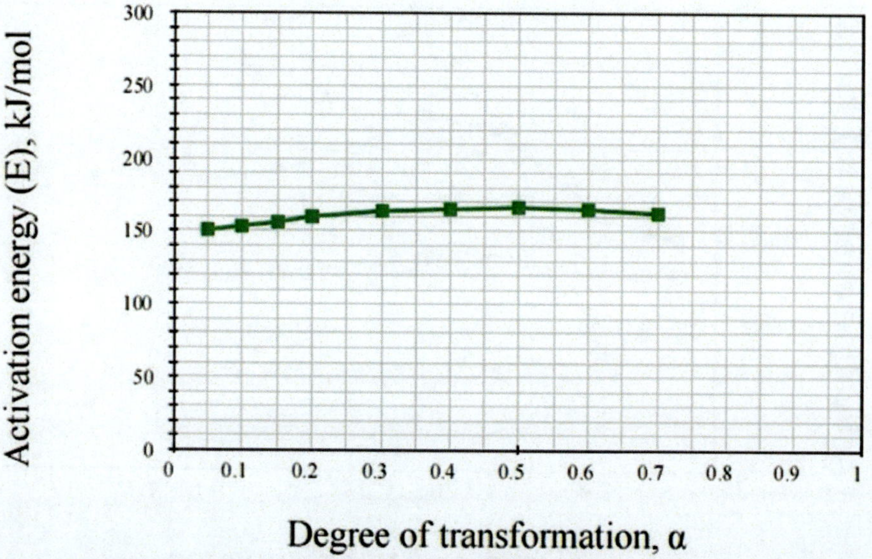

B

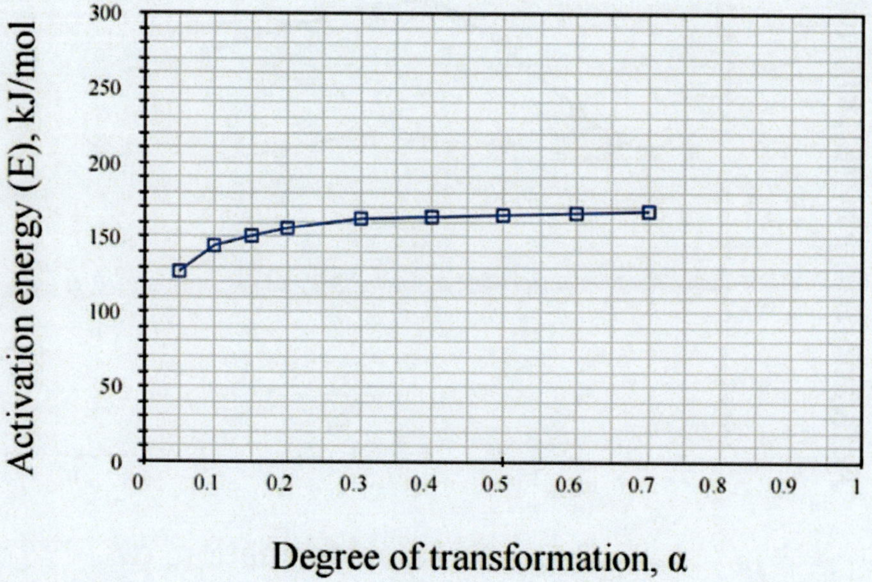

Fig. 2.11 Influence of degree of transformation during pyrolysis of pine (**a**) and spruce (**b**) wood samples on the change in activation energy of the process

tabulated characteristics should be attributed to the mechanism controlled by reactions at the interface. As can be seen from the table, the order of the decomposition reaction for all samples is practically equal to 1.

A)

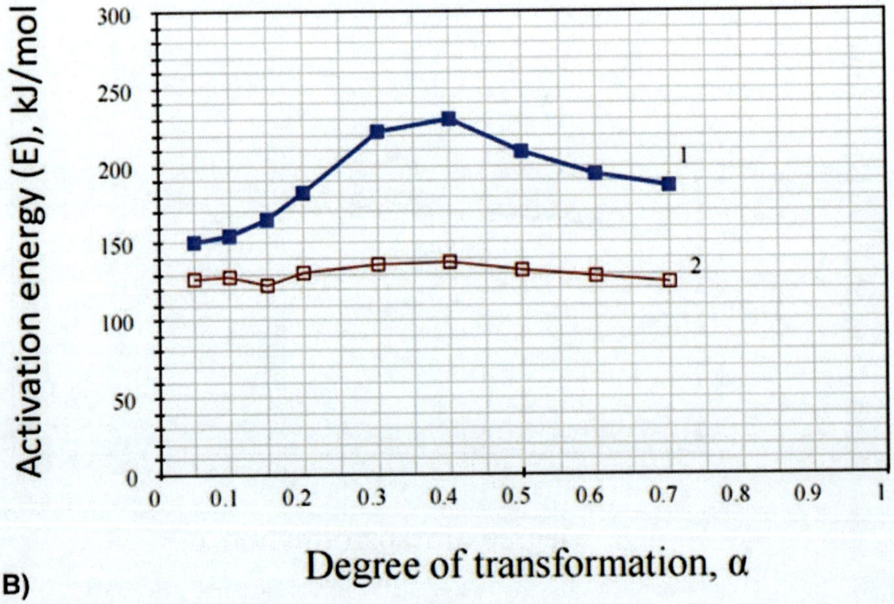

B)

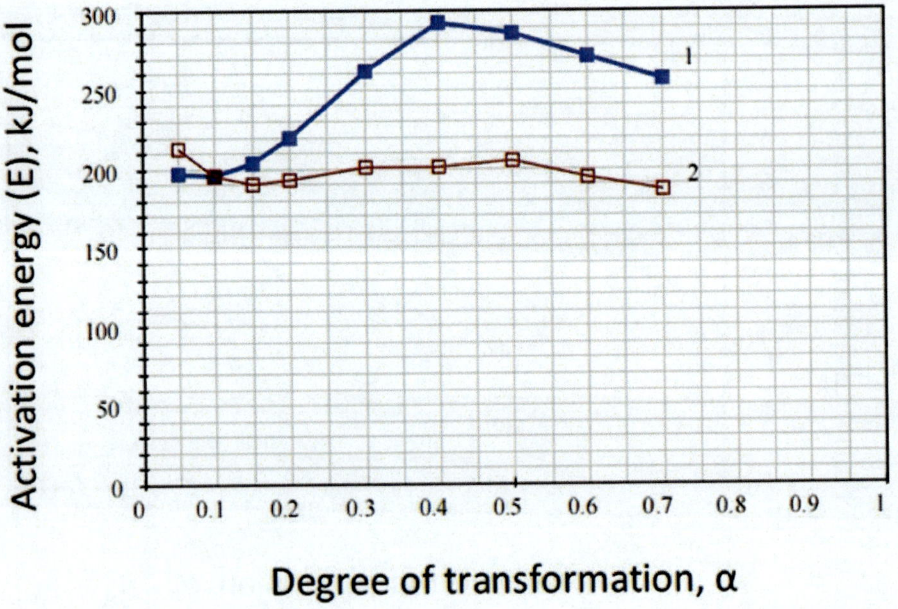

Fig. 2.12 Influence of degree of transformation during pyrolysis (1) and thermal-oxidative decomposition (2) of birch wood samples (**a**) and oak wood samples (**b**) on the change in activation energy of the process

Table 2.9 Experimental values of reduced rate and α_{max} in thermal decomposition of tropical wood

Index	Pine thongkaribe	Spruce vansham	Eucalyptus bachdan	Acacia Keo tai tuon	Acacia keo lai
$\alpha_{макс}$	0.626	0.754	0.633	0.624	0.64
$V_{0.75}/V_{0.5}$	1.007	1.209	1.015	1.002	1.005
$V_{0.8}/V_{0.5}$	0.932	1.13	0.933	0.939	–
n	0.940	0.928	1.01	1.108	0.963

Table 2.10 Kinetic parameters and mechanism of thermal decomposition of tropical wood

Wood species	Interval α	Mechanism, type $g(\alpha)$	E, J/mol	lgA, (min^{-1})	n
Thongkaribe pine	0.35–0.80	R $[-\ln(1-\alpha)]^n$	119510	9.380	1
Wansham spruce	0.35–0.80	B $1-(1-\alpha)^n$	70513	4.936	1
Eucalyptus bachdan	0.38–0.80	R $[-\ln(1-\alpha)]^n$	105170	8.272	1
Acacia Keo tai tuon	0.34–0.80	R $[-\ln(1-\alpha)]^n$	109570	8.662	1
Acacia keo lai	<0.47 <0.87	R $[-\ln(1-\alpha)]^n$	70200 159300	5.058 12.774	1 1

Gorbachev equation in the logarithmic form of the graphical relationship $\lg[g(\alpha)/T^2]$ from $1/T$ makes it possible to determine the kinetic parameters of wood decomposition (Table 2.10).

If the first order of the reaction is the same, the change of R type of the decomposition mechanism in conifers to B type is accompanied by the expected decrease in the activation energy by almost 1.5 times. The reason for the different formal mechanism of high-temperature thermal decomposition in two conifers remains unclear. It may be related to the changes in the morphological structure of wood after the first stage of decomposition, which refers to the hidden peak on the DTG curves.

By the example of keo lai wood, we can see that the activation energy of the process increases as the degree of decomposition increases.

In order to quickly diagnose the formal pyrolysis mechanism of wood samples, the experimentally found value of the parameter α_{max} at the reaction order $n = 1$, compared to the tabulated values [16], was used.

Thus, it was found that at the first low-temperature stage, the pyrolysis process of wood samples is carried out by a diffusion-controlled mechanism. At the main stage, mass loss is controlled by the mechanism of formation of nuclei (nucleation)—active destruction centers by the law of chance.

Table 2.11 Kinetic parameters of thermooxidative decomposition of tropical wood

Wood species	$E_{eff.}$, kJ/mol	lgA, (min^{-1})	n
Thongkaribe pine	164.0	13.97	1
Wansham spruce	159.6	13.60	1
Eucalyptus bachdan	164.7	14.48	1
Acacia Keo tai tuon	166.8	14.73	1
Acacia keo lai	163.7	14.24	1

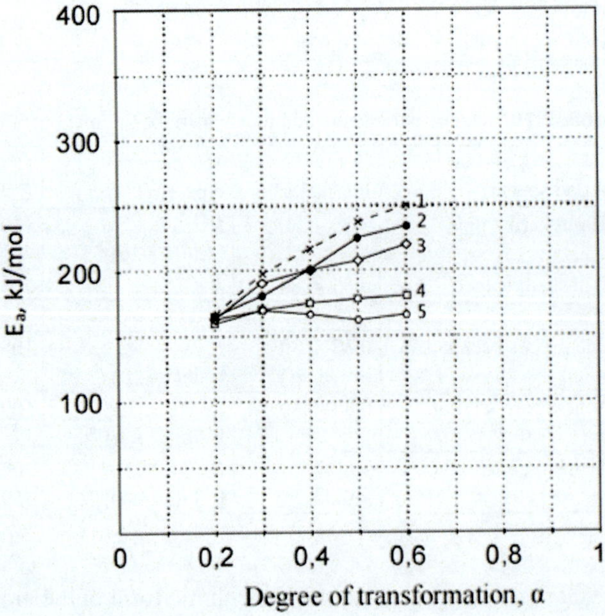

Fig. 2.13 Dependence of effective activation energy of thermooxidative wood decomposition on the degree of decomposition: 1—Eucalyptus bachdan; 2—acacia Keo tai tuon: 3—acacia keo lai; 4—spruce van sham; 5—thongkaribe pine

Oxidative decomposition of wood proceeds in two stages. The first stage is associated with the degradation of wood components, the second stage with the oxidation of the resulting coke. Both stages of decomposition are exothermic, with release of significant amount of heat (Fig. 2.3b). When heated to 460 °C, almost complete loss of mass of the spruce vancham sample with heat release in the amount of 5352 J/g is observed [13]. Kinetic parameters of thermooxidative decomposition of wood obtained by heating samples at a rate of 5–20 g/min at a decomposition rate of 20% are quite similar for different species (Table 2.11).

The effective activation energy is 159.6–166.8 kJ/mol, and the logarithmic values of the pre-exponential multipliers, lgA, vary between 13.60–14.73 (min^{-1}). As the degree of decomposition increased up to 40–60%, the values of E_{eff} increased, especially noticeably during thermooxidative decomposition of hardwood (Fig. 2.13) [13].

Apparently, the values of E_{eff} at large degrees of wood decomposition reflect the individual participation of wood components in this process and the contribution of energy oxidation of coke residue. It can be concluded that the carbon residues formed during decomposition of coniferous wood are more prone to oxidation than the coke of hardwood [13].

TG and DTG analysis of wood of different species in an air atmosphere at a heating rate of 5–20 deg/min leads to the conclusion that the oxidative decomposition of the solid phase corresponds to the mechanism of diffusion-controlled process.

2.3 Fire Hazard of Wood

One of the important factors determining the fire hazard of wood is its ability to initiate and develop a combustion process when heated in air. The onset of combustion is only the initial stage of the process, which can take place in both flaming and smoldering modes, which differ fundamentally in terms of the mechanism and kinetics of chemical reactions and the importance of certain physical factors [10].

Figure 2.14 shows a schematic diagram of physical and chemical transformations during flame combustion of wood [10].

The occurrence of smoldering or flaming combustion is a critical condition for the transition from a slow decomposition reaction of wood material to a self-accelerating exothermic reaction capable of sustaining itself and spreading in space due to feedback [10, 18].

Numerous studies of the conditions under which smoldering combustion occurs and of the factors that influence the rate of spread of the smoldering zone, the transition of smoldering to a flaming process, or the attenuation of smoldering have mainly been conducted on materials with high open porosity and gas permeability. These are materials like polymer foams, powdered materials including coal

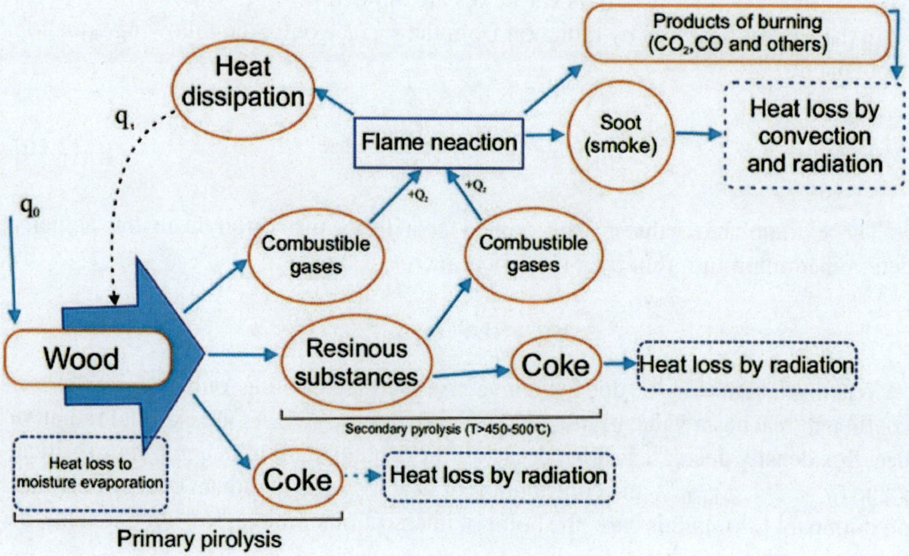

Fig. 2.14 Scheme of the chemical and physical processes in flame burning of wood [10]

as well as sawdust and wood shavings, cellulose fabrics, and heat insulation materials [19–23].

A characteristic feature of such materials is their rather high ratio of total surface area to occupied volume. Studies of wood smoldering itself are very limited.

The process of flame wood combustion arises from self-ignition and ignition of the wood material. Self-ignition is a typical process that takes place in the gas phase. It can occur when a concentration of combustible volatile wood decomposition products in the gas phase near the wood surface exceeds the lower flammable gas ignition concentration limit. The temperature of the combustible gases produced must be high enough to accelerate the oxidation reaction and cause combustion [24–26].

The critical heat flux, qe.cr, below which flame combustion cannot occur, is one of the important parameters for (self-) ignition of materials. Early studies on the self-ignition of different wood species provide data showing a large scatter in the values of $q_{e.cr}''$ (from 4.3 to 117 kW/m^2). To a large extent, this variation is explained by different exposure times to external heat fluxes of different intensities [27].

The initiation of flame combustion of wood when it is heated by an external radiation heat flux in the presence of an additional local energy pulse is different from the self-ignition conditions. Once wood decomposes, combustible volatile degradation products are released into the environment, mixing with air to form a boundary layer of combustible mixture at the wood surface. In the presence of an additional pilot energy source, it is actually sufficient to reach the appropriate flammability limit of the resulting combustible gases [10].

Theoretical analysis of the effect on the ignition delay time of materials of heat losses due to radiation from the surface alone or to convective cooling alone has led to the conclusion that this factor affects the numerical value of q_{cr}'', determined by the intersection of the line of experimental data ($t_{ign}{}^{-1/2} - q_e$) with the abscissa axis when sufficiently high external heat fluxes are applied [28].

In the case of heat loss by radiation from the surface only, the following equation is obtained:

$$t_{ign}^{-1/2} = \frac{2}{\sqrt{\pi k\rho c}} \frac{q_e'' - q_{cr.\,min}''}{T_p - T_0}. \quad (2.10)$$

The extrapolated value of the critical heat flux must differ from the actually determined minimum flux by a numerical factor of 0.64:

$$q_{cr}'' = 0.64\, q_{cr.\,min}''. \quad (2.11)$$

When all heat loss is due to convective surface cooling only, the numerical coefficient reaches a value of π/4 and $q_{cr}'' = 0.785 \cdot q_{cr.min}''$. As the external radiative heat flux density decreases and approaches the actually determined minimum critical value ($q_e < 1.1 \cdot q_{cr.min}''$), the contribution of heat loss due to radiant energy becomes predominant [28]. In this case, the point of intersection of the regression line with the abscissa axis in the coordinates $t_{ign}^{-1/2} - q_e''$ must correspond to the value of $q_{cr}'' = q_{cr.min}''$ according to the equation

$$t_{ign}^{-1/2} = \frac{\pi}{\sqrt{\pi k\rho c}} \frac{q_e'' - q_{cr.min}''}{T_p - T_0}. \tag{2.12}$$

Slope of the regression line of the experimental points $f(t_{ign}^{-1/2} - q_e'')$ is different from the slope of the line obtained at high values of q_e''.

Experimental results were compared with theoretical values calculated on the basis of a one-dimensional integral model of ignition applied to carbonized materials [29]. According to the developed model, the value of q_{cr}'' corresponding to the point of intersection with the abscissa axis of the regression line in the coordinates $t_{ign}^{-1/2} - q_e''$ at $q_e'' > 20$ kW/m^2 must differ from the value determined at $q_e'' < 20$ kW/m^2 by the numerical coefficient $\beta = q_{cr.cross}''/q_{cr.min}'' = 0.76$.

Table 2.12 shows some experimental and theoretically predicted flammability values for wood [29].

The time to ignition of wood at an external heat flux approaching the critical flux, as shown in Table 2.12, is quite considerable. However, in the case of parallel fiber orientation with respect to the direction of heat flow, the ignition delay time is considerably reduced compared to that for perpendicular orientation. This factor affects not only the value of q_{cr}'' but also the surface temperature of wood samples during ignition. The observed effects are due to anisotropy of thermophysical properties of wood, in particular, the values of thermal inertia, $k\rho c$, in the direction parallel or perpendicular to the axis of the wood fibers.

In the analysis of flame propagation over the surface of solid materials, the concept of a continuously repeating ignition process is used, in which the flame itself serves both as a source of heating of the solid material and as a local source of ignition of the resulting combustible decomposition products. The main feature of the spreading combustion process of wood (as well as other combustible material) is the ability to sustain itself through the feedback of heat flow from the reaction zone to the adjacent layer of the original substrate, even in the absence of an initial external heating source. A great number of experimental and theoretical studies on the spreading of the combustion process have been devoted to these aspects, starting from the first works of Semenov N.N., Zeldovich J.B., Frank-Kamenetsky D.A., and the followers of their scientific school.

The main quantitative characteristic of flame or flameless combustion of various materials is the rate of development of the process. It is estimated as the distance covered by the flame (smoldering) front per unit time. It should be emphasized that materials that form non-volatile coke during thermal decomposition differ from non-carbonizing materials in their ability to spread flame across the surface. In addition to wood, the former includes a considerable number of natural and synthetic polymeric materials [30].

Ignition in carbonizing materials is sometimes followed by a rapid run of flame across the surface, followed by instant charring of the near-surface layer and attenuation [31]. By this time, a small part (~10%) of the material mass is consumed,

Table 2.12 Measured and predicted pilot radiation ignition rates for softwood and hardwood

Wood species	Orientation fibers	$\rho_{med.}$, kg/m^3	$W_{med.}$, %	t_{ign}, hour: min: s	$q_{cr.min}''$, kW/m^2 (measuring/theoretical)	q_{cr}'', kW/m^2 (extrapolated)	T_{ign}, °C (theoretical)
Sequoia	⍰	354	8.6	0:36:10	13/15.5	11.7	375
	⍰	328	7.4	0:23:36	9/5.9	4.5	204
Douglas Fir	⍰	502	7.4	1:33:00	12/16.0	12.2	384
	⍰	455	8.5	0:39:55	9/8.4	6.4	258
Oak	⍰	753	5.1	–	–/10.8	8.2	304
	⍰	678	5.2	–	–/9.2	7.0	275
Maple	⍰	741	4.8	1:10:00	12/13.9	10.6	354
	⍰	742	4.8	0:44:40	8/3.8	2.9	150

but the formation of coke changes the conditions of heat transfer and mass transfer and increases the resistance of the material to flame propagation.

Experimental studies of flame propagation on the surface of wood-based materials using different test methods have shown that not only the conditions of external heat exposure but also the type of wood affect the characteristics of flame propagation [10].

Internationally, building materials have traditionally been classified for their ability to propagate flames through the ceiling of rooms during direct exposure to fire by means of the large-scale ASTM E84 method, developed as early as the mid-XXth century and often referred to as the tunnel method. The fire hazard rating of building materials using this method is legislated in various American building codes. It is believed that large-scale methods for testing the flammability of materials better reflect their behavior in actual fire situations.

In ASTM E84, a 7.32 m long and 51 cm wide specimen is mounted on the ceiling of a test chamber and ignited at one end with a propane torch flame, covering nearly 1.37 m of the specimen length. Once a steady flame has developed, determine how far and fast the flame spreads across the underside of the material with forced air flow. Forced air flow is created by the action of the chimney ventilation (1.2 m/s). The flame propagation index value for this method is calculated from the position of the flame front as a function of time or the maximum movement of the flame front over the surface of the test piece in 10 min. Red oak with Flame Spread Index = 100 serves as a reference. Building materials are divided into three classes according to the flame spread index: Class I (or A) with an index of 0–25; Class II (or B) 26–75; and Class III (or C) 76–200 and higher.

Table 2.13 shows the ASTM E84 flame propagation index values for untreated wood specimens 19–25 mm thick [32]. Flame propagation index values for asbestos cement board and dry plaster are given for comparison.

Although most wood types are classified as Class III, the most dangerous class in terms of their ability to spread flames across the surface of ceiling structures in a fire, some softwood samples, as shown in the table above, have lower flame spread index values and are classified as Class II. It can be assumed that this is due to the absence or low content of energy-intensive extractives in the chemical composition of samples of these wood types [32].

The flame propagation process is considered stationary due to the adoption of a moving coordinate system. The adopted oxidizer velocity profile makes it possible to exclude the equations of motion from the analysis and simplify the solution of the problem. Formulas connecting the flame propagation velocity over the surface with the thermophysical properties of the material and its thickness, the velocity of the flowing air flow, the corresponding temperatures of the flame, and the decomposition of the condensed fuel are obtained. The flame propagation velocity over thermally thin materials is expressed by equation [33]:

Table 2.13 Influence of coniferous and hardwood species on flame spread index

Wood species	Flame spread index	Flame spread class	Wood species	Flame spread index	Flame spread class
Red cedar	69	II	Birch	80–110	III
Yellow cedar	78	III	Red oak	100	III
Cypress	145–150	III	White oak	100	III
Douglas fir	70–100	II–III	Maple	104	III
Silver spruce	69	II	Lime	130–140	III
White spruce	65	II	Poplar	170–185	III
White pine	75–82	III	Alder	104	III
Pine	98	III	Walnut	130–140	III
Ponderosa pine	105–230	III	Gum wood	140–155	III
Larch	55–74	II	Asbestos cement	0	I
Sequoia	70	II	Gypsum plasterboard	10–15	I

$$V_{\mathrm{FS}} \approx \frac{\sqrt{2}k_{\mathrm{g}}(T_{\mathrm{f}} - T_{\mathrm{V}})}{\rho_{\mathrm{s}}c_{\mathrm{ps}}k_{\mathrm{s}}(T_{\mathrm{V}} - T_{0})}, \tag{2.10}$$

where FS—flame spreading, k_{g}—thermal conductivity of the gas phase and τ—the thickness of the material sample.

Speed of flame propagation through thermally thick materials [16]:

$$V_{\mathrm{FS}} = \frac{V_a \rho_g c_{pg} k_g (T_{\mathrm{f}} - T_{\mathrm{V}})^2}{\rho_s c_{ps} k_s (T_{\mathrm{V}} - T_0)^2}, \tag{2.11}$$

where V_a—velocity of the inflowing air flow and the subscripts g and s refer to the gas and solid phases, respectively.

It follows from these formulas that the flame propagation velocity is thickness-dependent in the case of thermally thin samples and thickness-independent, but directly proportional to the rate of oxidant flow in the case of thermally thick materials. Later, the $\sqrt{2}$ multiplier in the formula for the flame propagation velocity in thermally thin materials was refined. It was shown that it should be equal to a value of $\pi/4$ [34].

One important characteristic of the ability of materials to propagate flame across the surface is the value of the minimum (critical) external heat flux below which flame propagation across the surface is impossible. In a study of the surface flame propagation rate of wood samples of various species from New Zealand using the ASTM E1321 method, known as the Lateral Ignition and Flame Transport (LIFT) method, it was found that the value of $q_{\mathrm{e,min}}$ was the lowest (2.2 kW/m^2) for wood of large-fruited cypress (bulk density 514 kg/m^3). Samples of beech and pine (Radiata Pine) with a thickness of 20 mm had $q_{e,min}$ = 4.4–4.5 kW/m^2. Dacrydium cupressinum with a high bulk density of 660 kg/m^3 had $q_{e,min}$ = 7.6 kW/m^2 and showed significant charring before reaching the limiting distance of flame propagation [35]. Studies of the critical phenomena in diffusion combustion of various polymeric materials lead to the conclusion that the limits of flame propagation are determined primarily by the ability of the material to provide the flame with sufficient combustible material and heat loss from different zones of the burning material [10].

Many fire hazard indicators are directly related or related in one way or another to the rate of heat release from burning substances, materials, structural elements, and other substances (e.g., the rate of flame spread, smoke, and toxic carbon monoxide formation). The development of fires in different types of rooms is determined by the nature of the dynamics of heat release intensity. It is natural to pay more attention to the determination of different heat release characteristics in different countries of the world.

The results obtained for heat release make it possible to calculate the effective heat of combustion of the material under study if the mass loss of the sample under the same conditions is known [10].

The presence of two peaks in the graphical dependence of the rate of heat release on the time of the experiment, often observed in the case of wood samples and other carbonizing polymeric materials, means a transition from the flame combustion mode to the heterogeneous combustion mode of the formed coke layer. Its emergence and increase in thickness after reaching the maximum of the first peak leads to the isolation of the underlying layers of the parent material and a decrease in the rate of decomposition.

The short period of stable heat release between the two peaks, in other words the stable combustion period, seems to be caused by a peculiar equilibrium between the endothermic decomposition of the parent material and the exothermic oxidation process of the coke residue. Therefore, when the charred layer spreads over the entire thickness of the material, the latter process begins to dominate until the complete cessation of heterogeneous combustion as a result of the exhaustion of the combustible mass.

Indeed, in a calorimetric study of the rate of heat release during combustion in air of very thick wood samples (about 60 mm), it was found that within 5 min of the experiment, the second peak in the heat release curves was absent.

A very important characteristic of the heat release rate during material combustion is the time of reaching the maximum heat release rate during the first stage of flame burning or the rate of heat release rate rise. It is an indicator used in the SBI method, introduced in EN-13823, called the FIGRA-Fire Growth Rate.

Laboratory calorimetric methods for evaluating the heat release characteristics of building material combustion, developed on the basis of oxygen consumption or temperature measurement, differ from each other not only by the type of system (open or closed chamber). They differ in terms of the size of the samples to be tested, their orientation, the air flow rate, the method of measuring the oxygen concentration, and the type and location of the local ignition source. These experimental differences affect the results of heat release rate measurements. On the same material samples, comparative tests were therefore carried out on the OSU calorimeter with the conventional temperature measurement and OSU modified for oxygen consumption [36]. With the second method, the recorded heat release rate was about 15–30% higher than with the standard method.

In the same way, the heat release rates obtained from the oxygen consumption of vertical and horizontal samples were compared using three types of calorimeters [37]. In addition to the abovementioned cone and OSU calorimeters, a device developed at the Swedish Forest Products Research Laboratory (STFI) was used in [37].

The STFI calorimeter is characterized by a vertical 150 × 150 mm sample arrangement in an open chamber. Oxygen concentration was measured in the flow of combustion products in the chimney. In a cone calorimeter, 100 × 100 mm samples were tested in a horizontal and vertical configuration.

The values of the average combustion heat release rate of the wood specimens were obtained higher in the horizontal orientation of the specimens compared to the vertical arrangement.

The structure of wood and charred layers developing at a heat flux of 50 kW/m^2 was studied by electron microscopy. In the structure of soft wood species (poplar and spruce), many cracks occur after heat exposure, penetrating across the fibers into the carbonized area. No cracks were detected in the structure of oak and larch, which are considered hardwoods. Coke layers were dense and hard, with enhanced thermomechanical properties. This circumstance, according to the author [38], has the most significant influence on the time of reaching the peak value of the heat release rate in different wood species. In particular, the FIGRA value for poplar was 0.68 kW/m^2·s, for spruce 0.50 kW/m^2·s, and for oak and larch 0.36 and 0.38 kW/m^2·s, respectively. However, the average heat release rate for the entire 5-min test period was twice as high for those wood varieties that formed a dense and hard surface coke layer.

Thus, the structural characteristics of the wood and, presumably, its chemical composition should have a greater influence on the heat release characteristics of combustion than simply whether it is a hardwood or softwood species.

One of the most important studies is to establish the relationship between the chemical composition of wood of different species and species and its smoke-producing capacity and the toxicity of the products of combustion.

Most wood species contain lignans, which are dimers of coniferous and synaptic alcohols. Lignans, like lignans, act as a matrix that supports and strengthens the skeletal structure of the plant made of cellulose macromolecules. Having an aromatic structure, these components largely determine not only the intensity of heat release during wood combustion, but also its smoke-forming capacity. In pine wood smoke gases, lignans are found in significant quantities and unchanged state. Smoke from burning oak wood contains oxidized dimer derivatives of synaptic alcohol [39]. All these compounds are useful biomarkers of the origin of woody plants. The presence of incomplete combustion products of lignin and lignans in flue gases provides additional information about the type of original burning biomass. Coniferous wood lignin flue gases are enriched in phenolic derivatives with the basic structure of coniferous alcohol.

Substances derived from synaptic alcohol are found in trace amounts. Smoke fumes from hardwood lignins are enriched in compounds that retain two methoxyl groups in the phenolic cycles of the synalcohol [39].

The main characteristics of wood combustion smoke are soot particle concentration and particle volume fraction in the aerosol, shape, and particle size distribution.

They determine the optical properties of smoke, the scattering and absorption of light by smoke, and the visibility in smoke. Visibility is determined by the distance, *l*, at which an object is clearly visible to an observer under room lighting conditions. Visibility depends not only on the smoke-producing capacity of the material when burning but also on the illuminance in the room, whether the object will emit or reflect light.

In practice, the smoke-forming capacity of materials is determined by the maximum value of the optical density of smoke per unit specimen area, D_s^{max}, or per unit initial specimen mass, D_m^{max}, or per unit mass loss of the specimen during the test period, $D_{\Delta m}^{max}$. In contrast to the latter two values, D_s^{max} is a dimensionless value used in the ASTM E-662 method. In addition, D_m^{max} is the recommended value. This value and $D_{\Delta m}^{max}$ are used in con-calorimetric tests of materials (ISO 5660-1 and ISO 5659, respectively). D_m^{max} and $D_{\Delta m}^{max}$ have the advantage of showing the correlation between the smoke generation process and the amount of material and its composition. In the new European standard EN 13823, which focuses on the SBI method, the smoke-generating capacity of building materials must be evaluated in addition to the heat release properties of the combustion process. The smoke generation rate and the total smoke emission in the test period—total smoke emission in 600 s—are taken into account. Important is the introduction of a new index called SMOGRA (Smoke index Growth Rate) similar to FIGRA.

Unfortunately, there is no detailed information on the influence of wood species and species on smoke production. However, there is a general trend: the smoke-forming capacity of wood in flame combustion when exposed to an external radiation heat flux is much lower compared to smoke production in decomposition and smoldering mode. According to [40], indices of smoke-forming ability of pine timber under external heat flux of 25 kW/m^2 change in the following way: in flame combustion $D_s = 54–68\ m^2 \cdot m^{-2}$; $D_m = 4–6\ m^2/kg$; and $D_{\Delta m} = 7–10\ m^2/kg$; in smoldering regime, these indices have values 339–477 $m^2 \cdot m^{-2}$; 30–39 m^2/kg; and 137–150 m^2/kg, respectively. In pine wood flame combustion, maximum smoke optical density $D_s^{max} = 20\ m^2 \cdot m^{-2}$ is reached already after 2 min after ignition. In smoldering mode, this process takes up to 20 min.

The combination of heavy smoke and toxicity of the products of combustion in the event of a fire not only poses a major threat to people inside the buildings but also makes it difficult to rescue people and extinguish the fire. Statistics show that fatalities in fires are caused by smoke in escape routes and increased toxicity of combustion products in 50% of cases [41]. Up to 80% of deaths are due to inhalation of toxic smoke [42]. Toxic substances are the main hazards in the flue gases released from burning wood.

Although a large number of toxic compounds have been found in wood-burning flue gases, methods of assessing fire hazards by toxicity of combustion products based on an analytical approach have focused mainly on the presence of CO, CO_2, and loss of oxygen concentration.

Carbon monoxide (CO) was found to be the largest contributor to the toxicity of wood burning products. In [43], it was found that in smoldering combustion mode, the yield of CO was 70–240 times higher than in flame combustion mode for spruce wood.

Similarly to the data on smoke-forming capacity, the information on the effect of wood species and species on the indicators of toxicity of combustion products, patterns of yield of toxic combustion products when exposed to external heat fluxes of various intensities, is limited. It was found that, when exposed to external radiation heat flux, the ability to form toxic combustion products is much lower compared to the release of various toxicants in flameless (smoldering) combustion mode.

With data from small-scale methods for testing the fire hazard characteristics of wood-based materials, it is possible to estimate the rate of fire hazard buildup during the initial stage of fire development in a room and the time to reach a human-critical situation. A mathematical description of the growth process of such fire hazards in rooms as the average values of temperature, concentration of toxic products of combustion, optical density of smoke, and reduction of oxygen concentration is presented in the monograph [44].

Currently, there is very limited information on the fire hazard properties of wood used in modeling the development of fire inside buildings and structures with wooden structures, assessing the growth of fire hazards, and predicting the fire hazard and fire resistance of building structures made of wood of different species. When estimating the fire hazard of wooden buildings, the physical-chemical and fire-hazardous indices of wood itself are actually not taken into account, having wide numerical discrepancies of these indices depending on numerous external and internal factors.

Despite the huge number of experimental works carried out in the field of wood fire hazard research, it is often difficult to compare the results obtained not only because of the use of different research methods but also because of different wood samples often without mentioning the species or any description of their characteristics. Apparently, this way of working is caused by the outdated approach to wood as an average generalized substance, with little variation in its properties. This is evident from the current reference databases on the fire behavior of different building materials.

In addition, the long-term processes that take place in wood during low- temperature heat exposure have hardly been investigated.

The conditions and factors influencing the occurrence of wood smoldering combustion and its transition to flame are unclear. It is necessary to understand the extent to which the wood combustion process and the critical conditions of its existence will be affected by the relative content of not only the main components in wood but also the constituents of these components and their chemical nature. It is of great scientific and practical interest in this respect to analyze the changes in some important indicators of fire hazard of wood in the course of its natural and artificial aging.

Another problem facing modern researchers is the study of the effect of wood charring processes in fire conditions on the fire resistance of wooden load-bearing and enclosing structures. There is a general trend: conifers char at a higher rate than hardwoods, all other things being equal.

However, because of the great variety of wood-based materials that are used in timber construction, their fire behavior and resistance to fire—the quantification of which is dependent on many natural and functional factors—must be carefully studied. In doing so, serious consideration must be given to the quality and properties of the surface charred layer produced by burning wood and wood-based materials, including naturally and artificially aged wood, and to the possibility of purposeful regulation of the porous structure and properties of the charred layer. This will affect its thermal properties, the temperature distribution in the material in the immediate vicinity of the charring front, and thus the temperature dependence of the mechanical properties of the structure as a whole.

2.4 Fire Resistance and Fire Risk in Timber Structures

One of the most important requirements for solid-wood and laminated timber structures with building or structural functions is acceptable fire and structural safety during the foreseeable future, i.e., fire resistance. Factors affecting the fire resistance of timber structures include the type and species of wood, the type of wood (glued or solid), the presence of reinforcements, the geometric dimensions and cross-sectional configuration of the structure, operating conditions and duration, fire exposure conditions (fire), and others.

Changes in fire conditions in the strength and, for wood, in the geometric characteristics of sections cause a reduction in the load-bearing capacity of the elements and units of wooden structures, expressed as a temporary index—the fire-resistance limit of the structure.

The fire-resistance limit of a timber structure is determined by calculation or by experiment, taking into account the effects of the standard temperature dependence of fire development. The standard temperature dependence of the time course of a fire is described by the equation

$$T_f - T_0 = 345 \lg(8\tau + 1), \tag{2.12}$$

where T_f и T_0—is the current and initial temperature in the furnace chamber, °C, respectively, and τ—time, min.

The fire-resistance limit of timber structures may be determined by calculation based on the patterns of charring and heating of their sections in standard thermal

Table 2.14 Results of fire resistance of wooden beams with different cross-sections

Type of construction	Fire-resistance rating, min.
Wooden beams bend-stressed unprotected (three-sided fire exposure) with dimensions in mm:	
100 × 140	25
120 × 160	30
140 × 200	40
180 × 260	50
Wooden columns, unprotected, longitudinally bending stressed at flexibility $\lambda = 75$ (four-sided fire exposure) with dimensions in mm:	
120 × 120	10
160 × 160	15
200 × 200	20

conditions and taking into account the limit states for fire resistance. In this case, the fire-resistance limit of the element joints and support joints of timber structures, including the use of metallic and non-metallic reinforcement elements, shall not be less than the required fire-resistance limit of the structure as a whole.

The fire-resistance limit of timber structures is usually determined by the time taken for the load-bearing capacity of the section to decrease due to charring and heating up to the value of the effective load.

Experiments have shown that the fire resistance of glulam structures is higher than the fire resistance of solid natural timber structures. This can be explained by the better quality structure of the laminated elements and by the use of heat-resistant adhesives in the laminated timber industry. For example, a glued wooden beam with a section size of 100 × 140 mm has a fire-resistance limit of 35 min when tested, which is 10 min greater than the fire-resistance limit of solid wood with a similar section [45].

The most significant factors determining the fire resistance of timber structures are the size of the cross-section of the structure and the number of heating sides under fire conditions. The results of determining the fire-resistance limit of timber beams as a function of section size and number of heating sides under standard fire temperature conditions are shown in Table 2.14.

The results shown in Table 2.14 show that as the cross-sectional area of the structure decreases, the fire-resistance values decrease significantly. For a 100 × 140 mm cross-section structure, the fire-resistance limit is 25 min, and for a 180 × 260 mm cross-section structure, the limit is doubled to 50 min. The increase in the number of heating sides also reduces the fire-resistance limit of the construction.

The fire resistance and fire hazard of timber structures is greatly influenced by the configuration of the cross-section. For example, in elements with rectangular cross-sections, corners of sections are more intensively charred in a "standard fire," rounding of which is observed 10–15 min after the beginning of the carbonization of the wood.

Ribs and similar projecting parts of elements with a large ratio of the surface area perceiving heating from an external source to the volume of heated material ignite faster than flat or convex elements with a large radius of curvature [46].

Under fire conditions, the fire resistance of complex timber structures is mainly determined by the reduction in the load-bearing capacity of their timber elements and the knotty joints of these elements.

The load-bearing capacity of wooden structures decreases due to charring of wood and reduction in the size of their operating cross-section capable of absorbing the existing loads, as well as due to changes in wood strength in the uncharred part of the cross-section [47]. Changing the load-carrying capacity of joints in fire is influenced both by wood charring and by a reduction in the strength of steel elements used in wooden structures—steel dowels, reinforcing linings, and shoes.

A characteristic of the charring process is the rate at which the wood is charred. For different species of wood, charring rate may vary from 0.6 to 1.1 mm/min and depends on many factors, in particular, on the species, volume weight, and moisture content of wood, number of sides of heating structure, duration and features of temperature regime, heating rate, section size, surface roughness, and others [47].

With increasing bulk (apparent density), moisture content of wood and cross-section size of wooden element, charring rate decreases, and increasing temperature of heating medium in fire, air inflow, number of sides of heating section, and surface roughness of their planes, charring rate of wood increases. During prolonged temperature exposure, the charring rate of wood decreases.

Compared to glued wood, the charring rate of solid wood is higher. An experimental study [45] of 100 × 140 mm and 150 × 250 mm softwood beams exposed to standard fire conditions showed that the average values of the rate of charring in the direction of the core of the section are as follows:

- For solid wood beams:
  The charring rate from the side is 0.65 mm/min. Charring rate at the bottom is 0.95 mm/min.
- For glulam beams:
  The side charring rate is 0.55 mm/min. Charring rate at the bottom is 0.85 mm/min.

The charring rate is not linear; it varies during the individual phases of wood combustion.

It has been shown that the charring rate of wood is significantly dependent on the type of support structure. A number of European countries have adopted standards that set the average charring rate for a particular type of timber structure [48].

For example, Italian Ministry of the Interior Circular No. 91/61 states the following average charring rates depending on the type of supporting timber structure: exterior beams forming the vault and laterals 0.8 mm/min, interior beams forming the vault 1.1 mm/min, pilasters and columns 0.7 mm/min, and other horizontal components 1.1 mm/min. A more recent Italian national standard, UNI

9504 “Analytical evaluation of fire performance of wood structural members,” sets charring rates of 0.9 mm/min for solid wood and 0.7 mm/min for laminated wood.

Experimental studies performed under standardized conditions in European Community countries on numerous unprotected wooden elements and constructions have shown that the average charring rate of wood pulp under fire exposure is about 1 mm/min and varies from a minimum of 0.7 mm/min for dense and wet wood to a maximum of 1.2 mm/min for light and dry wood [48].

The charring process is inevitably accompanied by a reduction in the cross-sectional dimensions of the timber structure. In addition to the reduction of the cross-sectional dimensions as a result of charring, a reduction in the strength and elastic properties of the timber can be observed when heat is applied to the components of wooden structures. The uneven distribution of temperature over the cross-section causes the values of mechanical and thermophysical characteristics to change unequally at different points in a given cross-section. The temperature distribution from the charring front to the center of the section of solid glued wooden elements from 300 °C to the initial temperature changes according to the hyperbolic law [47]. The mechanical properties change linearly with temperature.

Measurements of charring rates of eight different types of wood at standard fire temperatures according to ASTM E 119 have led the authors to the following relationship:

$$\tau = m{x_c}^{1,23}, \tag{2.13}$$

where τ—thermal exposure time, min; x_c—thickness of charred layer, mm; and m—charring rate coefficient. The coefficient is a function of apparent density of dry wood, ρ, kg/m^3, moisture content in the sample (in relative proportion to dry wood), and absolute contraction factor of charred layer, f_c.

The charring rate coefficient, m, is calculated according to the formula

$$m = 0.000564\rho + 1.21\mathrm{u} + 0.532f_c - 0.147,\ \mathrm{min/mm}^{1,23}. \tag{2.14}$$

Here f_c is a dimensionless parameter for the reduction of charred layer thickness. It is defined as the ratio of the thickness of the charred layer at the end of fire exposure to the initial thickness of the layer of wood that should be charred. The authors [49] believe that the fc factor is primarily a function of the lignin content in the chemical composition of the wood.

For dry spruce wood with $\rho = 460$ kg/m^3, the charring rate factor $m = 0.47$ min/mm^{1,23}, and for 9% moisture content, $m = 0.58$ min/mm^{1,23}.

The charring rate of hardwood under standard fire conditions is 10–20% lower compared to softwood specimens. The above formula refers to standard temperature test conditions for timber structures in unloaded condition. It is known that the mechanical stress applied to the specimen “helps” the thermal movement to destroy the material. It has been shown, for example, to reduce the fire-resistance limit of 3 m pine beams of 120 × 410 mm section from 65 to ~20 min with an increase in the load applied in the middle of the span from 2300 to 9200 kg under standard fire temperature conditions [50].

It can be noted that, under the same mechanical loading conditions, larch wood beams showed almost 25% higher fire resistance than pine wood beams. The charring rate of larch wood was also lower. In general, the charring rate of wood decreased with increasing duration of standard heat exposure and thickness of charred layer [50].

Under standard temperature conditions, the actual fire-resistance limit of wooden structures, $FR_{ac.}$, is determined by taking into account the total time from the beginning of thermal effect on the wood in fire to the beginning of charring, τ_o, and the time from the beginning of charring to the onset of the limiting state in fire, τ_{cr}:

$$FR_{ac.} = \tau_o + \tau_{cr}, \text{ min.} \tag{2.15}$$

The latter is determined on the basis of the charring rate of the wood:

$$\tau_{cr} = (\delta_{cr} - \delta)/v, \tag{2.16}$$

where v—charring rate, mm/min; δ_{cr}—thickness of charred layer at limiting condition, mm; δ—thickness of wood layer behind the charred layer, which practically does not resist the load acting on a structural element. It is usually considered that $\delta = 5$–7 mm [51].

It is assumed that this very layer of wood heated from 175 °C to 300 °C should not be taken into account in calculations of the fire-resistance limit of timber structures due to a significant reduction and loss of strength properties.

Calculation methods for the fire-resistance limit of building structures, including wood and other polymer composite materials, are based on the numerical solution of the system of heat transfer equations and statics using modern software thermal engineering complexes. Mathematical modeling and calculation of fire resistance of building structures is based on knowledge of macrokinetic parameters of decomposition and thermophysical, thermochemical, and mechanical characteristics in a wide temperature range of materials of which this or that structure is created. In the presence of flame retardants, relevant data on the characteristics of the flame retardant material is also needed [52, 53].

Using the equations of energy conservation in the pyrolysis front and on the surface of the charred layer during wood burning (or generally carbonized polymer composites), the authors of [53] obtained an analytical expression for the time dependence, τ, of the charred layer thickness build-up, δ_k, taking into account its shrinkage.

This was possible by assuming a quasi-stationary combustion process, a relatively small thickness of the pyrolysis zone, the predominant radiant heat transfer between the flame and the surface of the combustible material, and the use of a preconceived temperature change profile in the decomposing material exponentially. The shrinkage coefficient, β_y, of the charred layer anbd the yield of carbonized residue, K, resulting from the pyrolysis of the material were assumed constant.

The following equation was obtained:

$$\delta_k = \sqrt{2a_k\tau + b_k^2} - b_k \tag{2.17}$$

here b_k is the parameter taking into account the conductive heat conductivity of the charred layer and the radiant heat transfer from the flame at the temperature T_f:

$$b_\kappa = k_\kappa / A_{\text{eff}}\, \sigma T_f^{\,3}, \tag{2.18}$$

where $A_{\text{eff}} = 1/[(1/\varepsilon_f)\,(1/\varepsilon_w - 1) + 1/\varepsilon_w]$—effective function of radiation parameters of flame and combustible surface; ε_f and ε_w—emissivity of flame and combustible surface.

Parameter $a_k = (1 - \beta_y)\, k_\kappa\, (T_f - T_\kappa)/\rho[(1 - K)\, L_v + c(T_\kappa - T_o)]$ takes into account the ratio of heat transferred from flame to pyrolysis front and heat spent for fuel gasification.

Applied to spruce douglas-fir combustion in flame with temperature $T_f = 1200$ °K and the assumed wood properties: $\rho = 450$ kg/m^3, $c = 2800$ J/kg·K, $k_k = 0.5$ W/m· K, $\beta_y = 0.2$, $K = 0.2$, $L_v = 1820$ kJ/kg, and $T_k = 623$ K calculations showed that the charred layer thickness in 20 min of burning period reached almost 18.5 mm. After the beginning of charring of spruce wood, the mass pyrolysis rate during this period decreased from 12 to 4 g/m^2·s [53].

Relatively recently, a wood pyrolysis model has been proposed for engineering estimation of the charring rate of wood structures under fire conditions [54].

The one-dimensional model called CROW (Charring Rate of Wood) is based on the application of the energy conservation equation in individual k-phase zones with appropriate heat transfer boundary conditions:

$$\rho\, c_p\, \partial T/\partial t = \partial/\partial\, (k \partial T/\partial x) - (\Delta h_v + \Delta h_w)\, \rho_0\, \partial u/\partial t \tag{2.19}$$

here x is the length coordinate, m; ρ_0 is the density of dry wood, kg/m^3; u is the moisture content; ρ is the density of the original wood, partially charred and coke; c_p is the heat capacity of the respective layers; k is the thermal conductivity; and Δh_v and Δh_w are the heat of water evaporation and wetting, J/kg.

The pyrolysis reaction is assumed to be thermoneutral, with water evaporating at 100 °C, partially retained in the sample mass. The properties of the original wood are considered at $T \leq 200$ °C and the coke layer at $T \geq 800$ °C.

As in [49], four main factors affecting wood pyrolysis were taken into account: apparent density of dry wood, moisture content, lignin content, and regression of the char layer. It was proposed to calculate the limiting depth of charring of wooden structures under standard temperature and load conditions by solving equation

$$S_f = \left((b - 2x_c) \cdot (d - x_c)^2\right)/6 - 0.215x^2\,(d - x_c), \tag{2.20}$$

where S_f—sectional modulus of the timber at the limit of loss of bearing capacity of the structure at four-sided heating and rounding of sectional angles during charring,

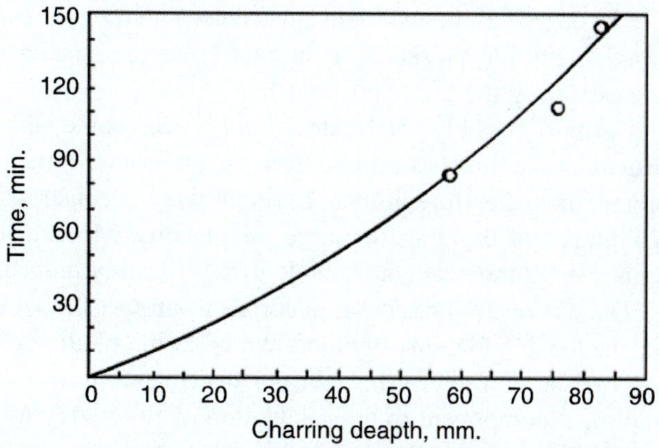

Fig. 2.15 Predicted and experimentally found charring depths in glulam timber structures for loss of load-bearing capacity: —— calculation; ○—experiment

m^3; *b*—initial width of the timber section, m; x_c—depth of charring, m; and *d*—initial depth of the section, m.

Value:

$$S_f = M / k_f k_{med.} F_{all.}, \tag{2.21}$$

where *M*—maximum induced stress moment, kN·m; k_f—coefficient of reduction of strength when section is heated: $k_f = 1 - P/200A$, where *P* is the section perimeter, m, *A*—section area, m^2; $k_{med.}$—coefficient of transfer from the allowable stress value to the average value of stress at failure; and $F_{all.}$—allowable stress, kPa.

The results of ASTM E 119 fire-resistance testing of glulam timber beams of Douglas fir 4.57 m length and 222 × 419 mm overall cross section, under different loads: 27%, 44%, and 91% of maximum use, have been reviewed. Permissible values of stress limit $F_{all.} = 16550$ kPa and modulus of elasticity $E = 11$ GPa have been determined. The moment of resistance was estimated to be 302 kN-m, and the induced moments under different loads were 25.7, 41.6, and 88.2 kN·m, respectively.

The fire-resistance limit of laminated timber structures in this case was found to be 147, 114, and 85 min.

Figure 2.15 shows the calculated dependence of the charring depth of timber-laminated structures on the test time compared with experimental data at the load-bearing capacity loss limit [54].

Determination of the fire hazard class of timber structures is necessary to establish the extent to which they are involved in the development of a fire, in the formation of fire hazards. The fire hazard class of a structure depends on the fire hazard of the materials of which it is made.

In fire tests, the thermal damage degree and the presence of thermal effects in standard fire temperature conditions are evaluated as the main indicators of fire hazard of wooden structures.

The degree of thermal damage is determined by the high intensity of the charring process, and the presence of thermal effects is characterized by a significant heat release during the fire exposure (fire).

Despite possibly high structurally achievable fire-resistance limits, timber structures are fire hazardous. The use of timber structures without adequate fire protection is therefore limited. Especially in the construction of public and industrial buildings and buildings of large dimensions. Modern fire protection methods for timber structures make it possible to significantly reduce their fire risk.

The use of fire protection materials to protect timber structures can significantly reduce the flammability of wood in a fire. First of all, the time from the beginning of fire exposure (fire) to the ignition or self-ignition of wood, which serves as an important component of the calculation methodology for determining the fire resistance of wooden structures, can be significantly increased.

Under standard fire temperature conditions, the time to onset of autoignition (τ_0) and charring of solid wood is 3–4 min. With fire protection, it can be extended to 100 min or more.

2.5 Main Methods and Types of Fire Protection for Timber Structures

To prevent the occurrence and spread of fire in buildings and constructions with load-bearing and enclosing structures as well as wood finishes and cladding, fire retardants are widely used, which are distinguished according to the method of application (treatment) and the mechanism of fire protection.

The fire protection of natural timber building structures can be provided by constructive means similar to the protection of concrete, reinforced concrete, and metal structures by means of board and roll materials and plasters of various types. Their purpose is to slow down the heating of the protected wood surface to the critical temperature at which active pyrolysis begins and the wood ignites. They should also slow down the rate of wood charring and the heating of the remaining unchanged part of the wood behind the charring front to the limit temperature corresponding to the loss of physical and mechanical properties of wood. The effectiveness of wood fire protection by the above structural methods is primarily determined by the resistance to fire and the insulating capacity of the fire-retardant board and roll materials. Structural measures can significantly increase the fire-resistance limit of timber structures. However, the required thickness of such fire-retardant materials is considerable. It usually exceeds 10–15 mm [55, 56].

Currently, methods such as surface and deep impregnation with special compositions containing flame retardants as well as the application of flame-retardant coatings are popular for fire protection of natural wood. Particularly attractive is the use of intumescent coatings. With a relatively small thickness, they exhibit high fire protection efficiency, allow the use of modern mechanized methods of application to the protected object, and preserve the original wood texture and its aesthetic appearance [57].

As a rule, fire-retardant impregnation compositions and coatings for wood are multicomponent systems. Each component of the system performs its functional role in providing the overall fire-retardant effect and the required performance properties.

In most cases, impregnating compositions are saline solutions in aqueous medium or in organic solvents. Surface impregnation is carried out by repeated application of the composition to the protected surface of wooden elements of constructions or facing materials by hand or mechanized method. The surface impregnation is less effective than deep impregnation at its cheapness and manufacturability, but allows by means of wetted surface with compositions to protect the mounted wooden constructions directly on building objects. It is extremely difficult to obtain reliable fire protection for wood in this way. This is largely due to the poor penetration of salts in the wood. High consumption is required. Deeper penetration of the flame-retardant solution into the surface layers of wood is ensured by impregnation methods such as the use of hot-cold baths as well as processing in industrial apparatuses—autoclaves in the mode of sequential pressure increase or alternating vacuum and increased pressure.

The hot-cold bath method has found industrial application as the main impregnation method for wood building parts at a number of large woodworking factories [58].

Deep impregnation of wood with flame-retardant saline solutions is based on the peculiarities of the capillary-vascular structure of wood. A number of technological difficulties, the need for large consumption of flame retardants, and the need for special equipment for deep impregnation of wood with fire-retardant compositions limit the practical use of this direction of fire protection of wood.

At present, a large number of fire-retardant impregnating compositions for wood, which differ from each other by different set and quantitative combination of low molecular weight inorganic substances and derivatives of organic compounds, exhibiting properties of fire retardants, have been developed. Scientific and practical research in this field continues to evolve with the aim of identifying even more effective fire-retardant formulations for wood and improving the performance of the final products.

Various derivatives of phosphoric and phosphonic acids (mono- and diammonium phosphates or their mixtures (ammophos); urea-, melamine-, and amido-phosphates; and amidomethyl phosphonates) should be included among the substances frequently used in formulations of fire-retardant impregnating compositions. Boric acid, ammonium and sodium tetraborates, ammonium salts of sulfuric and hydrochloric acids, chlorides of alkaline earth metals, and metals of variable valence, sodium, and potassium carbonates—this is not a complete list of inorganic substances used as components of impregnating fire-retardant compositions for wood [55].

An important direction in the creation of modern flame retardants for wood is the development of a new generation of complex flame-retardant impregnation compositions with a partial or complete replacement of salt components [59, 60]. Thus, a new-generation biocide based on water-soluble esters of phosphoric acid with a wide range of bioprotective action against mold and wood-destroying fungi was found to be quite effective. Impregnation of wood with this composition in combination with oligomeric organosiloxanes, in particular, with oligoethylhydridesiloxane, allows not only to provide fire bio-protection of wood but also to increase moisture and water resistance of fire-resistant material while maintaining high mechanical properties of the original wood.

A new generation of water-soluble oligomeric flame retardants containing phosphorus and boron elements in their macromolecules is proposed for fire protection of wood and other organic materials. Their synthesis is based on the reaction of interaction of dimethyl phosphite with boric acid at different molar ratios of reagents and temperature. Treatment of pine wood with an aqueous composition of P, B—containing flame retardant, can increase the oxygen index of wood from 20–23 to 53 [61].

Among the currently used flame retardants for wood, a special place is occupied by blowing-type coatings, whose protective properties are manifested when exposed to high temperatures and fire. The development of this type of fire protection for wood has received considerable attention worldwide [62, 63].

The bloating-type coatings necessarily contain important components: 1—film-forming binders, which are simultaneously the source of formation of the carbon skeleton during decomposition; 2—catalysts of reactions of formation of the carbon skeleton; and 3—foaming (bloating) agents. To enhance the overall flame-retardant effect, various additives are introduced which can influence technological, heat-protective, and other operational properties of the coatings. These are fillers with flame-retardant properties, thickeners, pigments, stabilizers, etc.

Thermoplastic polymers are recommended as binders—source of carbon framework formation: polyvinyl acetate, polyvinyl alcohol, chlorinated polyvinyl chloride, chlorinated rubber, latexes of vinyl chloride copolymers with vinylidene chloride, chlorinated polyethylene, and polyurethanes.

Among thermosetting polymers, urea-formaldehyde, phenol-formaldehyde, dicyandiamide-formaldehyde, and epoxy resins are often used, which form a small number of cross-links during curing. The polymer must be in a plastic, viscous-fluid state in order for the swelling of the paint coating to take place. The rapid curing of thermosetting polymers interferes with the formation and release of non-flammable vapors and gases as a result of the thermal decomposition of the foaming agents. Disruption of this synchronicity prevents the formation of a foamed coating during heating.

The additional addition of low molecular weight substances such as sugars, polyatomic alcohols (e.g., pentaerythritol), and multi-atomic phenols to organic binders increases the overall yield of the carbonized product.

The catalysis of reactions of carbon framework formation during thermal decomposition of organic binders is carried out in the presence of substances which lead to strong acidic agents of dehydration, dehydrochlorination, cyclization of formed fragments with unsaturated double bonds, and cross-linking reactions in the system. Such substances include the abovementioned ammonium salts of phosphoric and polyphosphoric acids, sulphates, and other compounds used in impregnation fire-retardant compositions for wood.

These same derivatives of various acids are also effective foaming agents of coatings, releasing large amounts of non-combustible gases (NH_3, CO_2, HCl, H_2O). In the formulation of foaming, coatings can be introduced in an independent form of high-temperature gas-forming agents: dicyandiamide, melamine, and melem.

These foaming agents are typical chemical gas formers, which form non-combustible vapors and gases as a result of chemical decomposition reactions when the substances are heated. Physical agents forming blowing flame retardants include vermiculite and thermally expandable graphite. Vermiculites are layered mineral compounds which expand and increase in volume many times when heated. Thermoexpandable graphites are interlayer compounds of natural graphite inclusions. The graphite particles separate and expand when heated.

Depending on the method of interlayer formation and the intensity of heat treatment, 300–400 times swelling of the substance can be achieved.

Bloating coefficient of coatings with thermoexpandable graphites also depends on their quantitative content in the total composition of coating. Coatings based on chlorosulfonated polyethylene, rubber, and thermally expandable graphite have a bloating coefficient of 44.

Experimental studies of the effect of flame-retardant coatings of the bloating type on the fire-resistance limits and fire hazard class of wooden structures are very limited. Of great interest are the results of experimental work on fire protection of laminated timber structures with transparent bloating coatings. The experimental results obtained confirm the high effectiveness of fire protection of wooden laminated structures with organic polymer-based intumescent coatings consistent with the thermal engineering calculations of their fire resistance [64].

Flame-retardant coatings are characterized by a complex fire behavior. Mathematical modeling of the performance of flame-retardant coatings of the bloating type and calculation methods is presented in [64, 65]. The calculation of optimum thicknesses of such coatings for effective protection of building structures requires knowledge of many fire protection characteristics. These parameters are determined by experiment and calculation based on the determination of the temperature fields in the material to be protected.

In world practice, to evaluate the effectiveness of the fire protection effect of impregnating compositions and coatings for wood, the most commonly used results of complex studies with the help of laboratory con-calorimeters (ISO 5660, ASTM E1354). The method makes it possible to determine the mass loss rate, heat release rates (maximum and average heat release rate, total heat release in a certain period of time), effective heat of combustion and smoke production capacity, toxic gas release, and other indicators. Thus, for example, high flame-retardant efficacy has been shown for a blowing coating based on urea-dicyandiamide-formaldehyde resin in combination with monoammonium phosphate and dextrin. At a radiant heat flux of 35 kW/m^2, no ignition, mass loss, or heat release was observed during a 30 min test of a coated pine wood specimen. After ignition, the maximum heat release rate was almost half that of the original wood. The time to reach the peak heat release rate, which characterizes the onset of intensive charring of the coating, also increased [66].

It was found experimentally, that under standard temperature conditions, the time to the beginning of autoignition and charring of solid wood, τ_0, by fireproof semi-rigid mineral wool board (GOST 9573-89) with thickness of 50 mm is 30 min. Typical modern board materials used for fire protection of wooden structures include cement-silicate binder-based boards, vermiculite-based liquid glass, non-combustible basalt fiber, and mineral wool boards.

Thermal calculations of the fire resistance of laminated timber structures with the abovementioned structural fire retardants do show an increase in τ_o values and a reduction in the average charring rate of wood over a given period of thermal exposure [65]. For example, the use of vermiculite boards with a thickness of 40 mm results in an increase of τ_o value to 106.7 min, which makes it possible to reduce the average charring rate of wooden laminated structures from 0.73 in an unprotected structure to 0.3 mm/min.

Structural wood fireproofing methods are correspondingly effective with thicker materials. It is interesting to note that the calculations of temperature fields in laminated wooden structures made by the authors of [65] showed not only that in most cases the layer thickness $\delta = 5$ mm, limited by the isotherms 175 and 300 °C and unable to resist the applied load, was exceeded. A tendency to an increase in the size of this layer with increasing thickness of fire protection and an increase in thermal exposure time was revealed.

Thus, various methods have been proposed to protect timber structures from the effects of fire, including both structural fire protection and surface treatment with fire-retardant compositions. The question remains how the type of wood, its structural characteristics, and the conditions and duration of use of timber structures affect the effectiveness of fire protection. It is not clear to what extent flame retardants reduce the fire risk of wood and timber structures. The development and use of flame retardants capable of reducing the intensity of the charring process of wood, which determines the fire resistance of wooden structures, remains uncertain.

References

1. Nikitin, V. M. (1951). *Chemistry of wood and cellulose* (p. 495). Goslesbumizdat.
2. Kachaeva, N. Y. (2004). *Proceedings of the IV international symposium "structure, properties and quality of wood – 2004"*, St. Petersburg, Vol. 2, p. 569.
3. Baldvin, S. F. (1987). *Biomass stoves: Engineering design, development and dissemination, volunteers in technical assistance (VITA)*, Arlington, p. 178.
4. White, R. H. (1987). Effect of lignin content and extractives on the higher heating value of wood. *Wood and Fiber Science, 19*(4), 446–452.
5. Monakhov, V. T. (1979). *Methods for research on fire hazard of substances* (p. 424).
6. Marlair, G., Cwiklinski, C., & Tewarson, A. (1999). An analysis of some practical methods for estimating heat of combustion in fire safety studies. In *Proceedings of the 8th Internal Conference on Fire Science and Engineering, Interflame'99*. Interscience Communication, Vol. 1, pp. 201–212.
7. Van Krevelen, D. W. (1976). *Properties and chemical structure of polymers* (p. 415).
8. Aseeva, R. M., Serkov, B. B., Sivenkov, A. B., & Thanh, B. D. (2004). *Heat and smoke release at the combustion of the different wood varieties. Fire science and technology 6th Asia*

- *Okeania symposium Daegu, Korea* (pp. 203–211). Korean Institute of Fire Science and Engineering., March 17 (Wed)–20 (Sat).

9. Aseeva, R. M., Serkov, B. B., Sivenkov, A. B., & Thanh, B. D. (2004). Characteristics of heat and smoke emission during burning of different wood. In *Proceedings of the IVth International Symposium RKSD: Structure, Properties and Quality of Wood*. St. Petersburg State Forestry Academy named after Kirov, pp. 347–350.
10. Sivenkov, A. B. (2015). *Influence of physical and chemical characteristics of wood on its fire hazard and fire protection efficiency: Dissertation PhD thesis: 02.00.06 High-molecular compounds*. N.N. Semenov Institute of Chemical Physics, Russian Academy of Sciences, p. 289.
11. Vakeg, A. J. (1982, November). Wood fuel properties and fuel products from woods. In *Proceedings of Fuel Managements and Utilization Seminar*. Michigan State University, pp. 14–25.
12. Sivenkov, A. B. (2007). *Research of features of thermal and thermo-oxidative decomposition of wood of various species. Collection of scientific papers: Technology and equipment for wood processing* (pp. 115–124). MSULE.
13. Aseeva, R. M., Serkov, B. B., Sivenkov, A. B., & Thanh, B. D. (2005). Thermal decomposition of wood of tropical species. *Bulletin of the Moscow State Forest University - Forest Bulletin, 2*(38), 70–76.
14. Ilyin, A. V., Naganovsky, Y. K., Serkov, B. B., Sivenkov, A. B., & Uvarov, I. A. (2005). The behavior of some wood species under heat exposure in inert and oxidizing environments. *Bulletin of the Academy of State Fire Service of Russia., 3*, 56–65.
15. Shestak, J. (1987). *Theory of thermal analysis* (p. 456). Mir.
16. Rogers, F. E., & Ohlemiller, T. J. (1981). Pyrolysis kinetics of a polyurethane foam by thermogravimetry: A general kinetic method. *Journal of Macromolecular Science: Part A - Chemistry, 15*(1), 169–185.
17. Criado, J. M. (1878). Kinetic analysis of DTG data from master curves. *Thermochimica Acta, 24*, 186–189.
18. Frank-Kamenetsky, D. A. (1967). *Diffusion and heat transfer in chemical kinetics* (p. 492). Nauka.
19. Moussa, N. A., Toong, T. Y., & Garris, C. A. (1976). Mechanism of smouldering of cellulosic materials. In *Proceedings of the 16th Symposium (International) on Combustion*. The Combustion Institute, pp. 1447–1457.
20. Rogers, F. E., & Ohlemiller, T. J. (1980). Cellulosic insulation materials. I. General degradation Kinetics and reaction heats. *Combustion Science and Technology, 24*, 129–137.
21. Ohlemiller, T. J., & Rogers, F. E. (1980). Cellulosic insulation material II. Effect of additives on some smolder characteristics. *Combustion Science and Technology, 24*, 139–152.
22. Ohlemiller, T. J. (1985). Modeling of smoldering combustion propagation. *Progress in Energy and Combustion Science, 11*(4), 277–310.
23. Ohtani, H., Maejima, T., & Uehara, Y. (1991). In-situ heat release measurements of smoldering combustion of wood sawdust. In *Proceedings of 3rd Internal Symposium on Fire Safety Sciences*. Elsevier, pp. 557–564.
24. Babrauskas, V. (2001). Pyrophoric carbon and prolonged low-temperature ignition of wood. *Fire and Arson Investigator, 52*(2), 12–14.
25. Bowes, P. C., & Townshend, S. E. (1962). The ignition of combustible dust on hot surfaces. *British Journal of Applied Physics, 13*, 105.
26. Bilbao, R., Mastral, J. F., Aldea, M. E., Ceamanos, J., & Betran, M. (2001). Experimental and theoretical study of the ignition and smoldering wood including convective effects. *Combustion and Flame, 126*, 1363–1372.
27. Babrauskas, V. (2002). Ignition of wood. A review of the state of the art. *Journal of Fire Protection Engineering, 12*(3), 163–189.

28. Delichatsios, M. A., Panagiotou, T., & Kiley, F. (1991). The use of time to ignition data for characterising the thermal inertia and the minimum (critical) heat flux for ignition or pyrolysis. *Combustion and Flame, 84*, 323–332.
29. Spearpoint, M. J., & Quintiere, J. G. (2001). Predicting the piloted ignition of wood in the cone calorimeter using an integral model - effect of species, grain orientation and heat flux. *Fire Safety Journal, 36*(4), 391–415.
30. Mirzakhmedov, B. H., Berlin, A. A., Sivenkov, A. B., & Mukhamedgadiev, B. A. (2022). *Polymeric flame retardants: Synthesis, properties and application: Monograph* (p. 140). Tashkent Publishing House.
31. Newman, J. S., & Tewarson, A. (1991). Flame spread behavior of char-forming wall/ceiling insulating materials. In *Proceedings of 3rd International Symposium on Fire Safety Sciences, Edinburg, Scotland*, pp. 679–688.
32. White, R. H., & Dietenberger, M. A. (2002). Fire safety, Chapter 17. In *Wood handbook - wood as an engineering material.* US Department of Agriculture, Forest Products Laboratory.
33. De Ris, J. N. (1969). Spread of a laminar diffusion flame. In *Proceedings of 12th Symposium (International) on Combustion, Pittsburg*. The Combustion Institute, pp. 241–252.
34. Wichman, I. S. (1992). Theory of opposed-flow flame spread. *Progress in Energy and Combustion Science, 18*, 553–593.
35. Merryweather, G., & Spearpoint, M. J. (2010). Flame spread measurements on timber products using the ASTM E 1321 LIFT and a reduced cone calorimeter adaptation. *Flame and Materials, 34*(3), 109–136.
36. Krause, R. F., & Gann, R. G. (1980). Heat release rate measurements using oxygen consumption. *Journal of Fire & Flammability, 12*(2), 117–130.
37. Ostman, B. A.-L., & Svenson, I. G. (1985). Comparison of three test methods for measuring rate of heat release. *Fire and Materials, 9*(4), 176–184.
38. Pofit-Szczepanska, M. (1994). The influence of the natural structure of wood on the heat release rate during its combustion. In *Abstracts of works of 25th international symposium on combustion, California, Irvine, USA, Poster Session 4*, p. 324.
39. Simoneit, B. R. T., Rogge, W. F., Mazurek, M. A., Standley, L. J., Hildemann, L. M., & Gass, G. R. (1993). Lignin pyrolysis products, Lignans and resin acids as specific tracers of plant classes in emissions from biomass combustion. *Environmental Science & Technology, 27*, 2533–2541.
40. Romanenkov, I. G., & Zigern-Korn, V. N. (1984). *Fire resistance of building structures made from effective materials* (p. 240). Stroyizdat.
41. Drysdale, D. (1990). *Introduction to fire dynamics* (p. 424). Stroyizdat.
42. Gann, R. G., Babrauskas, V., Peacock, R. D., & Hall, J. R. (1994). Fire conditions for smoke toxicity measurements. *Fire and Materials, 18*, 193–199.
43. Babrauskas, V., Levin, B. C., Gann, R. G., Paabo, M., Harris, R. H., Peacock, R. D., & Yusa, S. (1991). *Toxic potency measurement for fire hazard analysis* (p. 107). NIST Special Publication 827.
44. Koshmarov, Y. A., & Rubtsov, V. V. (1999). *Processes of the growth of dangerous factors of fire in industrial premises and the calculation of critical duration of fire* (p. 89). MIPB MVD of Russia.
45. Zoufal, R., & Kaspar, Y. (1986). *Investigation of wooden structures with regard to fire safety* (pp. 70–88). Bulletin, Higher Specialised School named after Georgiy Dmitrov, Ministry of Internal Affairs of the Republic of Bulgaria.
46. Roitman, M. Y. (1975). *Fire Prevention in Building Industry* (p. 525). Higher Engineering School of Fire Engineering of the USSR Ministry of Internal Affairs.
47. Sivenkov, A. B., & Frolov, A. Y. (2011). *Tasks and exercises on fire resistance of timber structures* (p. 103). Academy of the Russian Ministry for Emergency Situations.
48. Gusev, A. I., Paznikova, S. N., & Kozhevnikova, N. S. (2008). Increasing the fire resistance of wooden structures. *Fire and Explosion Safety, 15*(3), 30–35.

49. White, R. H., & Nordheim, E. V. (1992). Charring rate of wood for ASTM E 119 exposure. *Fire Technology, 28*(1), 5–30.
50. Kharitonov, V. G., & Khmelidze, T. P. (1989). *Behaviour of laminated wooden beams under standard fire conditions. Coll. Fire resistance of building constructions and ensuring of fire safety of people and material values* (pp. 40–47). VNIIPO MVD USSR.
51. Roitman, V. M. (2000). *Engineering solutions for assessing fire resistance of designed and reconstructed buildings* (p. 382). Publishing House of Fire Safety and Science Association.
52. Geraskov G. V., Voinov, A. N., & Molchadsky, I. S. *Mathematical modelling of charring process in combustion of thermally thick material in view of influence of external thermal flow*, pp. 172–175.
53. Strakhov, V. L., Krutov, A. M., & Davydkin, N. F. (2000). *Fire protection of building constructions* (p. 433). TIMR.
54. Janssens, M. L. (2004). Modeling of the thermal degradation of structural wood members exposed to fire. *Fire and Materials, 28*, 199–207.
55. Almenbaev, M. M., & Sivenkov, A. B. (2021). *Fire hazard of wood and its fire protection. Monograph* (p. 180). Publishing House "World of Press".
56. Romanenkov, I. G., & Levites, F. A. (1991). *Fire protection of building constructions* (p. 320). Stroyizdat.
57. Strakhov, V. L., Krutov, A. M., & Davydkin, N. F. (2000). *Fire protection of building structures* (p. 433). TIMR.
58. Novitskiy, G. I., & Stogov, V. V. (1959). *Wood impregnating plants* (p. 316). Transzheldorizdat.
59. Pokrovskaya, E. N., Kobelev, A. A., & Naganovskiy, Y. K. (2009). Mechanism and efficiency of fire protection of phosphor- and organosilicon systems for wood. *Fire and Explosion Safety, 3*, 44–48.
60. Pokrovskaya, E. N., & Naganovskiy, Y. K. (2004). Firebioprotection of monuments of wooden architecture. *Fire and Explosion Safety, 6*, 33–36.
61. Bondarenko, S. N., Kablov, V. F., Keibal, N. A., & Krekaleva, T. V. (2009). *Synthesis and application of phosphorus containing oligomers. "Oligomers-2009"*. Theses of reports of X international conference on chemistry and physicochemistry of oligomers, Volgograd, VGTU, p. 199.
62. Sivenkov, A. B., Khasanova, G. S., & Sharipkhanov, S. D. (2013). Effective fire-protection mechanisms to reduce the fire hazard of wood (in Russian). *Izvestia of Southern Federal University, 8*(145), 76–79.
63. Fialkov, A. S. (1997). *Carbon, interlayer compounds and composites on its basis* (p. 718). Aspect Press.
64. Garaschenko, N. A. (2006). Results of fire tests of laminated timber constructions with bloating coatings. *Fire and Explosion Safety, 15*(2), 12–16.
65. Garaschenko, N. A., Garaschenko, A. N., & Rudzinskiy, V. P. (2006). Thermal calculations of a fire resistance of laminated timber constructions with a fire protection. *Mounting and Special Works in Building, 10*, 14–18.
66. Wladyka-Przybylak, M., & Kozlowski, R. (1999). The thermal characteristics of different intumescent coatings. *Fire and Materials, 23*, 33–43.

Check for updates

Influence of the Wood Ageing Process on its Physico-chemical and Fire Hazard Properties

3

3.1 Essence of Wood Aging Process: Critical Analysis of Works in the Field of Ageing of Wood-Based Materials and Structures

Aging of wood in natural conditions occurs already during biological development and life of wood plants under certain climatic conditions. The biological age of plants is reflected in all properties of the wood of felled trees. How the biological age of plants affects the natural aging of wood products in their subsequent use is not clear and requires special consideration.

It should be noted that at present the study of natural wood aging is mainly aimed at determining and analyzing the dynamics of changes in the macroscopic characteristics of the structure and properties of the material during long-term use. However, the true aging process of wood substance, mechanism and kinetics of chemical reactions of the component, limiting stages, and possibilities of effective regulation or inhibition are not adequately considered. Solving this problem requires the creation of new approaches and methods, the cost of a large collaborative effort of the scientific community.

The research carried out so far shows that significant physical and chemical changes occur in wood during long-term aging under natural conditions [1–4]. Recently, such studies have become increasingly relevant and practical significance to ensure the longevity of wood structures during the reconstruction of wooden architecture monuments [5–7].

The results of these studies can determine the optimal conservation conditions to increase the longevity of wood structures. At the same time, much attention has been paid by the researchers to study the impact of wood biodegraders on its longevity.

The longevity of wooden structures under various climatic conditions is affected by a large number of different factors. As with synthetic polymeric materials [8, 9], the most significant physical and chemical changes in this natural polymer composite during operation occur as a result of photochemical, hydrolytic, thermal oxidative

S. Andrey Borisovich et al., *Fire Hazard and Fire Resistance of Wooden Structures*,
https://doi.org/10.1007/978-3-031-24074-4_3

degradation, the effects of mechanical stresses, and other types of processes under the influence of solar radiation, heat, rainfall, wind, and hurricane loads.

Unfortunately, a clear relationship between the conditions and duration of natural aging of wood and the nature of transformations in the structure, changes in the chemical composition, and many properties of the material has not yet been established. There is almost no information about the effect of wood aging on fire behavior and fire hazard indicators.

The variety of physical and chemical processes that occur in wood during its use does not allow all the transformations that occur in it to be reproduced artificially with high accuracy, nor does it allow the nature of changes in properties to be predicted. To assess the effect of natural aging on the fire risk of timber cladding and finishing materials and elements of timber building structures, it is preferable to examine samples of wood from existing old timber buildings and monuments of wooden architecture in an experimental way. This approach requires precise knowledge of the operational age of the object and the natural and climatic conditions in which it was located.

The study of the peculiarities of physical and chemical transformations occurring in the wood complex over time and leading to significant changes in its properties is a traditional subject of research in wood science.

So far, no clear correlation between the conditions and duration of exploitation of wood with its chemical and elemental composition, as well as with its structure, has been established. There are few research results associated with a comparative analysis of the abovementioned physicochemical parameters of different wood species depending on the climatic-geographical zones of their growth. The use of these results has, for example, valuable practical value for the production of wood and cellulose-based materials.

Great contribution to the development of the study of wood and wooden structures of long-term natural aging was made by a pleiad of such scientists as P.P. Erinsh, S.N. Gorshin, Y.A. Varfolomeev, M.V. Kisternaya, L.G. Shapovalova, B.N. Ugolyev, L.M. Perelygin, I.I. Pischik, E.N. Pokrovskaya, L.M. Kovalchuk, Kovalchuk L.M., Kazanskaya S.Y., Vikhrov Y.V., Golman L.P., Moskaleva V.E., Isaeva T.I., Brukhanova N.K., Fefilov V.V., Burkovskaya Y.I., Nikitin M.K., Oshkaev A.Kh, Kozlov V.L., Krutov V.I., Terentyev V.Y., Gusev B.P., Okamoto K., Fukada E., Keskulla T., Fengel D., Wegener G., Vitinen N., and many other scientists.

According to the results of these scientific schools:

- It has been established that the process of wood ageing is based on the process of hydrolysis.
- The kinetic parameters of chemical reactions occurring in long-term natural aging wood.
- The structure and destruction of lignocarbohydrate matrix of naturally aged wood have been studied.
- The basis for ensuring durability of wooden structures has been formed.

- Some properties (physical, mechanical, acoustic, piezoelectric, etc. of long-term natural ageing wood have been studied.
- The complex non-linear regularities of physical-chemical and mechanical properties changes during the long natural ageing of wood have been revealed.
- There have been developed the efficient preparations (antiseptics (biocides), fungicides etc.) for increasing the durability of the wooden constructions.
- Various technical constructive solutions have been developed to increase the longevity of wood objects.

Studies of the wood aging process show that significant physical and chemical changes occur in wood that affect its performance characteristics. It was found [7, 10] that the effect of wood age on these characteristics is cyclical and consists of two symmetrical branches: ascending and descending. The time cycle is 200 years for conifers and 260 years for deciduous wood.

Aging of polymeric materials can be regarded as a set of physical and chemical processes occurring in the material during storage, processing, and operation and leading to a change in its properties. In this case, there is a decrease of polymerization degree or molecular weight of the polymer. The process of aging of polymer materials is a destructive phenomenon, caused by the action of various factors that lead to structural and chemical changes in the polymer. Depending on which of the factors prevails, there is thermal, photochemical, radiation, mechanochemical, hydrolytic, biological aging. Depending on where chemical bonds are broken, degradation in the main and side chains of the polymer macromolecule is distinguished [11].

Chemical and physical factors of aging of polymeric materials are known. Chemical transformations usually proceed by the mechanism of chain reactions with the formation of active centers—free radicals, ions, and electronically excited particles at three main stages: initiation, process development, and death of active centers. The most common types of chemical transformations in the aging of polymers are processes that lead to the breaking of chemical bonds, cross-linking, etc.

Physical transformations can include physical processing of the material, prolonged exposure to stress, and removal of individual components of the material, leading to changes in its mechanical properties, density, volume, and the occurrence of mechanical stresses [12].

The mechanism of the processes occurring during aging of polymers determines the change in time of the index Y of macroscopic material properties depending on the condition X [12]:

$$Y = f(X, t), \tag{3.1}$$

where $f(X,t)$ is called the aging function.

The change in physicochemical properties of polymeric material when exposed to various factors over time has a complex nature associated with changes in reactivity, orientation, and reaction rate of molecular groups and complexes [12].

It is known [5] that destruction of polymers is the result of large thermal fluctuations that lead to a breakdown of the chemical bonds of the polymer chain. Changes in the chemical bonds of the polymer chain reduce the activation energy of polymer degradation.

The longevity of polymer can be described by the following formula [5]:

$$\tau = \tau_0 \exp(U_0/kt), \tag{3.2}$$

where U_0—is the bond breaking energy.

The natural aging of wood is generally a more complex and diverse process compared to polymers, as wood is a multicomponent polymer composite consisting predominantly of hydroxyl-containing polymers. The least reactive in the wood complex is the high molecular weight polysaccharide cellulose [13].

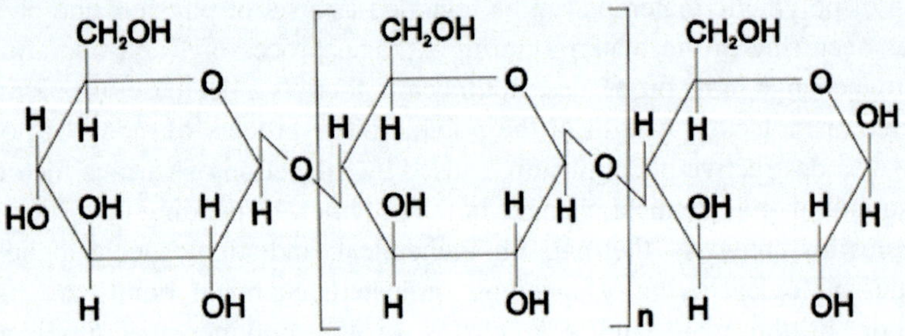

The ability of cellulose to react in various ways is determined by its functional groups, which are involved in all reactions. These include the three alcoholic hydroxyl groups of the anhydroglucopyranose links and the oxygen bridges between the chain links. The reaction depends on the accessibility of the reagent to the functional groups. Many cellulose modification reactions are heterogeneous due to the fact that cellulose is insoluble or partially soluble in the reaction medium. Cellulose solvents (dimethylformamide ether, aqueous solutions and hydrophilic solvents, water vapor) play an important role in the degree of reaction. Conducting chemical modification of cellulose by hydroxyl groups allows to obtain products with versatile properties and a wide range of applications (methyl cellulose, butyl cellulose, nitrate cellulose, triacetate cellulose, carboxymethyl cellulose, and many others) [13].

Some cellulose and wood degradation is accompanied by a hydrolysis process. For example, hydrolytic degradation or exposure of wood to microorganisms causes hydrolysis of cellulose and wood and oxidation of the hydrolysis products to form simpler polysaccharides [14]:

$$(C_6H_{10}O_5)_x + xH_2O \rightarrow xC_6H_{12}O_6$$

The end product of cellulose hydrolysis is glucose. The decomposition of cellulose by microorganisms in the presence of air oxygen leads to its oxidation to carbon dioxide and water.

Hemicelluloses are less resistant to oxidative processes. Hemicelluloses are polymeric carbohydrates with five and six carbon atoms in the main chain. Their general formulas are as follows: pentosans ($(C_5H_8O_4)_n$) and hexosans ($(C_6H_{10}O_5)_n$). Hemicelluloses in coniferous woods consist of hexosans and pentosans in equal proportions, while deciduous woods consist of pentosans by 90% [13]. During hydrolysis, pentosans are split into pentoses [14]:

$$(C_5H_8O_4)_x + xH_2O \rightarrow xC_5H_{10}O_5$$

The products of hydrolysis of hexosans are glucose and other hexoses (galactose, fructose). Hemicelluloses are more easily hydrolyzed than cellulose, and unlike cellulose, they are soluble in alkalis [14].

When heated with acids, pentose is converted into cyclic aldehyde—furfural. Three water molecules are detached from each pentose molecule [14].

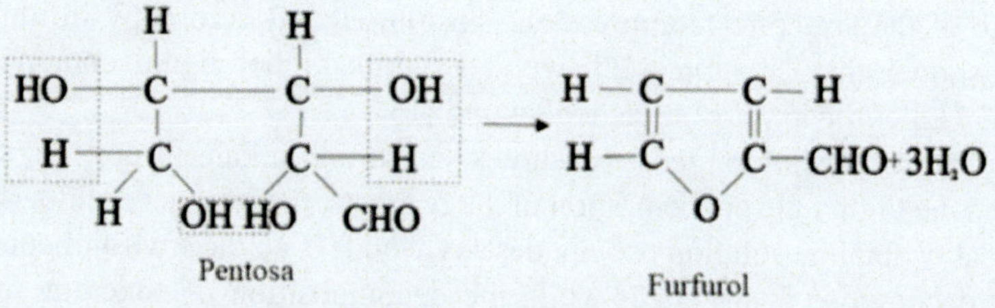

Apparently, in wood during its natural aging, not only destructive and reducing processes at the molecular level occur but also the formation of new chemical complexes through the interaction of hydroxyl, acetyl, aldehyde, carboxyl, and other groups. The greatest changes as a result of oxidative processes are observed in the lignocarbohydrate complex. The aging process that occurs with wood is a result of the action of a combination of climatic, mechanical, and biological factors, accompanied by various types of degradation (photochemical, enzymatic, direct photolysis, photosensitized, and others).

When the surface of wood is exposed to sunlight, the most destructive is the UV component with a wavelength of the spectrum below 380 nm. The wood darkens rapidly and the fibers peel off. This destroys the wood at the molecular level with active degradation of the lignin.

Lignin is less resistant to chemical attack than cellulose. The presence of a significant number of methoxyl groups in the lignin molecule explains why dry distillation of wood produces methyl alcohol. Dry distillation of pure cellulose does not produce methyl alcohol. When lignin is fused with alkali and oxidized with ozone, aromatic series compounds are formed [14].

During mild selective oxidation of lignin with nitrobenzene in an alkaline medium, a cleavage of the bond between α and β-carbon atoms in the macromolecule links containing the benzyl alcohol group or C=C bond is observed, yielding a mixture of hydroxyaromatic aldehydes and acids.

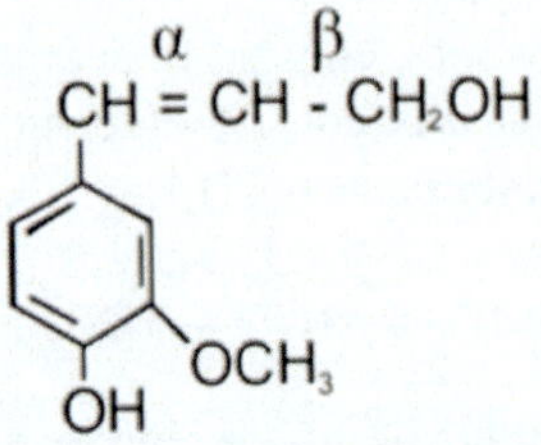

The oxidation of lignin by the action of O_2 in an alkaline environment of H_2O_2, ClO_2, and hypochlorites in order to remove its residues in cellulose bleaching is important. Lignin in wood is bound to polysaccharides, mainly hemicelluloses, as suggested by ester, glycoside, and simple benzyl ester bonds [13].

The content of the main components in wood also varies depending on the type of stress state of the structure (compression, tension, etc.). Microscopy in ultraviolet light has shown that "compressed" wood exhibits higher lignin content in the secondary wall compared to "stretched" wood [13].

During hydrolytic prolonged exposure to acids and alkalis, wood degradation occurs, and the higher the concentration of the reagents, the stronger their destructive effect. Weakly alkaline solutions do not destroy wood. In acidic environments, wood begins to decompose at pH < 2, while the decomposition of concrete and steel begins at pH < 4. Conifers are more resistant to sulfuric, nitric, hydrochloric, and acetic acids and caustic soda than deciduous trees, and larch is the most resistant of conifers. Timber is less resistant in sea water than in river water. In water of high bacteriological aggressiveness, the resistance of wood is low, which is why it is not used in sewage networks [15]. The only case of high resistance of wood to seawater is the use of coniferous larch. The increase in the strength properties of larch wood over time when exposed to sea water for a long time is truly phenomenal. An example of larch longevity and durability is its use as support piles on which Venice stands. The most characteristic difference of larch from other species of wood is its content of up to 15% of arabinogalactan, a water-soluble larch gum or gummie of larch [16].

The elemental composition of wood changes along with its chemical composition due to the influence of environmental factors. Thus, during rotting of wood, significant changes in its elemental composition are observed. In the destructive type, depending on the stage of rot, the carbon content of wood increases from 49–50% to 61%, and in the corrosive type, it decreases to 41%. Thus, the calorific value of wood changes along with it [13].

3.2 Effect of the Natural Aging Process of Wood on Its Physical and Physicochemical Properties

Wood is a reproducible material of plant origin. Wood cell walls are made up of approximately 99% organic compounds. Mineral substances usually account for no more than 1%. When wood is burned, it produces ash [13]. The organic substances in wood are divided into three parts: 1—carbohydrate, 2—aromatic, and 3—extractive substances.

During exploitation, wood undergoes, first of all, hydrolytic destruction under the influence of moisture, which is accelerated in the presence of acidic substances [15]. During hydrolysis, cellulose partially breaks down its macromolecules to form water-soluble substances, which reduces the strength of wood. The cellulose content in old wood, compared to modern wood, decreases insignificantly. Lignin, in turn, acts as a cellulose microfibril binding agent, thus ensuring, together with it, the strength of the wood. The content of lignin increases somewhat during the life of the wood due to the reduction of extractive substances and carbohydrate components. Easily soluble substances are found in the thickness of the cell walls of wood. In high humidity and in the presence of acidic substances, they are washed away intensively. The rate of polysaccharide leaching from the surface zone of long-term exploited wood depends largely on the operating conditions [15].

Hydrolysable hemicelluloses change the most during wood aging. Under the influence of air and moisture, they change into oligosaccharides which are soluble in hot water. While during hot water extraction of fresh wood only traces of reducing substances are found in the extract, from old wood up to 9% of carbohydrates in relation to absolutely dry wood are transferred into the extract [15].

It is known [13] that prolonged exposure of wood to oxygen, light (photochemical degradation), and enzymes leads to breaking of polymer chains, primarily hemicelluloses, which have a low (compared to cellulose) degree of polymerization and therefore are the most vulnerable component. At the moment of chain breaking, the oxidation mechanism is activated. The speed of this process depends on the operating conditions. According to these, the material had to be divided into two groups: 1—wood of interiors and monuments and 2—wood of archaeological finds. It is also known that the content of cellulose, lignin, substances extracted by organic solvents, and ash increases with increasing age of wood, while the content of pentosans decreases [13].

Table 3.1 shows the chemical composition of wood from the X to XIII centuries found during excavations in Novgorod and Brest [14].

The results of the study showed that the amount of carbohydrate components in archaeological wood decreases significantly. Pine wood retained significantly more pentosans than hardwood, which is explained by their greater resistance to degradation [15].

To obtain additional information about the peculiarities of the structure of aged wood, its infrared spectra were studied, and spectra of the same species of modern wood were also taken for comparison (Fig. 3.1) [15]. The results showed (Fig. 3.1) [15] that the spectra of modern and archaeological wood samples are different. In the

Table 3.1 Chemical composition (%) of wood from X to XIII centuries

Name of wood component	Wood species		
	Ash	Birch	Pine
Ash substances	6.29	2.71/0.14	7.88/0.17
Cold water extractable substances	6.20	4.94	3.93
Substances extractable with water at 90 °C	7.60	5.42/1.41	5.14/3.19
Substances extractable with an alcohol and benzene mixture	2.88	3.82	4.88
Lignin	51.69	46.71/19.74	44.97/24.68
Cellulose	14.53	16.49/35.38	31.53/44.10
Pentosans	17.93	15.13/24.57	6.36/7.60
Easily hydrolysable polysaccharides	18.52	16.95/26.54	11.62/17.84
Polysaccharides that are difficult to hydrolyze	15.48	18.16/39.40	33.21/47.66

Note. The denominator indicates the amount of modern wood components

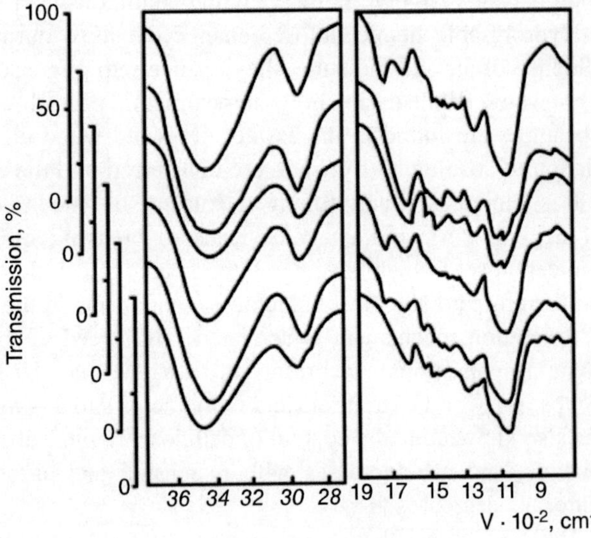

Fig. 3.1 Infrared spectra of wood samples—modern (**a**) and archaeological (**b**): 1—birch; 2—ash; 3—pine

spectra of archaeological wood of all species, there is no band in the range 1740–1765 cm^{-1} and the band 1380 cm^{-1}, which are present in the spectra of modern wood. In addition, there is a change in band intensity in the region of 1220–1270 cm^{-1}. The bands around 1750, 1380, and 1240 cm^{-1} are known to be characteristic of acetyl groups [15], the presence of which is assumed in hemicelluloses. The absence of them indicates that, over time, hemicelluloses are degraded in archaeological wood, leading to the loss of acetyl groups.

Table 3.2 Chemical composition of seasoned wood

Characteristics of wood samples	Content in wood, %				
	Ash	Extractives	Cellulose	Lignin	Light hydrolysable substances
Fresh spruce	0.26	3.96	54.47	26.29	17.76
Spruce 3 years old	0.28	7.5	54.40	25.40	18.50
Spruce 50–70 years old	0.25	7.28	53.97	25.01	17.80
Spruce 100–200 year old	0.48	10.63	50.78	24.64	14.16
Spruce 200 years old	0.24	10.57	51.72	26.12	13.34
Spruce 200–300 years old	0.48	11.3	52.34	24.90	13.16
Spruce 300–400 year old	0.47	15.22	50.77	24.26	12.11
Spruce 400 years aging	0.39	13.18	52.78	25.03	12.36
Spruce 500–700 year old	0.42	21.36	49.45	23.97	10.35
Fresh maple	0.56	8.19	46.53	24.18	22.51
Maple with 3 years aging	0.39	7.59	45.99	24.60	22.79
Maple 70 year old maple	0.51	8.83	46.53	23.97	19.80
100 year old maple	0.52	7.3	47.71	21.91	18.98

Note. Lightly hydrolyzed hemicelluloses were determined after the samples had been treated with hot water for 3 h in a boiling water bath

In the spectra of archaeological wood samples, an increase in the intensity of bands 1610, 1520, 1470, 1430, 1270, 1140, and 840 cm^{-1} (relative to bands in the range 1000–1100 cm^{-1}) characteristic of lignin is observed, indicating a higher relative content of lignin in archaeological wood [15].

In the 1970s of the twentieth century, a group of scientists carried out research on old and fresh wood of two species: spruce and maple [16]. The results of chemical analysis of wood samples are presented in Table 3.2.

According to the chemical composition of aged wood, we can conclude that easily hydrolysable hemicelluloses are the most amenable to change over time (losses over 700 years reach 8%). This is due to the fact that under the influence of water and air, they transform into oligosaccharides, which in turn dissolve in hot water. Cellulose and lignin are less variable over time. Their loss is around 5% per 700 years [16].

Figure 3.2 shows a comparative analysis of changes in the ratio of chemical components of wood in the monuments of wooden architecture of different ages [5].

The data presented in Fig. 3.2 shows a complex pattern of changes in cellulose and lignin over time. The analysis of the data shows a sinusoidal dependence of changes in cellulose and lignin content. For pentosans, there is a steady decline in their content in wood with increasing age.

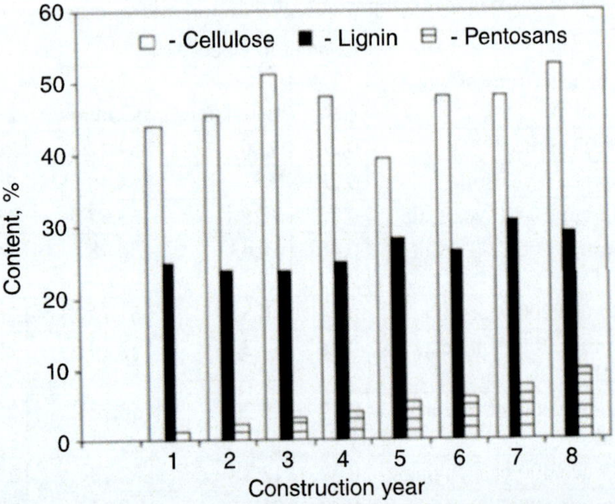

Fig. 3.2 Ratio of wood components in different monuments: 1—1493, 2—1600, 3—1699, 4—1700, 5—1750, 6—1790, 7—1870, 8—2000

Prolonged external exposure of wood to atmospheric oxygen, solar radiation, wind, precipitation, and increased air temperature leads to destructive chemical reactions, primarily of the carbohydrate part of the material. Hemicelluloses, which have a lower degree of polymerization than cellulose, are particularly vulnerable.

Of interest are the results of studies of the chemical composition of wood used in building structures of different durations. In [17], samples of wood from building constructions with different operating conditions were taken. The outer layer of wood up to 2–4 mm was analyzed. The results of the analysis of the chemical composition of wood are given in Table 3.3.

From the data presented in Table 3.3, we can see that the strongest changes occur in the wood under the direct influence of exploitation factors (changes in temperature and humidity, atmospheric, precipitation, biopollution). The greatest changes in the chemical composition took place in the cross-piece of the church, which is located in the direct impact of the abovementioned factors of wood aging.

The volume weight of the wood varies within a very wide range. Among domestic species the highest volumetric weight has boxwood (960 kg/m^3) and iron birch (970 kg/m^3); the lowest density has Siberian fir (375 kg/m^3) and white willow (425 kg/m^3). As the research results show, during the exploitation of wood a considerable change in its volumetric mass can occur. Some data on volumetric weight of pine and oak wood are presented in [18] Table 3.4.

The obtained average values of volumetric weight of wood, dated to the corresponding age, show their significant scatter for different monuments of wooden architecture. Analysis of the data allowed us to obtain dependences reflecting the complex nature of volumetric mass change of different wood species over time (Fig. 3.3).

Table 3.3 Chemical analysis of timber of building structures of different lifetimes

Lifetime, years	Design element and operating conditions	Lignin	Cellulose	Easily hydrolysable polysaccharides	Extractable substances	Ash content
0	–	26.5	54.3	17.8	10.7	–
85	Floor joist in the room	25.3	55.7	16.4	4.9	0.34
113	Eaves bracket—external structure	28.2	48.9	14.8	2.2	0.34
124	Top edge of wall underneath the boarding	28.7	49.9	16.7	2.1	0.25
124	As above	26.2	51.6	17.2	3.0	0.74
191	Staircase of poles under the roof	28.8	50.0	13.7	2.0	1.97
199	Internal structural element	27.9	48.9	16.8	4.6	0.67
223	Upper wall Wing under boarding	27.3	50.5	14.4	4.9	0.29
262	Ridge element	26.8	50.9	14.7	3.6	0.18
262	Church cross beam	29.6	48.5	9.8	0.8	0.89
351	Floor joist in the room	26.4	51.9	16.7	3.6	0.37

The most uniform decrease in volumetric mass over time is observed for oak and pine wood. A more complex change in volumetric mass associated with changes in their chemical composition is found in spruce and lime [18].

Studies of changes in wood volume mass over a shorter time period of up to 100 years show [19] that the volume mass of wood sampled at the level of the human chest increases in all the studied species almost over the entire time period (Fig. 3.4).

The average age of conifers at which the maximum values of wood volume weight are observed is about 100 years. For pine samples from southern taiga areas, this age averages 80 years. For spruce timber, the culmination of volumetric mass comes later in time [19].

In [5], refined data were obtained on changes in the volume weight of wood depending on the time of its exploitation (Fig. 3.5). The results show that a steady increase in wood volume mass is observed at time intervals of up to 100 years and

Table 3.4 Values for volume weight of seasoned wood

τ, years	ρ_0, g/cm^3	τ, years	ρ_0, g/cm^3	τ, years	ρ_0, g/cm^3	τ, years	ρ_0, g/cm^3
Pine							
1890–1894	0.499	1800	0.548	1717	0.555	1631	0.612
1882	0.362	1790	0.379	1717	0.482	1581	0.515
1869	0.423	1787	0.421	1717	0.411	1581	0.474
1857	0.564	1770	0.536	1714	0.500	1575	0.483
1850	0.474	1757	0.523	1702	0.441	1572	0.397
1850	0.477	1750	0.472	1699	0.454	1553	0.489
1850	0.519	1745	0.610	1689	0.568	1534	0.628
1846	0.493	1744	0.595	1687	0.403	1534	0.398
1845	0.509	1743	0.438	1672	0.518	1521	0.625
1844	0.502	1734	0.484	1670	0.569	1497	0.551
1844	0.458	1725	0.439	1668	0.505	1497	0.600
1830	0.490	1722	0.420	1656	0.533	1493	0.466
1820	0.452	1718	0.484	1649	0.433	1460	0.571
1801	0.475	1718	0.534	1646	0.486	1450	0.602
1801	0.566	1718	0.638	1645	0.426	1450	0.583
1801	0.451	1717	0.540	1645	0.618	1450	0.510
1800	0.420	1717	0.472	1645	0.638	1440	0.666
1387	0.315	1384	0.470	1384	0.455	–	–
Oak							
1930	0.579	1760	0.614	XVIII c.	0.769	1647	0.504
1920	0.717	1760	0.636	–//–	0.725	1632	0.709
1870	0.553	1710	0.648	–//–	0.600	1564	0.585
1870	0.744	1710–40	0.639	–//–	0.558	1527	0.569
1801	0.678	XVIII c.	0.646	–//–	0.643	End XVII c.	0.663
1773	0.662	XVIII c.	0.774	–//–	0.577	1228	0.449
1770	0.526	–//–	0.678	1651	0.616	–	–

from 200 to 300 years. Extreme time periods of 200 and 350 years are characterized by a maximum decrease in wood volume mass.

When analyzing samples of wooden structures, it was found that with increasing service life there is an increase in wood density due to removal of volatile low-boiling components from the wood material, shrinkage, and transformation of the morphological structure of the wood substrate [5] (Fig. 3.5). A similar increase in wood density over a service life of 100–150 years was noted by several authors in [10, 17].

For a more detailed study of the effect of natural wood aging on decomposition processes, heat and fire resistance, partly samples of deciduous and coniferous wood,

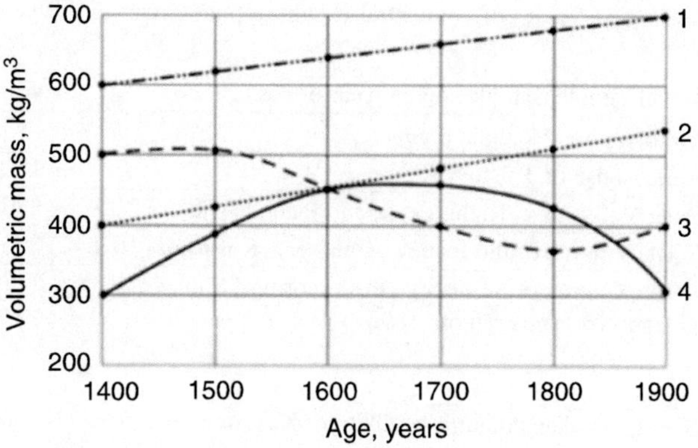

Fig. 3.3 Relation of volume weight to wood age obtained from a small number of samples: 1—oak; 2—pine; 3—spruce; 4—lime

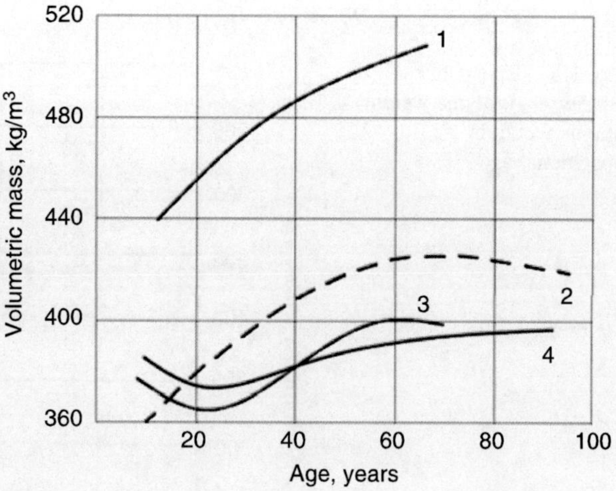

Fig. 3.4 Variation of wood volume weight with age (cut from samples taken at breast level): 1—birch; 2—pine; 3—aspen; 4—spruce

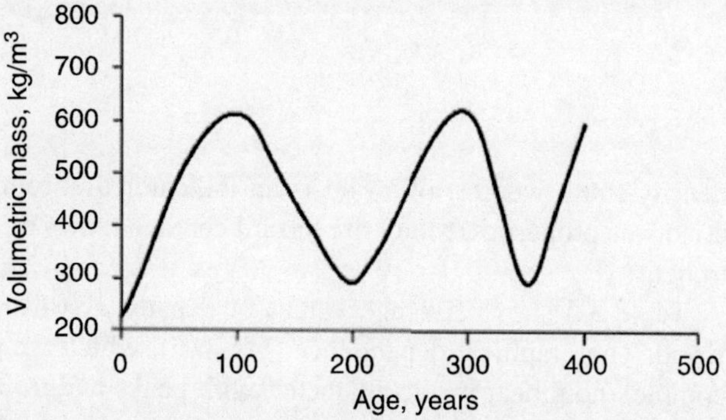

Fig. 3.5 Variation of wood volume weight as a function of operating time

taken from old religious buildings, were involved. Oak wood samples were taken from elements of molded building products. In total, the timber elements and products had a service life of between 2 and 370 years. To get an idea of the studied specific samples, we previously determined their density, elemental composition, and the content of the main components of wood; it has already been noted that the

Table 3.5 Values of volumetric weight of pine timber of different lifetimes

Wood sample, sample source, year of construction	Service life, years	Volumetric mass, ρ_{12}, kg/m^3
Modern pine, Vologda region	–	440
Pine, House of Tolstoy, 1830	180	410
Pine, Volkostrov, Kizhi, Peter, and Paul chapel	330	587
Sosna, Church of the Ipatiev Monastery, Kostroma, 1628	375	535
Sosna, Church of St. George the victorious, Shulevo village, Kostroma region, 1898	112	North—488 South—537 East—414 West—426
Sosna, wooden structure v. Nikitino, Kostroma region, 1876	130	North—505 East—591 West—561

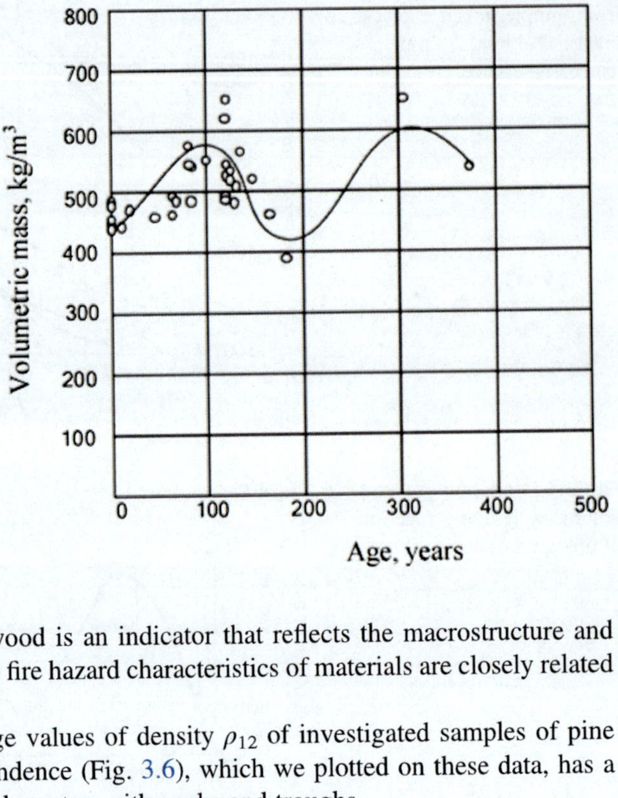

Fig. 3.6 Dependence of density ρ_{12} (volume weight) of pine wood on operational age

density (bulk weight) of wood is an indicator that reflects the macrostructure and affects its properties. Many fire hazard characteristics of materials are closely related to density.

Table 3.5 shows average values of density ρ_{12} of investigated samples of pine wood. The graphical dependence (Fig. 3.6), which we plotted on these data, has a complex nonlinear cyclic character, with peaks and troughs.

The maximum values of density are observed when the wood has been in service for about 100 and 330 years, and the minimum value is reached when pine wood has been in service for 190–200 years [20–23]. Such a nonlinear change in wood density as a result of its long-term exploitation is consistent with the works [5, 17].

According to I.I. Pischik [17], the cyclicity of natural aging processes in living and felled trees is interconnected with the double secular cycles of solar activity,

Table 3.6 Contents of the main chemical components of spruce wood depending on its operational age (Vologda region)

Sample location, operational age of the sample	Side of the world	Content of main chemical components, %	
		Cellulose	Lignin
Modern spruce (Vologda region)	–	54.5	25.4
Bar-built non-residential building (Lyabzunka village, Vologda region), spruce timber, 60 years old	North South	55.6 53.4	25.6 27.1
Uninhabited wooden structure (v. Levino, Vologda region), spruce wood, 90 years old	North South	50.6 49.0	27.1 28.2
Bar-built non-residential building (Demjanovo village, Vologda region), spruce, 110 years old	North South	50.8 48.6	26.9 28.1
Wooden residential building (Semigory village, Vologda region), spruce, 150 years old	North South	49.5 46.9	27.6 28.8

which leave their mark on the dendrological scale of woody plants. Decreasing values of ρ_{12} are explained by excretion of gaseous and water-soluble degradation products, while the volume of wood remains the same; the ascending branch of the curve is due to gradual shrinkage of the material.

What draws attention is the uneven density variation in different parts of the timber structures oriented towards the direction of the parts of the world. Particularly high values of density are observed on the southern and eastern parts of the world. This does not seem to be a coincidence. A response to the influence of sunlight is likely. Unfortunately, exact information about the biological age of the wood used for wooden buildings has been lost. Normally, 150–220-year-old pine was used for religious buildings [1], while 60–100-year-old pine was used for residential and farm buildings. For outbuildings, spruce wood was used more often.

The dependence of degradation processes in spruce wood on the location of the samples taken relative to the parts of light (the influence of sunlight) can be judged from the results of chemical analysis (Table 3.6) [23].

A simplified scheme for chemical analysis of wood was used [24]. The cellulose and lignin content was assessed. The remainder was a mixture of hemicelluloses and extractives.

The results of the study of the wood samples under consideration show that the cellulose content decreases with increasing service life up to 150 years, while the lignin content increases (Table 3.6). The southern part of the timber structure shows greater changes than its northern part. Under approximately equal conditions over 150 years of operation, the relative cellulose content in the elements of wooden structures on the southern side of the structure decreased by 13.9% and on the northern side by 9.1%.

As shown by analysis of replicas taken from the surface of the samples, electron microscopy and typical elements of coniferous wood structure, namely microfibrils in the form of ribbon-like fiber formations with fringed round-shaped pores, were found on the surface of the sample (Fig. 3.7).

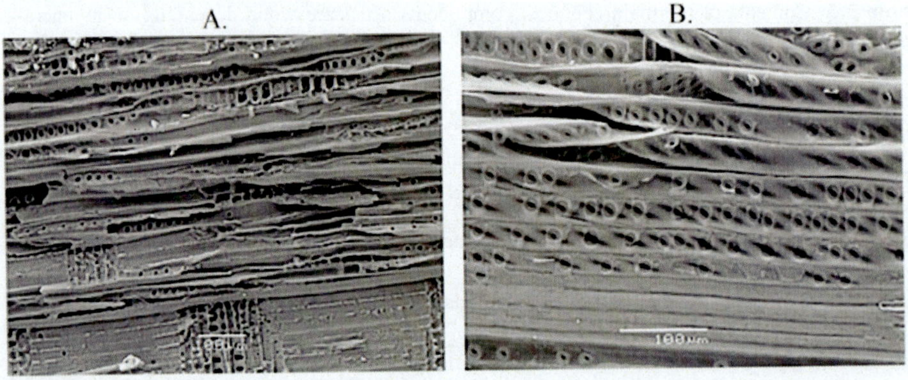

Fig. 3.7 Photo of the surface of modern wood. (**a**) 300× magnification and wood (150 years of use). (**b**) 150× magnification

After a long period of natural aging, a significant transformation of the wood's structure is noticeable. A plaque can be seen forming in the area of the fibers and pores. In some cases, the pores are partially or completely closed. This demonstrates some physical compaction of the wood substrate. A change in the physical and physical-chemical properties of long-term natural aging wood leads to a change in the strength characteristics of wooden structures. So in [25], it was found that the mechanical properties of larch wood from a wooden house built in Krasnoyarsk for 230 years of natural aging in the climatic conditions of Siberia significantly deteriorated compared to modern wood. The samples for the analysis were taken from the undamaged part of the lower walls of the house. Erosion was noted only in the surface layer of logs 2–3 cm thick. On average, the density of larch samples decreased by 8% (to 585 kg/m^3), compression strength along the fibers by 15% (to 45.3 MPa), ultimate bending strength in the tangential plane by 29% (to 68.1 MPa), and shear strength in the same plane by 28% (to 6.1 MPa). Anatomical analysis of tangential slices of old wood showed destruction of tracheids at their junctions and heart ray cells. The authors [26] consider the cause of such destruction to be the formation of large internal tensile stresses in the tangential direction when the logs periodically dry out to an air-dry state.

On wooden architecture objects in the Arkhangelsk region, standing from 85 years to 351 years, changes in pine wood density, ρ_{12}, strength under static bending in the tangential direction, and compression along the fibers were found [26]. Samples of healthy wood without defects were taken. Analysis of the chemical composition was carried out for externally undamaged outer surface layers with a thickness of 2–4 mm. In the process of aging, a nonlinear change in pine wood density is observed: a decrease in 140, 199 years from 490 to 420–430 kg/m^3 and further increase to 565 kg/m^3 (Fig. 3.8).

Almost synchronous changes in strength values were observed. At the same time, their spread decreased from 10–12% to 3–4%, indicating stabilization of wood condition during long-term operation of structural elements under load.

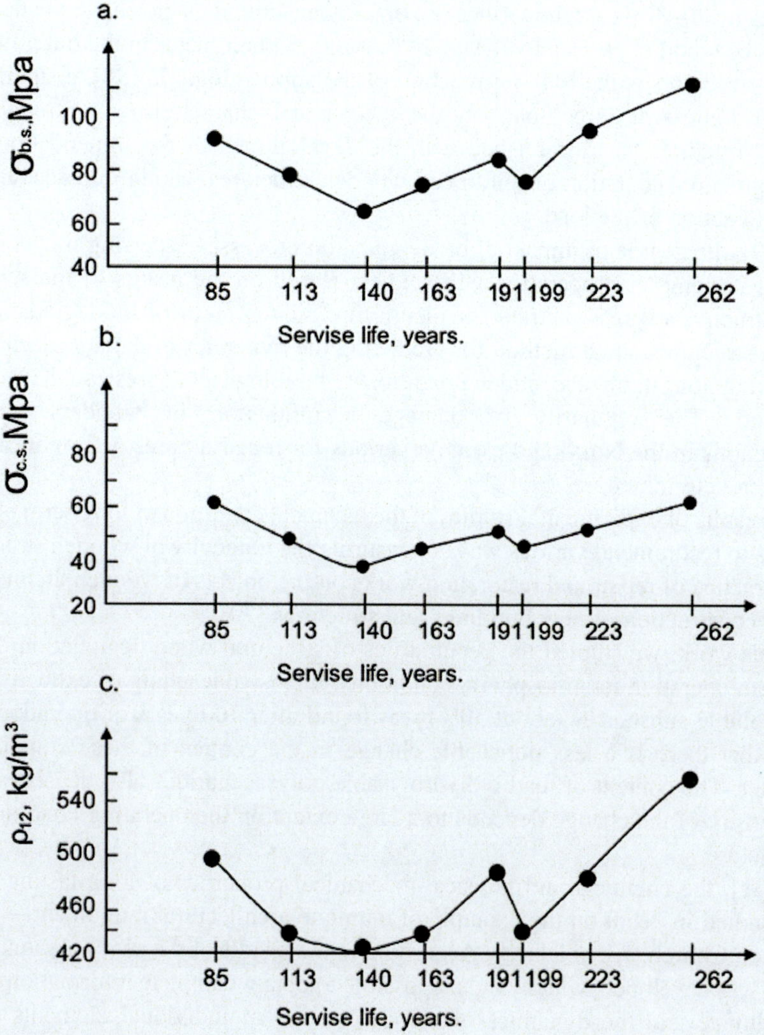

Fig. 3.8 Effect of the duration of natural aging of wood on bending strength (**a**), compression strength (**b**), and density ρ_{12} (**c**)

Few studies have been conducted on the effect of long-term operation on the performance of load-bearing and enclosing wooden structures, as well as increasing their durability [27–31].

In [27], a study of the load-carrying capacity of long-lived wooden structures was carried out, taking into account the damage occurred in them and refining the recommendation to increase their durability. As a result of the research, the main regularities of occurrence and development of mechanical and biological damages of wooden structures have been established. The author has found out that during exploitation of wooden structures up to 200 years, the strength of wood under compression and bending does not change essentially. The shear strength of wood decreases by 20–30%. As the service life of wooden structures increases, the flexural modulus of wood decreases by 4% to 24%. An important factor in the durability of wooden structures is the long-term action of the applied load. In [28], a significant effect of long-term force load on the mechanical characteristics of reinforced wooden structures was established with the development of recommendations for the design and calculation of reinforced wooden structures, taking into account the long-term action of the load.

In [29], the degree of impact of biodegradation processes occurring in wood as a result of its long-term operation was studied. Based on the results of the study, a non-destructive express method for identifying hidden areas of biodegradation of wooden structures and a method for predicting the dynamics of damage accumulation during long-term use under conditions of biological aggression have been developed. The regularity of damage accumulation in wooden building constructions in the North and extreme periods for repair and restoration measures have been determined.

The results of experimental studies of the technical condition of long-term objects allow us to recommend various ways of ensuring the longevity of wooden structures in the practice of repair and restoration works on the objects of wooden architecture and the construction of new buildings and structures [30].

In this work, we studied the peculiarities of structural wood degradation during long-term operation in atmospheric conditions. Intensive leaching of extractive and water-soluble substances (about 50%) was found after 100 years of operation. It is shown that there is a less noticeable change in the content of these components thereafter. The content of readily hydrolysable polysaccharides also decreases, and the intensity of the change depends to a large extent on the operating conditions of the structure.

In [31], the chemical and physical-mechanical properties of long-lasting wood were studied in detail on the example of a unique architectural monument—Transfiguration Church (Kizhi Island). According to the results of the study, a monitoring system was developed, which made it possible to obtain complete information on the microclimate and the dynamics of wood moisture of individual elements of the structure and assess the activity of the processes of biodegradation. The analysis of the obtained physical-mechanical parameters and chemical composition of the wood, which has been exploited for more than 280 years, allowed to conclude that it has sufficiently preserved its strength properties. The patterns of wood damage by wood staining and wood destroying fungi have been identified.

Hydrolytic and thermo-oxidative reactions of wood substance decomposition play an important role in the aging process. Aging reactions primarily affect oligomeric carbohydrate and extractive components of the natural polymer composite. The destruction of cellulose macromolecules in the amorphous part of the structure leads to the formation of water-soluble substances, their leaching from the material under the influence of rainfall, changes in the volume weight of the material, and its strength itself.

Thus, the service life of timber structures has a significant influence on the physical, physicochemical, mechanical, and other properties. Increase and decrease of quantitative indicators can have cyclic character. Few works are devoted to the peculiarities of influence of wood exploitation duration on its fire hazard properties and behavior of wooden constructions in fire conditions. To a greater extent, studies of fire hazard and fire resistance of wooden structures are carried out in relation to wood without taking into account the conditions and duration of operation. Extensive research has been conducted on the effect of wood species on the peculiarities of its combustion process and fire hazard properties [32–36].

3.3 Peculiarities of Fire Hazard of Long-Term Natural Aging Wood

3.3.1 Peculiarities of Thermal and Thermo-Oxidative Decomposition of Long-Term Natural Aging Wood

It is known that the processes of natural aging occurring in wood largely determine its thermal stability. In [37], it was found that the least thermal stability of wood is observed in the first hundred years and after 300 years of its operation. In this work, the author suggested that during these time periods wood is the most fire-prone.

It was found that in time periods of 150–200 years the activation energy and activation entropy significantly increase, i.e., any chemical processes, including thermal degradation, become more difficult (Fig. 3.9) [37].

This paper presents, for the first time, data on the thermal decomposition of wood as a function of operating time over 500 years. Many of the world's monuments of wooden architecture have a similar age of operation. Despite this, studies of the effect of physical and chemical transformations occurring in wood over time on its fire properties are limited.

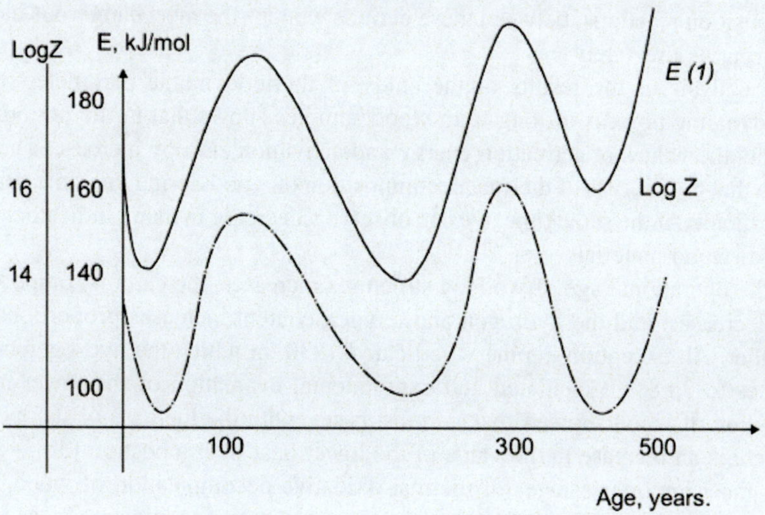

Fig. 3.9 Dependence of kinetic parameters of wood thermal decomposition on operation time: 1—activation energy kJ/mol; 2—LogZ

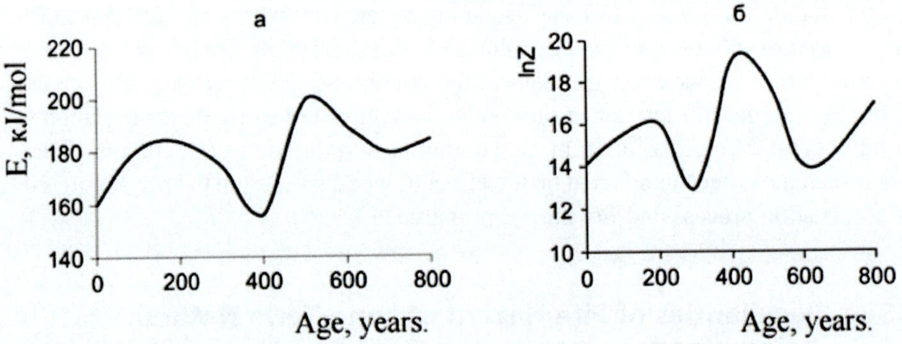

Fig. 3.10 Variation of thermodynamic parameters of the thermodynamic process of wood samples as a function of operation time: (**a**) *E*—activation energy; (**b**)—ln*Z*—activation entropy

Thermodynamic parameters of the thermodynamic process of wood decomposition (activation energy and activation entropy) indirectly reflect the physical and chemical changes occurring with it over time (Fig. 3.10).

The first attempts to find out the peculiarities of long-term wood behavior in fire conditions were related to the study of thermal stability of samples by the example of wooden architecture monuments. So the scientific team headed by Professor E.N. Pokrovskaya carried out a series of works devoted to the study of the dynamics of changes in the thermal properties of wood in the process of aging [5, 10]. As a result, it was found [5] that the change in the thermal stability of wood is not linear and the minimum activation energy and activation entropy are characteristic for samples aged 100 and 300 years and reach their maximum values in the period of 150–200 years (Fig. 3.11).

In this work, wood samples from structures with a service life of up to 525 years were investigated. According to the results obtained, the "step" of the curves is about 200 years, i.e., every 200 years, there are peaks of energy difficulties of thermal decomposition reactions; between these periods wood is the most dangerous in terms of ensuring its fire safety.

The analysis of the results of the study of thermodynamic parameters of the thermodynamic process of long-term wood samples shows that in the period up to 200 years the values of activation energy and activation entropy increase. This may indicate that the process of thermal decomposition of wood during this time period is more difficult. At the same time, we can observe a decrease in many indicators of fire hazard of wood material.

As the operational age of wooden structures increases, the carbon component in wood decreases, and the hydrogen and oxygen content increases proportionally to each other. An exception is pine wood dated 1830, in which the hydrogen content decreases to 3.88%. Calculated and experimental evaluation of the lower heat of combustion of wood showed that, in most cases, with the increase of its operating age, there is an increase in the value of the lower heat of combustion [38].

The most important stages of thermal-oxidative decomposition of wood, determining the features of its combustion, are the main decomposition stage, as well as the stage of formation and oxidation of the coke layer. In this regard, the kinetic

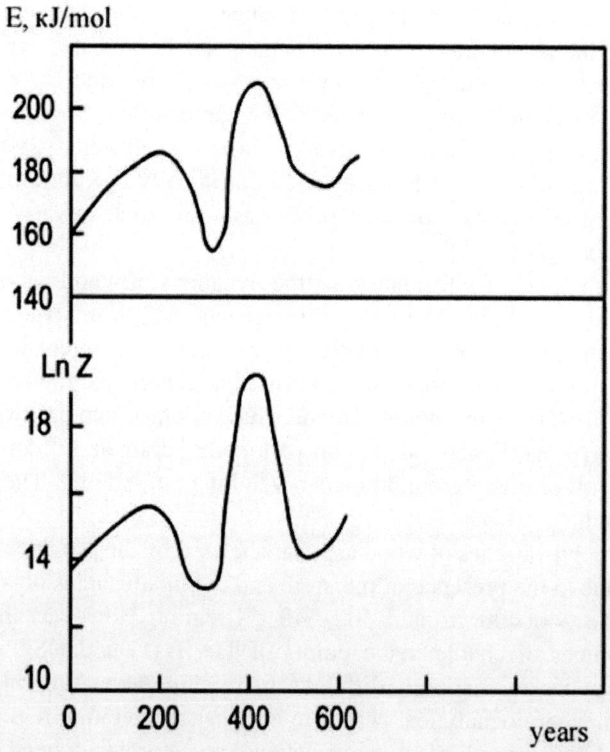

Fig. 3.11 Dependence of kinetic parameters of wood thermal decomposition on operation time: 1—activation energy (*E*), kJ/mol; 2—pre-exponential multiplier (ln*Z*), medium—air, heating rate 20 °C/min

parameters of the wood decomposition process, in particular, the change in the effective activation energy at the above stages, are particularly revealing. At the main stage of thermo-oxidative decomposition of the studied wood samples, insignificant changes in activation energy are observed for the degree of wood transformation from 0% to 80% for both modern and artificially aged wood. This indicates, first of all, the similarity of chemical reactions of wood components oxidation and the important contribution of polysaccharide cellulose component. For deciduous and coniferous varieties of wood the values of activation energy at the section of degrees of transformation from 10% to 80% are practically identical. The difference in the decomposition rate of wood samples observed on TGA curves is due to different values of pre-exponential multipliers (the role of entropic factors in the decomposition kinetics) in the Arrhenius equation.

As the degree of transformation increases, the effective activation energy increases. In the case of oak wood, it becomes much higher than in spruce wood. This is related to the initial activation energy of the oxidation process of the formed coke and, apparently, to its structural features. The quantity and quality of the formed carbonized product have a significant influence on the nature of the course of flame and smoldering combustion of wood materials. In this regard, the most important stage determining the energy of the combustion process of wood substance as a whole is the oxidation of the formed coke.

Unfortunately, the kinetic parameters for the main thermal decomposition stage and coke oxidation stage of long-term natural aging wood have not been evaluated in the works conducted. However, there are studies aimed at establishing the effect of artificial wood aging on the change in its thermal stability [38]. It was found that the maximum value of activation energy of oxidation of coke of modern birch wood (540 kJ/mol) was manifested at the degree of coke transformation of 60%, while for coke of aged wood the activation energy is reduced to 290 kJ/mol at the same degree of transformation. For the coke of artificially aged birch wood, the effective activation energy of oxidation processes was much lower than that for modern birch wood [38].

Important information on the dynamics of wood decomposition when heated in air can be obtained using thermal methods of analysis. In this case, to study the patterns of thermal-oxidative degradation of different wood species, as well as the carbon layer on its surface, we used an automated modular thermal analysis system TGA/DSC1 by Mettler Toledo (Switzerland). The unit was provided with a module for programmed calculation of kinetic parameters. Samples were prepared in the form of finely ground powder weighing 1.0–8.5 mg. The heating rate was 5–20 °C/min.

All varieties of wood are characterized by the presence of a low-temperature stage due to the presence of moisture and two main stages of weight loss of the sample in the 240–400 °C and 400–540 °C region. However, the position and maximum values of characteristic points of TG, DTG and DSC curves for different wood species are markedly different. Increasing the heating rate leads to a natural shift of the thermoanalytical curves to a higher temperature region. A comparative analysis of the curves leads to the conclusion that conifers when heated to 360–400 °C begin to decompose earlier and at a higher rate than deciduous varieties.

Thermal oxidative decomposition of wood is a process proceeding with considerable heat release at all its main stages. The presence of exopics on DSC curves serves as a confirmation of this fact. Decomposition of wood in air is accompanied by its charring. The greatest heat release during thermo-oxidative decomposition is associated with oxidation of the carbonized product at high temperature.

As a result of the natural aging of wood, a significant change in its thermooxidative stability occurs [21, 39, 40].

By the example of oak wood samples taken from the boards of icons of Holy Trinity Sergius Lavra (Moscow region), the effect of the duration of object exploitation on the process of thermooxidative decomposition of wood substance was shown (Fig. 3.12).

It can be seen that there is a significant decrease in thermal oxidative stability of oak wood samples, especially noticeable by DSC and DTG curves. During the natural aging of oak wood in the period of operation of the product for more than 460 years, all temperature parameters decrease: the temperature of the beginning of decomposition, the temperature of maximum mass loss rate, etc. [39].

Consideration of thermoanalytical curves of wood leads to an important conclusion that as a result of natural aging of the material during long-term operation there are transformations that favor the conditions for the reaction of decomposition, carbonization of wood and oxidation of the charred product. This situation is largely due to an increase in the content of lignin, a high-energy aromatic component of wood, as a result of aging, and changes in the porous structure.

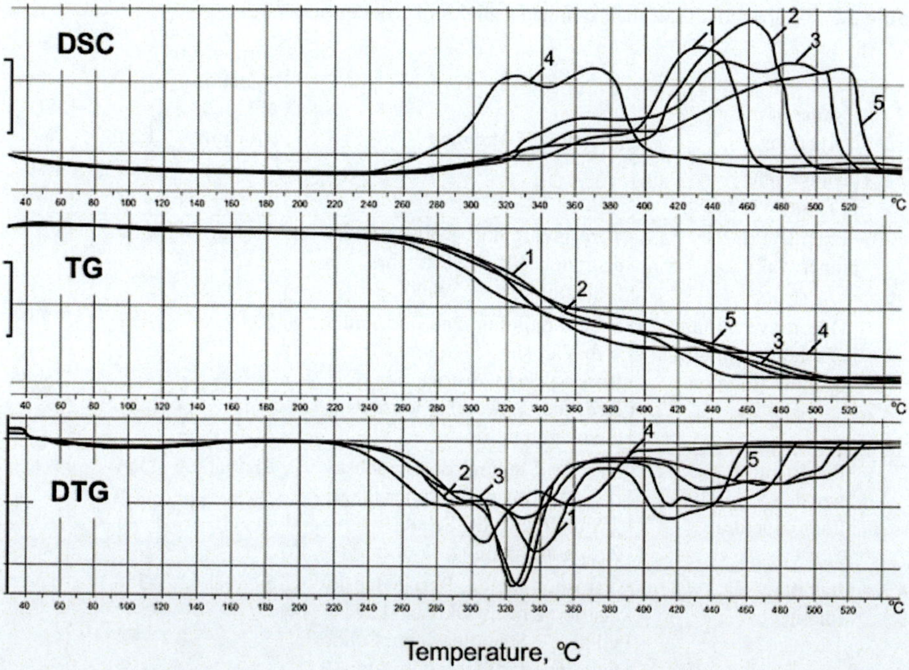

Fig. 3.12 TG, DTG and DSC curves of oak samples of different age: 1—1650 years; 2—1901 years; 3—artificially aged for 80 years; 4—1540 years; 5—modern wood (medium—air, heating rate 20 °C/min)

A similar pattern of changes in the thermo-oxidative stability of wood was observed by the authors of [23] studying linden wood samples from the frames and boards of old Romanian monastery icons. As the life span of the samples increased from 6 to 200 years, the thermo-oxidative stability decreased, the activation energy and the pre-exponential multiplier of the decomposition process decreased by three orders of magnitude compared to the reference linden sample. However, the 99-year-old wood was more stable than the reference and the bicentennial (1802) sample. The effective decomposition activation energy of this wood sample was estimated to be 105.2 kJ/mol at $\ln Z = 20.8$ (the reference had 85.5 kJ/mol and 16.6, respectively).

Especially large range in timber lifetime and its influence on thermal stability of wooden structure elements was traced in work [41]. The authors had samples of pine wood, which corresponded to a service life in the range up to 525 years. The cyclic nonlinear nature of changes in the values of mass loss as well as the kinetic parameters of the wood decomposition process was found.

Minima on the curves $E_{eff} = f(\alpha)$ corresponded to the operating time of wooden structures about 100, 300 and 500 years, the maximums were in the region of 200 and 400 years. The values of $\lg Z$ also changed almost synchronously. At the points of maximum they increased by 2–5 orders of magnitude. Wood is most prone to destructive processes in the first 50–100 years of operation. Cycles repeat after

Table 3.7 Characteristics of the examined samples of wooden structures

N arr.	Name of the structure	Location of sampling	Term of operation of structure	*p*, kg/m³	*W*, %
0	Native pine (Yaroslavl region)	–	–	459	12.0
1	Fragment of the inner wall	Yaroslavl region, Borisoglebsky district, c. Krasny Oktyabr, 2 Shkolnikov St. School, 2.	81 year	550	9.5
2	Decorative framing of a house window opening (log house)	School building, 2nd floor, inner wall Yaroslavl region, Borisoglebsky district, c. Kedskoe, a dwelling house near the church	200 years	482	12.0
3	Fragment of a wooden log in the floor structure	Yaroslavl region, Borisoglebsky district, Krasny Oktyabr village, 2 Shkolovaya St. School, 2. School building, floor 2	81 year	554	9.1
4	Internal wall fragment	Yaroslavl region, Borisoglebsky district, Krasny Oktyabrskaya village, 2 Schoolnaya str. School building, 2nd floor, corridor	81 year	558	12.2

200 years. According to the authors [41] it is during the periods of the lowest stability that one should expect an increase in the fire hazard of wood.

To determine the features of thermal and thermo-oxidative decomposition of long-term natural aging wood (up to 200 years) depending on the service life of objects for one climatic region we took samples from objects located in the Yaroslavl region, Borisoglebsky district (Kedskoe village and Krasny Oktyabr village).

The buildings were built in 1935 and 1816 from wooden pine logs of a diameter from 220 to 240 mm. Elements of wooden structures were preserved in satisfactory condition, without mechanical and biological damage. The characteristics of the samples under study, including sampling locations, density (p) and moisture content (W) of the samples are presented in Table 3.7.

Decomposition of wood when heated in an inert environment or in the presence of air oxygen provides the formation of combustible volatile products and is one of the main stages of the combustion process of the material. Thermal analysis of wood samples in a nitrogen atmosphere was carried out with a subsequent change of the medium to air at a temperature of 500 °C at a rate of 20 deg./min. Air induction leads to almost complete mass loss of samples at a high rate (Fig. 3.13).

When considering TG and DTG decomposition curves of modern and naturally aged wood samples, we can note the presence of two main decomposition regions: low-temperature (140–400 °C) and high-temperature (above 400 °C).

The first area reflects the processes of thermal degradation of wood substance and the formation of volatile combustible products. Small mass losses when heating samples up to 150 °C can be attributed to the evaporation of moisture. TG and DTG

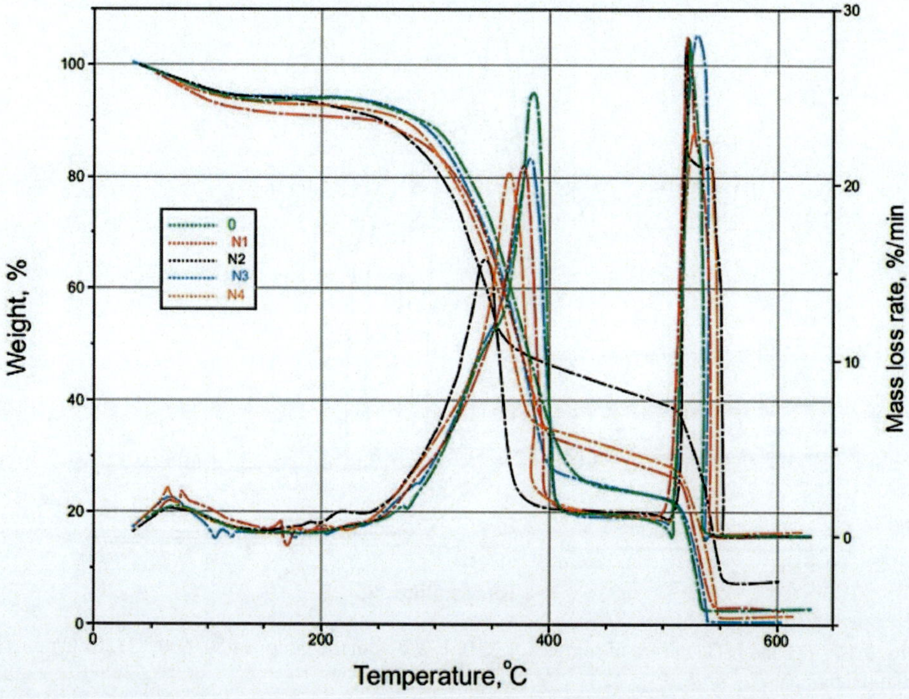

Fig. 3.13 TG and DTG curves of wood samples № 0, 1, 2, 3, 4 (20 °C/min, atmosphere—nitrogen to 500 °C further air)

Table 3.8 Results of thermal analysis of wood samples of different life cycles (medium—nitrogen)

Characteristics	Number of samples (Table 3.7)				
	«0»	«1»	«2»	«3»	«4»
Mass loss % (in the range 30–150 °C)	5.8	8.7	6.4	5.8	7.1
Mass loss, % (within interval 150–450 °C)	70.1	61.4	51.8	70.1	61.6
Maximum DTG (*T*, °C/A, %·min^{-1}) (in the interval 150–450 °C)	384/ 20.9	376/ 20.9	342/ 15.8	381/ 21.4	364/ 20.5
Oxidation rate of coal, %·min^{-1} (in interval 450–600 °C)	28.6	28.5	21.2	28.4	23.0

curves show that at the initial stage (temperature range 100–250 °C) there is more intensive mass loss of long-term wood samples compared to modern wood samples (Table 3.8).

In the temperature range of 150–450 °C the carbonization process obviously begins, which is reflected in the decrease of the maximum temperature and pyrolysis rate in the main section of the obtained thermograms (temperature range up to 450 °C).

Under conditions of thermal decomposition in an inert environment for long-term natural aging wood samples, a decrease in weight loss values in the temperature

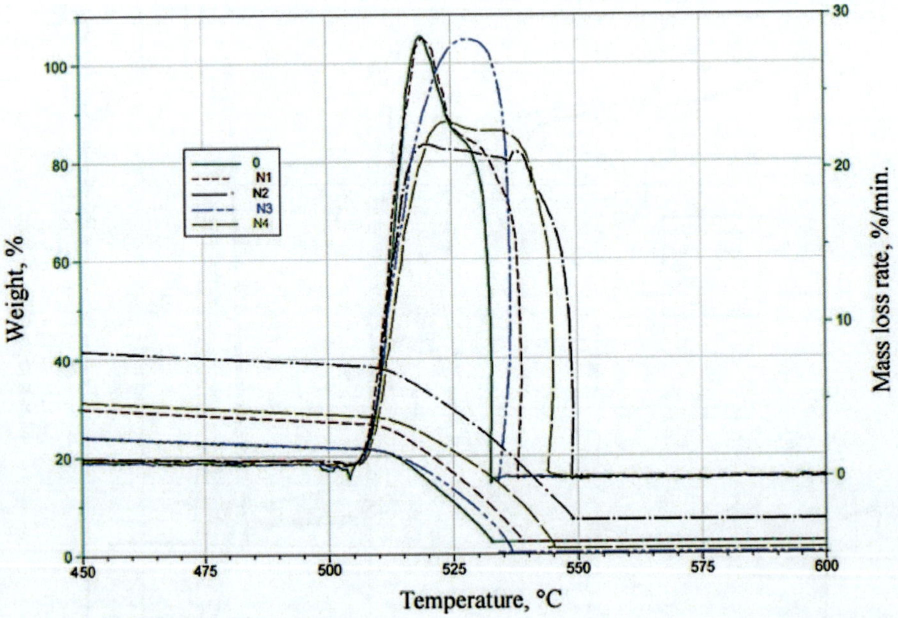

Fig. 3.14 TG and DTG curves of samples №№ 0, 1, 2, 3, 4 in the interval 450–600 °C (20 °C/min, coal oxidation)

range of 150–450 °C has been established in comparison with modern samples. The lowest values of weight loss were recorded for wood specimens with a service life of 200 years. This is apparently due to an abnormally low temperature of the beginning of the carbonization process of the examined wood sample (maximum DTG is observed at 342 °C). It should be noted that the decrease in the temperature threshold of the beginning of the carbonization process is characteristic of all the examined samples of long-term natural aging.

The second area of wood decomposition reflects the processes of active carbonization and oxidation of the charred product (Fig. 3.14).

Figure 3.14 shows that the most intensive oxidation process proceeds with respect to carbon residue of modern wood and wood with operating life of 80 years. It can be noted a higher anomalous duration of this process for pine wood with operating life of 200 years as compared to other samples (temperature interval 500–550 °C).

To clarify the features of thermal decomposition of wood samples under conditions close to the real fire, it was necessary to carry out thermal analysis of samples in the environment of air oxygen oxidizer. It is known that the thermal oxidative decomposition of wood is a process proceeding with significant heat release at all its main stages. Decomposition of wood in the air is accompanied by its charring. The greatest heat release during thermo-oxidative decomposition is associated with oxidation of the carbonized product at high temperature [5].

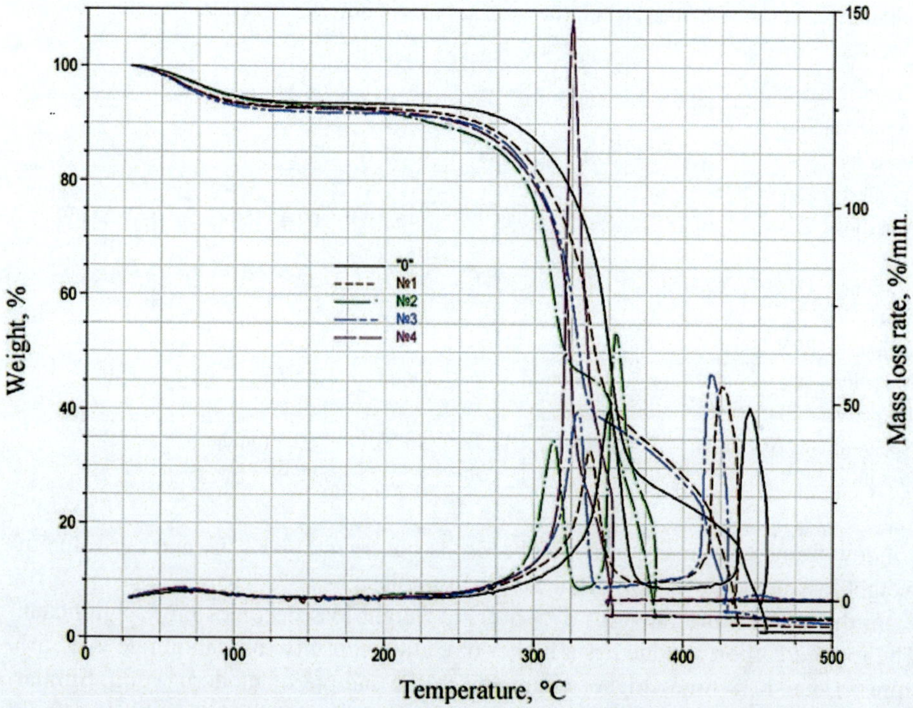

Fig. 3.15 TG and DTG curves of samples №№ 0, 1, 2, 3, 4 (20 °C/min, medium—air)

For the samples under study, a study of the effect of the duration of operation of wooden structures on the process of thermooxidative decomposition of wood substance was carried out.

According to the results of the study, a significant decrease in thermo-oxidative stability of pine wood samples with a service life of 200 years was found. This is especially noticeable by DSC and DTG curves (Fig. 3.15).

During the natural aging of pine wood in the period of operation up to 200 years, all temperature parameters decrease: the temperature of the beginning of decomposition, the temperature of the maximum rate of weight loss, etc. (Table 3.9).

The presence of a low-temperature stage caused by the presence of moisture and two main stages of weight loss of samples in the temperature ranges of 250–400 °C and 400–475 °C is characteristic for all wood samples under study. However, the position and maximum values of characteristic points of TG, DTG and DSC curves for different samples are markedly different (Table 3.9).

The obtained results show that with the increase of service life of wooden structure elements there are significant changes in the thermal stability of samples. These changes are associated with the fact that in the temperature range 150–375 °C there is a shift of the stage of maximum decomposition rate to the low-temperature region (DTG curves) for long-term natural aging samples, while the weight loss of the sample decreases. Thus, for wood with a service life of 200 years (decorative

Table 3.9 Results of thermal analysis of wood samples of different life cycles (medium—air)

Characteristics	Number of samples (Table 3.7) «0»	«1»	«2»	«3»	«4»
In interval 30–150 °C					
Mass loss, %	6.46	6.98	7.3	7.4	8.04
In interval 150–375 °C					
Mass loss, %	65.7	58.7	46.2	67.4 (150–340)	54.2 (150–330)
Maximum DTG *T*, °C/A, %·min^{-1}	350/48.2	337/38.0	312/40.5	325/47.1	329/47.7
In interval 375–500 °C					
Mass loss, %	26.5	32.0	42.7	21.7	35.0
In interval 400–500 °C					
Speed of coal oxidation, %·min^{-1}	48.8	54.3	67.5	30.5 (hidden peak)	57.3

window frame), the maximum weight loss is observed at 312 °C, and the value of weight loss is the lowest of all the studied samples.

In the temperature range of 375–500 °C weight loss becomes more significant. The formed carbon residue has a higher oxidation capacity (oxidation rate is 67.5% ·min^{-1}), i.e. 1.38 times higher compared to the sample of modern wood. Similar changes in thermal stability of wood samples were also established in [38].

This can be explained by the fact that as a result of natural aging of the material during prolonged use there are transformations that favor the conditions for the process of wood carbonization and oxidation of the charred product. This situation is largely due to the fact that the reactions occurring during natural aging of wood affect the oligomeric carbohydrate and extractive component of the natural polymer composite. Destruction of cellulose macromolecules in the amorphous part of the structure leads to formation of water-soluble substances, their leaching from the material under the influence of rainfall, changes in material density and its strength. At the same time, it should be noted that as a result of wood aging, the content of lignin, a high-energy aromatic component, may increase, which has a significant impact on the calorific value of carbonizate.

The greatest heat release during thermal-oxidative decomposition is established at the stage of oxidation of the carbonized product at high temperature (Fig. 3.16).

The results presented in air show an abnormal shift of the oxidation stage of carbon residue to the region of low temperatures for wood with a lifetime of 200 years.

For the oxidation stage of long natural aging wood, the DTG and DSC curves recorded a significant increase in the mass loss rate, the values of thermal effect and the heat of oxidation of carbon residue (Table 3.10).

Values of pyrolysis heat of wood in nitrogen medium with increasing service life show that the process of its thermal decomposition is facilitated in comparison with the modern wood sample. At the same time, the heat effect (J/g) and the carbon

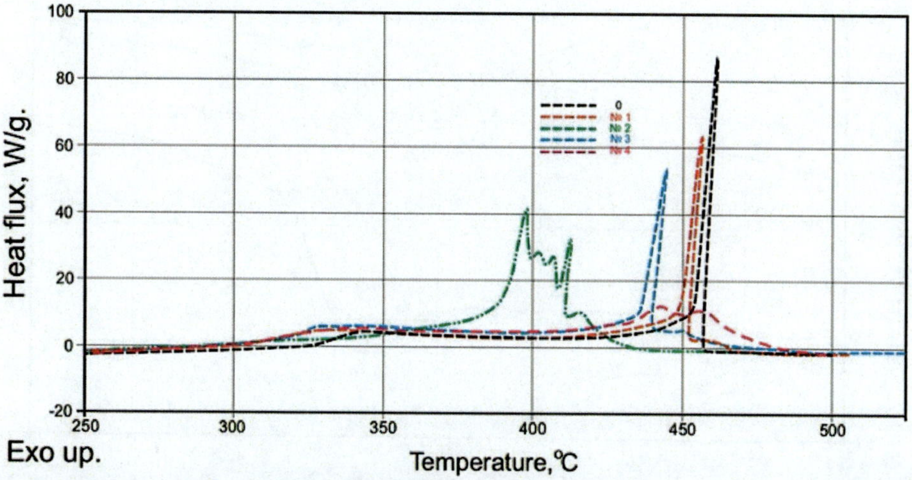

Fig. 3.16 DSC curves of samples №№ 0, 1, 2, 3, 4 (250–550 °C)

Table 3.10 Results of differential scanning calorimetry of the studied samples

Characteristics	Number of samples (Table 3.7)				
	«0»	«1»	«2»	«3»	«4»
Heat of pyrolysis, J/g (nitrogen)	−71.0	−46.1	−12.7	−33.4	−6.4
Thermal effect, J/g (air)	+5897	+6695	+7741	+7065	+7928
Heat of coal oxidation, J/g (air)	+3719	+4354	+5583	+4276	+4353

oxidation heat in their values for long-term natural aging wood significantly exceeds the modern samples.

In case of thermal analysis of the examined samples in inert medium (Fig. 3.17) we can make an important conclusion that during pyrolysis of long natural aging wood the endothermic effect is actually leveled.

Classical pronounced endothermic character of thermal decomposition process in conditions of nitrogen atmosphere has only DSC curve of modern wood sample with extremum at 370 °C. For the most representative sample with a lifetime of 200 years a slight visible thermal exothermic effect at 330 °C is observed. DSC curves in air environment (Fig. 3.18) confirm the highest calorific value of the oxidation stage of carbon layer of wood.

This stage actually accounts for about 54–70% of the total heat release during combustion of the studied wood samples. The maximum value of heat emitted during burning of the material in comparison with other samples was established for the sample with operating life of 200 years, with up to 70% in the mode of oxidation of carbon residue (1.3 times higher in comparison with the sample of modern wood). This important feature is due to the fact that during long-term use of wooden structures irreversible changes in the chemical composition and morphological structure of wood occur. Natural aging of wood is accompanied by destructive processes of hydrolysis with respect to the carbohydrate part of the wood composite, while the role of its aromatic component increases significantly.

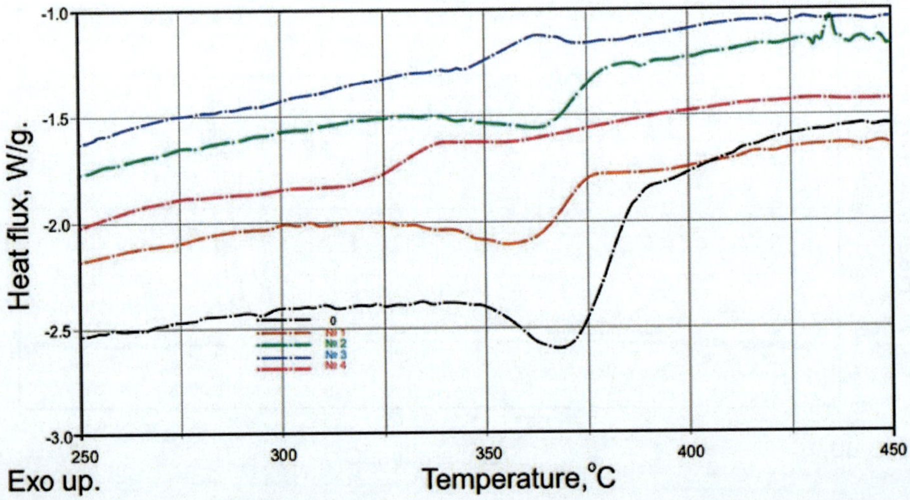

Fig. 3.17 DSC curves of samples №№ 0, 1, 2, 3, 4 (250–450 °C) (atmosphere—nitrogen)

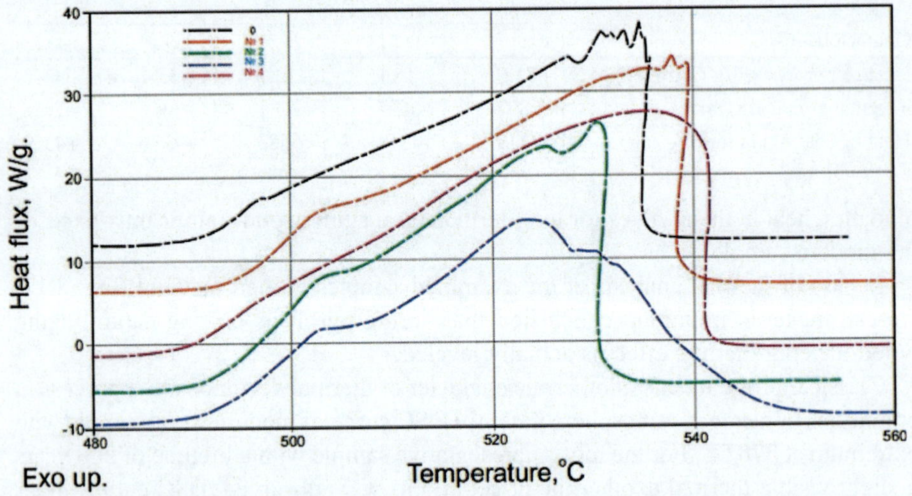

Fig. 3.18 DSC curves of samples №№ 0, 1, 2, 3, 4 (450–550 °C) (atmosphere—air)

Thus, multiple variations in the chemical composition and structural characteristics of wood depending on its service life affect the direction of chemical reactions, macrokinetic features of wood decomposition and charring, respectively, on the quantitative yield (thickness), structure and properties of the formed charcoal residue.

The results of the study revealed that at the main stage of thermal decomposition of long-life samples (300–400 °C) there is a marked decrease in the rate of mass loss. This is due to the earlier course of charring process of long natural aging wood compared to samples of modern wood, as well as the properties and structure of the resulting charcoal residue.

As a result of long-term natural aging, the thermal stability of wood, energy and oxidation capacity of the forming charcoal layer, peculiarities of its formation and structure significantly change. An important identified feature of the charcoal residue of wood of natural aging is the ability to oxidize at lower energy costs compared to modern wood, which has a pronounced exothermic character. In the process of oxidation of the charcoal layer, an overwhelming amount of heat is released during the entire exposure time of all the studied samples. At the same time, for wood with a lifetime of 200 years, the proportion of heat released at the stage of oxidation of carbon residue reaches the value of 70%, which is 1.3 times higher than for a modern wood sample.

Thermal transformations of wood components are extremely diverse. The processes of thermal decomposition of wood have a more complex character compared to its individual components. The change in the content and ratio of wood components during its operation largely determines its fire-technical characteristics. Under various conditions of fire exposure, including fire, it affects the burn rate, flammability and flame propagation over the surface of wood, formation of smoke and toxic products of combustion.

3.3.2 Fire Properties of Long Natural Aging Wood

The results of studies have shown that the content of the main components in wood determines its calorific value. Thus, in [38] it was found that the values of the lower heat of total combustion of wood depend to a greater extent not on the species (soft or hard), but on its chemical composition. The values of the lower heat of complete combustion of a particular kind of wood increase with the increase in the content of high-energy components in its composition, namely lignin and extractable substances. Certain changes occurring with the chemical composition in wood over time will affect not only the values of the heat of combustion, but ultimately the heat release characteristics of wood in general.

It was of interest to find out the effect of natural aging of wood samples on the lower heat of total combustion. For this purpose, dry wood samples of different operating ages were used. The main attention was paid to wood samples from existing objects of old buildings from different regions of Russia with a temperate climate (Vologda, Kostroma, and other regions).

Coniferous wood samples in the form of sawn wood were taken from the structures located in the villages Demjanovo, Levino, Lyabzunka, and Semigory (Vologda region), and in the villages Nikitino and Shulevo (Kostroma region). The operational age of the wooden buildings, according to the registered metrics, was 60–150 years. The structures were wooden residential and non-residential houses made of coniferous wood of spruce and pine. The samples were cut from the logs of the northern and southern parts of the structures, which are not subject to

Table 3.11 Influence of operating time of wooden structures on the lower heat of total combustion of wood [21]

Wood sample	Elemental composition, %			Q_{low}^{e}, kJ/g	Q_{low}^{cal}, kJ/g
	C, %	H, %	O, %		
Pine	52.14	5.91	41.95	19.6	19.1
Spruce	52.22	6.04	41.74	18.9	19.25
Oak	50.40	5.77	43.43	18.7	18.2
Oak, laurel, 1650	47.03	7.25	45.72	18.0	18.3
Pine. Tolstoy House, 1830.	49.38	3.88	46.74	15.2	15.52
Wooden uninhabited structure (Lyabzunka village), spruce, 60, north side)	50.3	6.44	43.26	18.4	19.21
Wooden non-residential building (v. Levino), spruce, 90 years old, north side)	48.3	7.11	44.6	18.15	18.6
Wooden non-residential building (Demjanovo village), spruce, 110 years old, north side of the building)	46.7	8.2	45.1	19.8	20.3
Wooden residential building (Semigory village), spruce, 150 years old, north side of the building)	50.5	8.47	41.03	20.1	20.7
The wooden building of the Church of St. George the victorious, Shulevo (year of construction—1898), pine	45.9	8.15	45.95	18.3	18.9
The wooden dwelling house,д. Nikitino (year of construction—1876), pine	48.8	8.34	42.86	19.9	20.4

biodegradative processes of rotting, at a distance of 1.5 meters from the ground surface [39].

Table 3.11 shows the obtained results of the study on the effect of the operating time of wooden structures on the lower heat of total combustion of wood.

As the duration of aging of wooden structures increases, the carbon component in wood decreases, and the hydrogen and oxygen content increase proportionally to each other. An exception is pine wood dated 1830, in which the hydrogen content decreased to 3.88%. Calculated and experimental evaluation of the lower heat of total combustion of samples shows that with the growth of the operational age of wood, there is a tendency to increase the values of the lower heat of total combustion. This is due to a decrease in the elemental composition of carbon and an increase in hydrogen, which four times exceeds carbon in its contribution to the heat of complete combustion. The observed effects are in good agreement with the results of the analysis of changes in the chemical composition of wood during aging.

Changes in the macrostructure of wood (volume weight of the material) and its chemical composition affect the characteristics of its fire hazard. In particular, a general pattern was confirmed [21]: at the same intensity of the external heat flux with increasing volume weight (density) of wood regardless of its type, the time delay to ignition and the value of the critical heat flux of ignition of samples increase. As the heat flux density increases, the ignition delay time decreases.

Table 3.12 Flammability values of spruce wood depending on the service life of wooden structures [41]

Name of wood sample, age	q_e, kW/m^2	$\tau_{ign.}$, s	$q_{cr.}$, kW/m^2	Maximum mass burn rate, g/m^2·s
Modern spruce (Vologda region)	30	20	11.5	–
	40	7		–
	50	3		30.9
Wooden non-residential building (Demjanovo village), spruce, 110 years old, south side	30	26	13.2	–
	40	12		–
	50	6		31.9
Spruce, (dwelling house, Vologda district, v. Semigory), 150 years old south side	30	21	12.5	–
	40	9		–
	50	4		35.7

The effect of the duration of natural aging of spruce wood components on ignition performance is shown in Table 3.12 [41].

It follows from Table 3.12 that the ignition delay time of each wood specimen depends on the intensity of external radiation heat flux. Under the same test conditions, spruce wood with 110 years of service life shows higher values of τB compared to the reference and the specimen with 150 years of service life.

The same trend is observed for the critical ignition heat flux density $q^{ign.}{}_{cr}$. However, despite some delay in ignition, wood aging leads to an increase in the maximum mass burnup rate, which determines the maximum heat release rate of the material in a fire. A similar result was also obtained when studying the flammability of pine wood with a service life of up to 110 years [23].

Changes in the chemical composition of wood during long-term natural aging should especially affect those indicators of fire hazard of materials that are directly related to the chemistry of the combustion process. Smoke-forming capacity and toxicity of combustion products can be attributed to such indicators.

When studying the smoke-forming ability of wood of long-term natural aging, it was found that the maximum mass density of smoke of the studied wood samples of different life cycles depends in a complex way on the density of the external radiation heat flux, q_e. The D_{max} index first increases with increasing intensity of thermal influence up to $q_e'' = 20$ kW/m^2, and then decreases. The extremum on the curves of dependence $D_m^{max} = f(q_e)$ is caused by self-ignition of samples. The position of the extremum corresponds to the value of the critical heat flux density, below which the flame process of wood combustion without an ignition source is not realized. Figure 3.19 shows the experimental results of estimating the smoke-forming capacity of pine wood depending on the duration of natural aging of wooden structures [21].

Smoke-forming capacity was determined according to GOST 12.1.044, item 4.18 (Russia) in the decomposition and smoldering mode with an external radiation heat flux of 20 kW/m^2. Smoke-forming capacity of elements of wooden structures of old buildings changes nonlinearly with the increase of service life. As can be seen from

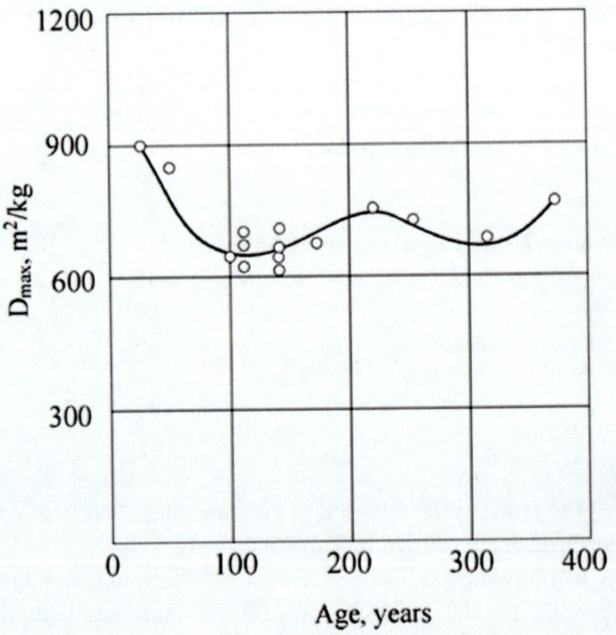

Fig. 3.19 Dependence of smoke generation coefficient (D_{max}) of pine wood on the operational age of wooden buildings

Fig. 3.19, pine samples have the lowest D_{max} coefficient when wooden buildings are operated for 100–150 and 300–330 years. This result generally agrees with the above presented dynamics of changes in the process of natural aging of other wood properties. Natural aging pine undergoes rather profound changes leading to transformation of the structure and content of the main wood components and, accordingly, to changes in the smoke coefficient. Separately, the main wood components are characterized by different smoke formation.

In [42] it was found that the greatest smoke-forming ability has cellulose (1000 m^2/kg). Smoke formation coefficient of aromatic component—lignin is 550 m^2/kg. Smoke-forming ability of wood in general should be determined by the smoke-forming ability of its individual components. It was shown [42] that the smoke-forming coefficient during wood combustion in smoldering mode is 700 m^2/kg. Depending on the operational age of pine wood, the maximum smoke-forming coefficient changes according to a complex sinusoidal curve, similar to what was found in the case of changes in the volume weight (apparent density) of wood. Wood has the lowest smoke generation coefficient after 150 and 300 years of use. This agrees with the predictions made in [10, 37]. A similar character of changes in smoke-forming capacity was established as a result of fire tests of spruce wood with the duration of operation up to 150 years.

A decrease in the smoke-forming ability of spruce wood as a result of its natural aging over 150 years is largely due to a change in the chemical composition of wood, in particular, an increase in lignin content. Lignin, due to its aromatic nature of the chemical structure, has an increased ability to carbonize and yield non-volatile coke

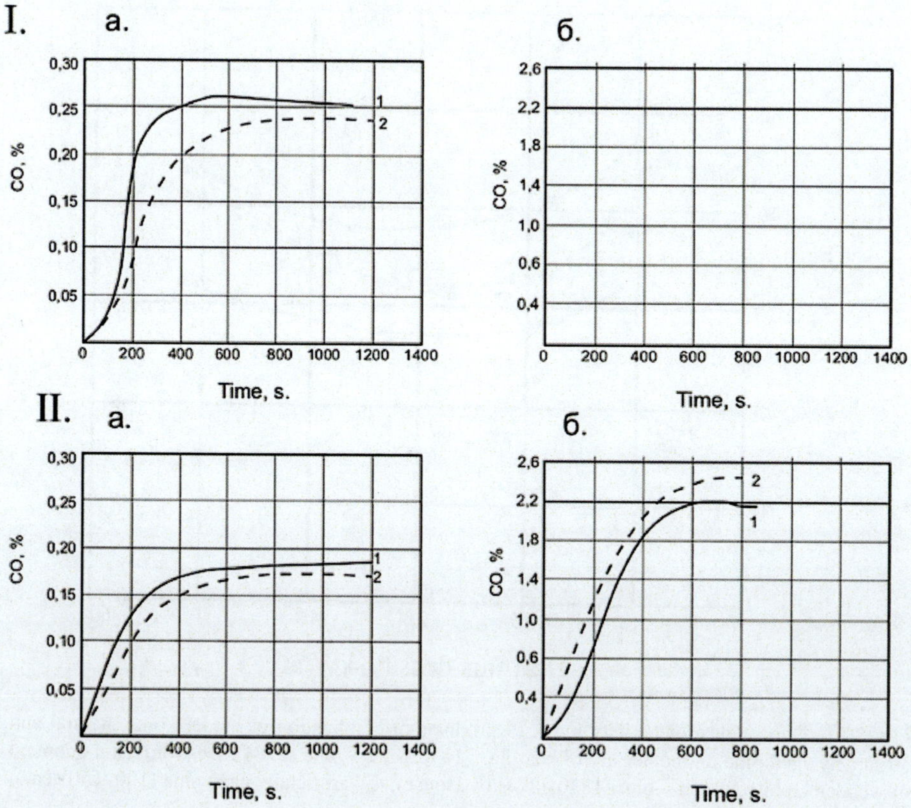

Fig. 3.20 Dependence of carbon monoxide yield (**a**) and carbon dioxide (**b**) in flameless combustion mode (I) (450 °C) and flame combustion mode (II) (500 °C) on test time of modern pine (1) and naturally aged wood (2)

residue during decomposition. This explains the reduced ability to produce smoke when burning spruce and pine after long-term use and aging for 150 years.

Toxicity of combustion products was evaluated by the yield of the most dangerous toxicants in the combustion of wood materials—carbon monoxide and carbon dioxide (CO and CO_2). Particular attention was paid to the "flame combustion—flameless combustion" boundary regime, in which the maximum yield of carbon monoxide is observed. This mode is realized at temperatures of 450–500 °C.

Figure 3.20 shows the dependences of CO and CO_2 yield on the test time of modern and naturally aged pine wood samples (150 years).

As the results of fire tests show, naturally aged wood samples have a lower carbon monoxide yield than modern wood. Accordingly, the time to reach critical human concentrations of carbon monoxide during combustion of long-lived wood is longer compared to modern wood.

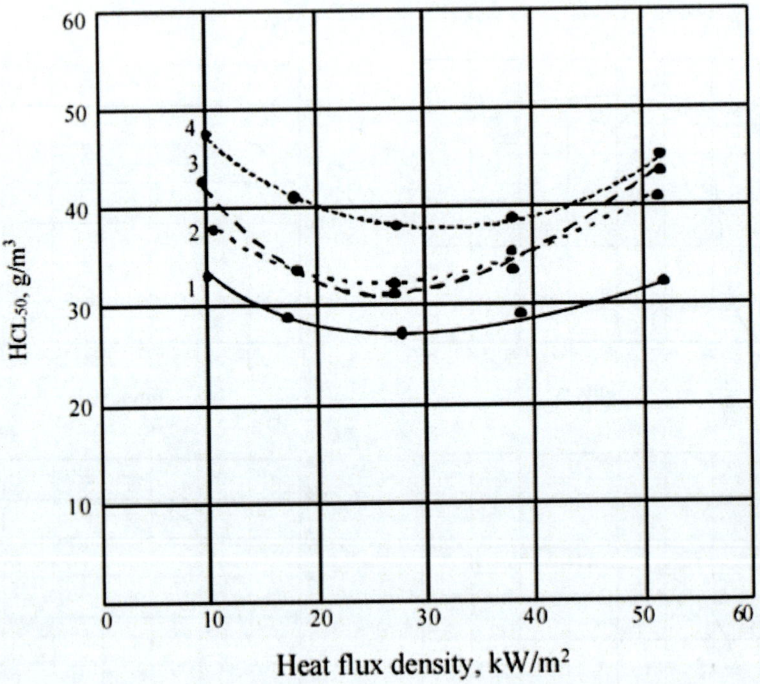

Fig. 3.21 Dependence of toxicity index of products of combustion of modern pine, natural and artificially aged pine on the density of heat flux: 1—pine; 2—pine (1784), Kostroma, a monument of wooden architecture; 3—pine, 1830. (Tolstoy House); 4—artificially aged pine (100–150 years)

Increasing the lignin content of wood increases charring and slows down the thermal decay of wood material. In the mode of flame combustion of wood, regardless of its service age, there is more complete combustion of combustible gases and intensification of carbon dioxide (CO_2) formation. At that, the highest CO_2 yield is observed when burning wood of long service life as compared to the wood of modern age.

Figure 3.21 shows the effect of external heat flux density on change in toxicity of combustion products of pine wood after long-term aging in tests according to GOST 12.1.044, item 4.20 (Russia) [21].

The results obtained indicate that the wood samples under study belong to the materials highly hazardous in terms of toxicity of combustion products (group T3): the toxicity index varies in the range from 27 to 38 g/m^3.

The curves have an extreme character due to the fact that at a certain temperature the self-ignition of the sample is observed, and the value of HCL_{50} increases, which means a decrease in the toxicity of the products of combustion. In the pre-ignition combustion mode, the highest carbon monoxide (CO) yield is observed. The pre-flame combustion mode for different wood species is in the range of external radiation heat flux from 10 to 30 kW/m^2.

The least toxicity of combustion products is in coniferous wood, which has been in operation for 100 to 150 years.

The results of the study together lead to the conclusion that the processes occurring during the natural aging of wood are quite long. They depend on the kind of wood, climatic local conditions of exploitation of wooden buildings and are caused by a complex combination of thermal and photo-oxidative, hydrolytic reactions of wood substance decomposition. Changes in the macrostructure and chemical composition of wood materials during aging cause changes in fire hazard characteristics. During very long natural aging of wooden structures, the observed transformations of wood substance have a non-linear nature in time, which affects the fire hazard characteristics of materials.

3.4 Influence of Accelerated Artificial Aging of Wood on Its Physico-Chemical and Fire Hazardous Properties

In engineering to determine the degree of influence of some or other factors on the change of operational properties of wood materials, the rate of negative processes and prediction of service life (longevity) of products usually use accelerated aging methods. The goals and objectives of the methods of accelerated artificial aging of wood are different. One of the directions of artificial aging of wood is giving building wooden constructions, beams, supporting poles, building facades and interior objects the appearance of antique products.

Antique styling of wooden structures and products has long been achieved by three main methods of wood surface treatment: chemical, thermal and mechanical, as well as a combination of these methods.

The task was to change only the appearance and quality while preserving the physical and mechanical strength characteristics of wood. The practice of imitating ancient, aged wood includes woodworking operations such as brushing, texturing, toning, and patination.

To emphasize the texture of the wood, the distribution of knots and fiber patterns, the surface of wooden products is burned with a blowtorch. After such treatment mechanically remove excess soot. With special brushes and other devices you can mechanically treat the surface of the wood (brushing), and then tint the roughly treated surface with alcoholic or water-soluble paints. To create the illusion of patina, lacquers and paints, special additives (such as fine-dispersed aluminum powder) are often used. Various mechanical methods of imitating wood damage by woodworms.

Well-known recipes for chemical imitation of an antique wood surface include treatment with ammonia alcohol; vinegar infused with rusty nails; aniline sulfate and alkali; a mixture of hydrogen peroxide and hydrochloric acid; and other chemical agents.

The second direction of really accelerated wood aging is to conduct the process in climatic apparatuses with predetermined conditions. This direction is realized usually in order to determine the efficiency of wood protection means from those or other external influences, service life (durability) of wooden laminated constructions elements, different materials and products. For comparison in the same conditions reference samples are tested (for example, from untreated wood). The conditions in climatic apparatuses for accelerated aging of samples of wood materials or products to a certain extent simulate the mode of their operation. The equipment used for accelerated aging of materials and products varies in terms of the volume of operation, the automatic control of the conditions created and the variants of conditions. Thermostats and chambers are used for accelerated tests of a wide variety of materials operated in different climatic zones and conditions:

- with programmable temperature and humidity;
- with programmable temperature and rate of change;
- with programmable temperature, humidity, intensity of ultraviolet or other types of irradiation;
- with adjustable temperature, environment and atmosphere: inert or aggressive gases, pure oxygen, salt spray, liquids, solvent vapors, etc.

Accelerated aging of materials and products is carried out in accordance with the standards adopted for each type of material or product (e.g [43, 44]. The revision of the standard [45] for wooden laminated structures includes accelerated testing methods, which allow to estimate the resistance of adhesive bonds under cyclic influences of variable temperatures and humidity, water resistance, heat and frost resistance. Based on the results of the accelerated tests it is possible to make a prediction about the service life of the products during storage and operation.

The conditions of artificial wood aging adopted in this work were aimed at accelerating the reactions of thermal oxidation and hydrolysis of oligo- and polysaccharides of wood substance.

The procedure of accelerated artificial aging made it possible to "age" hardwood samples by 50–80 years and softwood samples by 100–150 years in accordance with the scale of operational age, established by nondestructive methods of control by the nature of changes in physical, mechanical and other properties of wood [7].

Table 3.13 shows the data on the change in density and elemental composition after artificial aging of samples of 4 varieties of wood [21].

Table 3.13 Effect of accelerated artificial aging on the volume weight and elemental composition of wood

Sample	ρ, kg/m^3	*W*, %	*C*, %	*H*, %	*O*, %
Spruce	422	6.4	52.22	6.04	41.74
Spruce, 150 years old	430	7.2	48.81	11.2	40.0
Pine	448	6.0	52.14	5.91	41.95
Pine, 150 years	462	6.4	49.20	10.4	40.4
Oak	638	4.3	50.40	5.77	43.43
Oak, 80 years	626	5.1	48.9	6.43	44.7
Birch	567	5.5	50.58	5.78	43.64
Birch, 80 years	593	6.2	48.7	6.72	44.6

The small change in density, apparently caused by very little shrinkage and change in moisture content of the samples, is noteworthy. The changes in the elemental composition are more significant. All types of wood are characterized by a decrease in carbon content and an increase in hydrogen. Compared with artificially aged oak and birch wood, the hydrogen content in elemental composition of spruce and pine becomes almost two times higher. It is interesting to note that in coniferous wood samples after artificial aging there is a decrease in oxygen content, while in deciduous wood samples, on the contrary, there is an increase by 2.2–2.9%.

Aging of coniferous wood under natural conditions also leads to an increase in hydrogen content and a decrease in carbon in the elemental composition of wood substance (Table 3.13). However, too high enrichment of hydrogen elemental composition of artificially aged spruce wood compared to naturally aged for 90–150 years indicates different directions of physical and chemical transformations in wood in the process of aging.

Reduction of carbon content in artificially aged wood indicates the occurrence of degradation reactions causing release of low molecular weight carbon-containing products (e.g., carbon dioxide and monoxide, formaldehyde, methanol, formic acid, oxalic acid, and acetic acid). Hydrogen enrichment of the elemental composition of coniferous wood with a simultaneous decrease in oxygen content can, to a certain extent, be related to intermolecular condensation reactions of all wood components, especially with the participation of lignin. The possibility of relative accumulation in wood substrate (under the accepted conditions of artificial aging) of hydrocarbon derivatives from extractives—waxes, resin and saturated fatty acids C18–C22 is not excluded. They are unable to evaporate from wood at normal pressure and temperature of 160 °C. Decarboxylation of such acids should result in lower relative carbon content and hydrogen conservation.

In contrast to the accelerated artificial aging of wood by heating samples in the air at 160 °C, in obtaining the so-called thermowood, other processes take place. For example, as a result of heat treatment of spruce wood in a vapor-gas medium at 200 °C its elemental composition is enriched with carbon by 6–10%. Lignin content increases by almost 35%.

With the increase of spruce processing temperature up to 250 °C, lignin content in the material increases by almost 50%. Reactions of intra- and intermolecular dehydration of polysaccharides and condensation of lignin macromolecules lead to a decrease in hydrogen and oxygen content, respectively [38]. The results of analysis of the chemical composition of wood after artificial aging according to the accepted procedure are given in Table 3.14 [46].

As it follows from Table 3.14, as a result of artificial aging the cellulose content changes significantly. In the case of coniferous wood the cellulose content decreases by 30–32%. In hardwood, the decrease is only 21–27.5%.

It is known that lignin is the most oxidation-resistant component of wood. Therefore, an increase in the relative content of lignin in coniferous wood is quite understandable. A more complex picture is observed in artificial aging of oak and birch wood. The lignin content practically remains at the initial level in the case of birch, and in the case of artificial aging of oak wood, it even decreases by 8%. It is

Table 3.14 Chemical composition of wood of different species before and after artificial aging

Wood type	Lignin, %	Cellulose, %	Lignin[a], %	Cellulose[a], %
Spruce	28.6	44.2	36.1	29.8
Pine	27.5	43.3	36.4	30.3
Oak	29.6	41.0	27.0	32.4
Birch	21.0	41.0	21.2	29.7

Note: [a]Wood artificially aged

possible that the difference in the approximate duration of artificial aging of wood of different species is reflected in the rate of chemical reactions taking place.

Prolonged heat treatment of wood in the air at 160 °C already contributes to evaporation of volatile extractives (such as terpenes). Although mass loss during this period is insignificant, the wood substrate undergoes a noticeable transformation that affects the color of the wood. This is particularly characteristic of oak wood.

Reactions occurring in wood during accelerated artificial aging can be represented as follows. The initiation of the radical-chain process of thermo-oxidative degradation of wood by air oxygen occurs according to the law of chance. The rate of oxidation depends on the permeability and specific surface area of the porous structure of the material. This means that the surface and amorphous areas of the material's structural organization will be subject to changes first. Oxidative degradation of main chains of high-molecular-weight cellulose polysaccharides according to the radical mechanism and the law of the case is accompanied by the formation of smaller fragments in length compared to the original value. But they are still large enough, incapable of evaporation at normal atmospheric pressure. Therefore, they are difficult to remove from the matrix, belong to the group of non-cellulose polysaccharides together with hemicelluloses. Oligomeric polysaccharides—hexosans and pentosans—are partially hydrolyzed to sugars under the action of acids, products of oxidation of wood components. In turn, sugars are decomposed to form water and substances with low boiling points (furfural, hydroxymethylfurfural). The process becomes even more complicated when wood is exposed to a strong oxidizer—hydrogen peroxide. It can be assumed that oxidation of wood polysaccharides in hydrogen peroxide in aqueous medium in the presence of ash salts can cause partial cleavage of glucoside cycles by C2–C3 bonds at low temperatures with formation of polyoxy acids. Their formation and subsequent decarboxylation can explain the decrease of native cellulose content in the chemical composition and of carbon in the elemental composition of artificially aged wood.

Under the accepted conditions of accelerated artificial aging of wood, the processes of transformation of wood substrate proceed in the most thermodynamically advantageous directions. Autooxidation of lignin by air oxygen results in formation of phenoxyl and hydroxyl radicals. They attack hemicelluloses and cellulose, as well as macromolecules of lignin itself, and initiate decomposition reactions of the material. The acids formed accelerate hydrolytic reactions of cellulose degradation.

Thus, the process of accelerated artificial aging of wood is in full sense a closed cycle of complex conjugated heterogeneous reactions of degradation and structuring

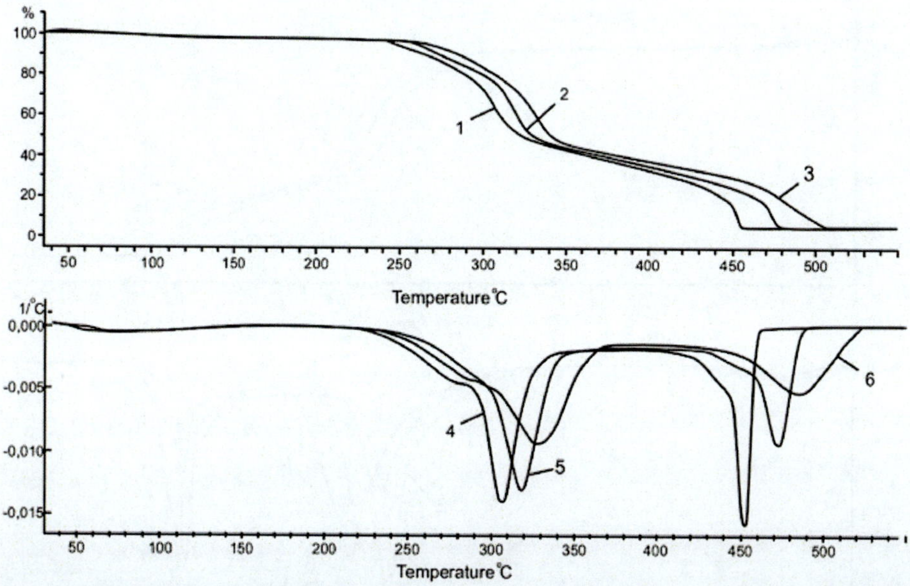

Fig. 3.22 TG (1, 2, 3) and DTG (4, 5, 6) curves of oak wood samples in an air atmosphere when heated at a rate of 5 °C/min (1, 4); 10 °C/min (2, 5) and 20 °C/min (3, 6)

of all components of the material. As a result of artificial aging of wood, removing some of its major components, the porosity of the material increases, increasing the possibility of contact of the wood matrix with air oxygen.

Cellulose is the main structural component of wood, providing its high physical and mechanical properties. Reducing the relative cellulose content in the wood matrix as a result of accelerated artificial aging should particularly affect such mechanical properties of wood as bending impact strength and flexural modulus, shear and shear strength. Strength and modulus of elasticity in compression along the fibers and hardness values should change to a lesser extent.

Thermal analysis of wood samples subjected to accelerated artificial aging reveals some changes in their behavior after aging. As an example Figs. 3.22, 3.23, 3.24, and 3.25 show TG and DTG curves of original samples and artificially aged oak and spruce wood when heated at different rates in an air atmosphere [40, 41].

When considering the TGA and DTG decomposition curves of modern and artificially aged wood samples, we can note the well-known pattern of the effect of heating rate that is common to all. Two main decomposition regions are characteristic: low-temperature (140–400 °C) and high-temperature (above 400 °C). The first area reflects the processes of thermal oxidative decomposition of wood substance and the formation of volatile combustible products. The second—the processes of carbonization and oxidation of the charred product. Small losses of mass when heating samples up to 140 °C are attributed to the evaporation of moisture.

Under equal heating conditions, weight loss in aged wood samples begins at a lower temperature compared to the original material. The effect of aging is especially noticeable when comparing the temperature peaks on the DTG curves of the

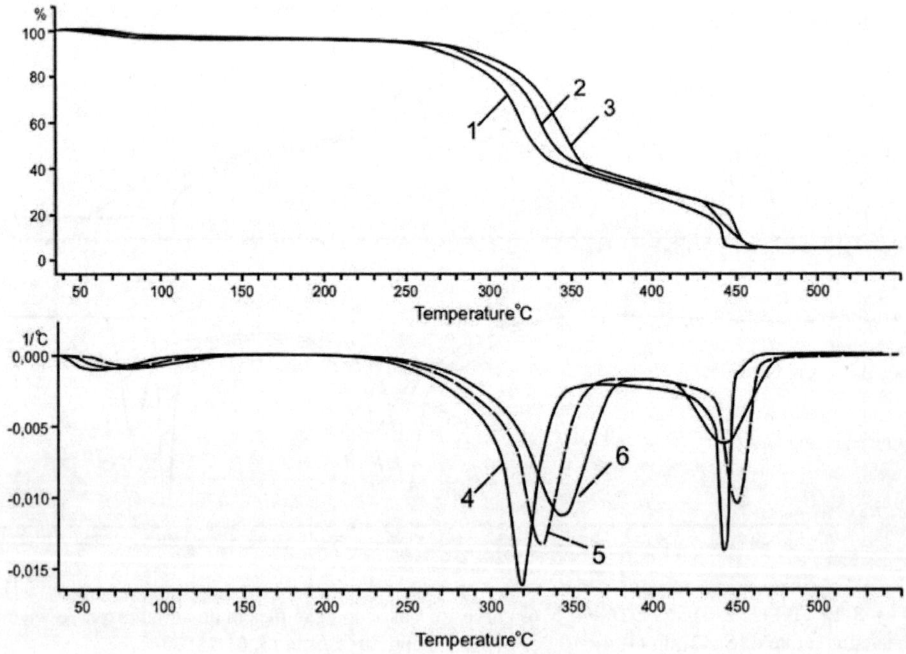

Fig. 3.23 TG (1, 2, 3) and DTG (4, 5, 6) curves of spruce wood samples in an air atmosphere when heated at a rate of 5 ° C/min (1, 4); 10 ° C/min (2, 5) and 20 °C/min (3, 6)

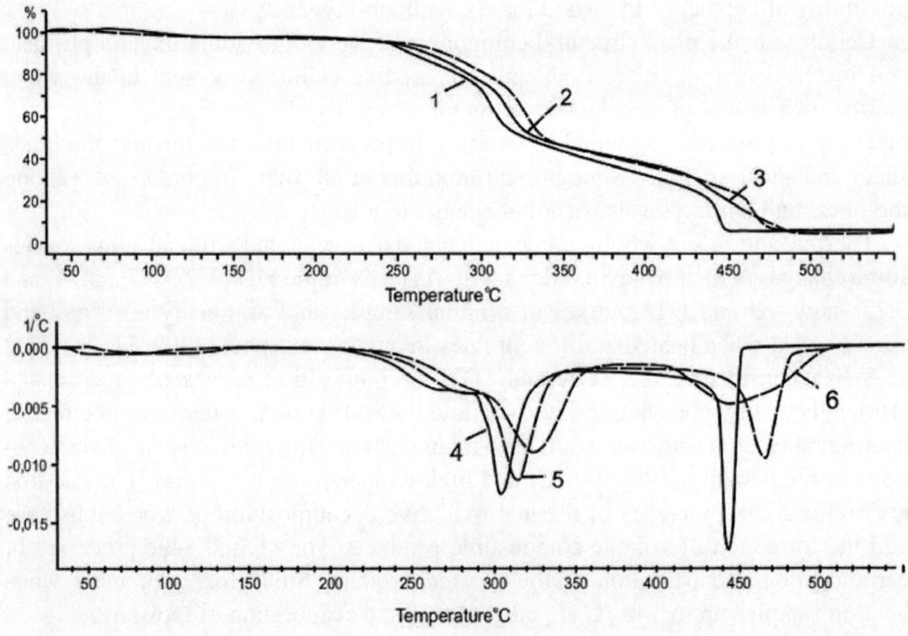

Fig. 3.24 TG (1, 2, 3) and DTG (4, 5, 6) curves of artificially aged oak wood in an air atmosphere when heated at a rate of 5 °C/min (1, 4); 10 °C/min (2, 5) and 20 °C/min (3, 6)

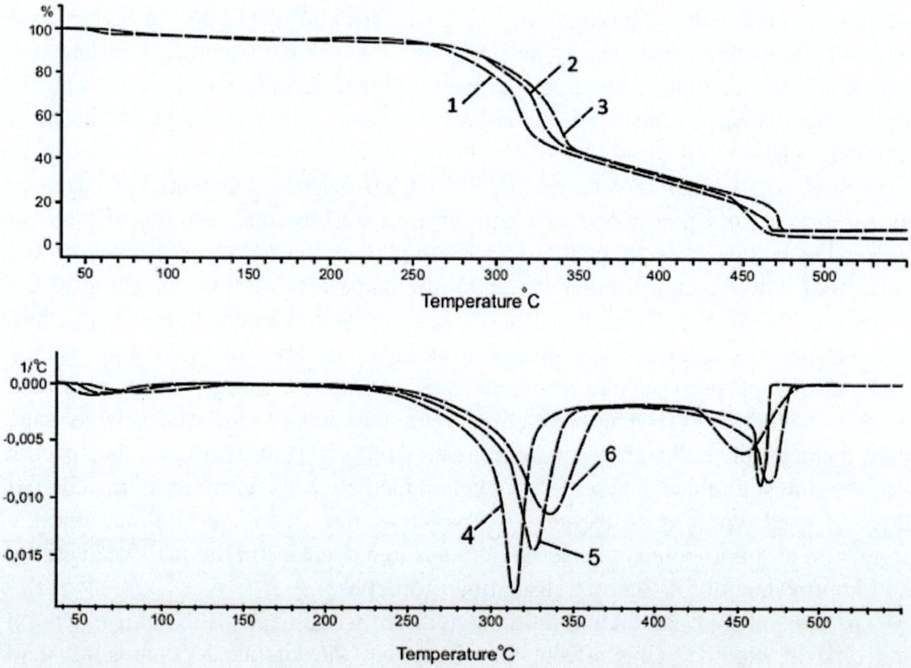

Fig. 3.25 TG (1, 2, 3) and DTG (4, 5, 6) curves of artificially aged spruce wood in an air atmosphere at heating rates: 5 °C/min (1, 4); 10 °C/min (2, 5) and 20 °C/min (3, 6)

corresponding samples. For artificially aged wood, the T_{max} value is several degrees lower compared to the T_{max} of the original wood sample.

At the first low-temperature decomposition stage, softwood and hardwood samples lose more than half of their weight. This stage is exothermic in nature. After artificial wood aging the amount of heat released at this stage increases by 1.5–1.8 times. The effective activation energy of thermo-oxidative process of wood decomposition depends on the degree of conversion and varies in a certain range from 10% to 90%.

The difference in the mass loss rate of artificially aged and modern wood samples observed from TGA curves is due to different values of pre-exponential multipliers in the Arrhenius equation. In other words, the kinetics of thermo-oxidative decomposition of wood is influenced by entropic, structural factors [39].

The second stage of thermo-oxidative decomposition of wood in the range of 400–550 °C is associated with active formation and oxidation of coke. Under dynamic heating conditions up to 500–550 °C the charred wood decomposition products are completely oxidized and lose their mass. Judging by DTG curves, oxidation of carbon residue of artificially aged samples of spruce goes at a higher rate compared to the coke of modern wood. Oak aging also leads to similar changes in the dynamics of mass loss at this stage, although not as pronounced as in coniferous wood. When the degree of oxidative burnout of charred oak products is up to 30–35%, the oxidation reaction of aged wood coke proceeds with lower activation energy as compared to the oxidation of coke of original wood. With a higher coke burnout the difference in E_{eff} is leveled. The coke oxidation process during this period proceeds with the activation energy of 200–215 kJ/mol. The same tendency in the change of E_{eff} is observed in oxidative decomposition of carbon residues of artificially aged birch wood [39].

A study of the porous structure of coke from original and artificially aged to 100–150 years old pine wood was carried out by isothermal sorption of benzene vapor. For comparative analysis we took samples of charred layers formed on the surface of wood samples under the action of flame after testing according to the method of GOST R 53292-2009 (Russia) [47], as well as in conditions of standard fire temperature regime. Total pore volume, average pore diameter and specific surface area of porous coke structure were calculated using obtained benzene sorption isotherms [48]. After accelerated artificial aging of pine wood, the average pore diameter of the charred layer decreased from 25.03 to 21.56 nm, the specific surface almost doubled. These results explain the increased reactivity of the charred layer of aged wood to air oxygen. At the same time, it should be noted that the formation of a fine-pored coke structure leads to a decrease in thermal conductivity and improvement of its thermal insulating properties.

The dynamics of natural and artificial aging of wood depends to some extent on the type of wood. Aging affects the change in the chemical composition and properties of the material as a whole. In this case of particular interest is the question of how accelerated artificial aging affects the various fire hazard characteristics of wood. The change in the chemical and elemental composition of softwood and hardwood after artificial aging affects the lower heat of total combustion of wood (Table 3.15) [46].

Accelerated artificial aging of wood, as well as prolonged natural aging, leads to an increase in the values of the lower heat of total combustion. Experimentally obtained values of Q_{low}^{calc} and calculated values of Q_{low}^{e} are close enough. Due to the increased content of lignin and enrichment of the elemental composition of hydrogen, artificially aged spruce wood up to 150 years has higher values of Q_{low}^{calc} compared to the sample of 150 years of operation in natural conditions.

The study of influence of artificial wood aging on mass burnout rate under action of radiation heat flux of constant density confirmed the known general regularities:

Table 3.15 Effect of accelerated artificial aging of wood on the lower heat of total combustion

Wood specimen	Wood density, kg/m^3	W, %	Q_{low}^{e}, kJ/g	Q_{low}^{calc}, kJ/g
Modern spruce	422	6.4	18.9	19.25
Spruce aged 150 years	430	7.2	22.7	23.6
Modern pine	448	6.0	19.6	19.1
Pine aged pine, 150 years	462	6.4	–	22.87
Oak modern	638	4.3	18.7	18.2
Oak aged 80 years	626	5.1	–	18.2
Birch modern	567	5.5	18.1	18.22
Birch aged 80 years	593	6.2	–	18.45

with increase of external heat flux density from 30 to 40 kW/m^2 mass burnout rate of samples increases, the maximum value of mass burnout rate increases and the rate of its achievement increases. Curves of mass burnout rate variation of artificially aged wood have more complicated character in comparison with modern wood. The peak values of mass loss rate corresponding to the beginning of active charring of artificially aged wood are much lower compared to modern wood. This means that artificially aged wood begins to char faster and the effect of the charcoal layer on the burnout of the sample appears earlier than in modern wood.

The standard setting of the flammability method does not allow recording the heat release during material combustion. For this purpose, we used a flow-through OSU calorimeter of HRR-3 brand of Atlas, USA. A wood sample of 150 × 150 × 10 mm in a vertical configuration was exposed to an external thermal radiation flux of 35 kW/m^2.

As an example, Figs. 3.26 and 3.27 show the results of fire tests of modern and artificially aged (100–150 years old) spruce samples with humidity of 6.4–7.2% [49].

Each figure shows the curves of 3 repeated experiments. It can be seen that the heat release rate curves have two pronounced peaks, which characterize the sequence of wood combustion stages.

At the first stage, heating, decomposition and ignition of combustible volatile products of material decomposition occur. After the beginning of intensive charring of wood (corresponds to the peak value of the heat release rate), the rate of heat release begins to decrease. The appearance of the second peak in the wood combustion rate curve is caused by the burnout of the carbon layer formed on the surface of the wood sample. The highest heat release rate is observed at the second stage.

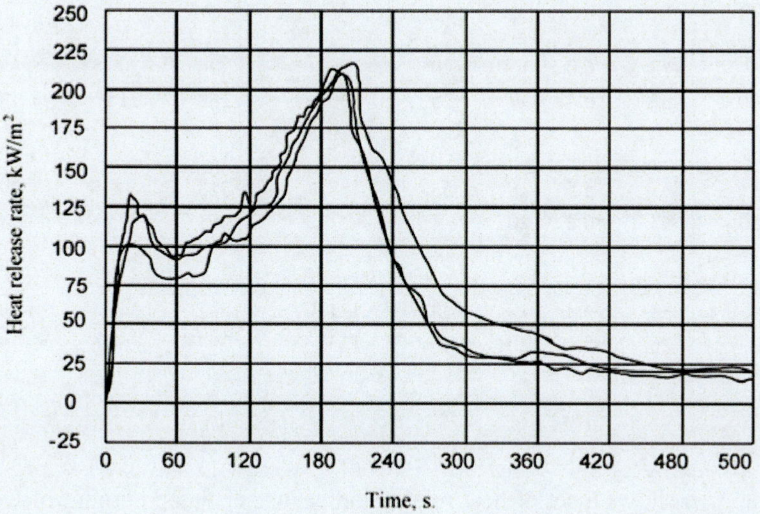

Fig. 3.26 Change in the rate of heat release during flame combustion of modern spruce wood as a function of time at $q_e = 35$ kW/m^2

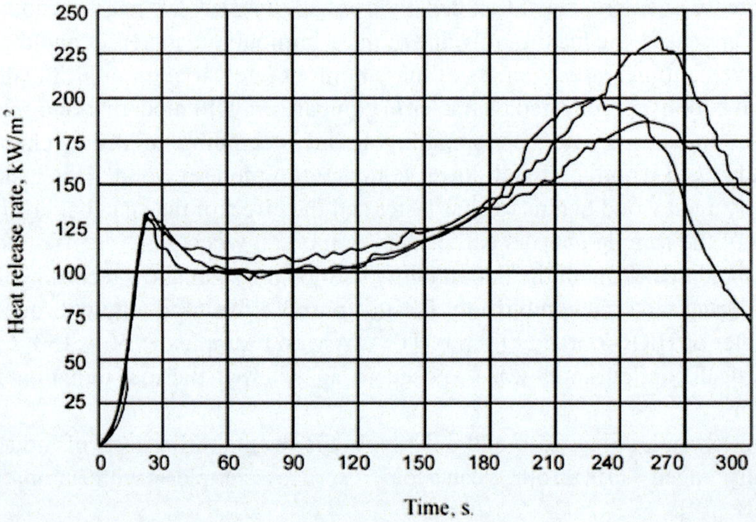

Fig. 3.27 Variation of heat release rate of artificially aged spruce wood (100–150 years old) as a function of time at a heat flux of 35 kW/m^2

Table 3.16 Effect of artificial aging of wood of different species on heat release characteristics

Wood sample	ρ, kg/m^3	W, %	τ_{ign}, s	τ1max, s	HDR1мmax, kW/m^2	τ2max, s	HDR2max, kW/m^2	THD2 min, kW·min/m^2
Spruce modern	422	6.4	<10	23	131.2	195	223.2	218.2
Spruce (100–150 years old)	430	7.2	15	26	135	245	207.9	190.7
Pine	448	6.0	<10	20	129.0	276	203.5	180.6
Pine (100–150 years old)	462	6.4	17	25	130	268	214	179.9
Birch	567	5.5	10	53	157.7	215	400.5	279.9
Birch (50–80 years old)	593	6.2	20	–	–	261	375	240
Oak	638	4.3	10	33	131.3	281	245.1	198.0
Oak (50–80 years old)	626	5.1	20	40	125	297	260.6	191.8

Table 3.16 shows a set of heat release indicators for flame burning of wood of different species before and after artificial aging at an external heat flux $q_e'' = 35$ kW/m^2 [49].

This table gives the average values of indicators for three repeated experiments. In particular, the following parameters were determined: the ignition delay time (τ_{ign}.); the maximum values of heat release rate at the first (CTB_{1max}) and second (CTB_{2max}) stages, as well as the time to reach them from the beginning of the test; total heat release for 2 min of burning process (THD_{2min}).

After artificial aging of wood as a result of removal of low molecular weight substances, part of easily hydrolysable hemicelluloses and extractive substances, as well as some increase in apparent density of samples, an increase in values of τv. is observed. However, then there is a decrease in the period from ignition to the peak value of heat release, τ_{1max}. This is especially indicative in the case of artificially aged spruce and pine wood specimens. At heat flux $q_e'' = 35$ kW/m^2 values of maximum heat release rate at the first stage of flame combustion, HDR_{1max}, of coniferous wood after artificial aging remain approximately at the same level, and in case of oak samples a slight decrease of HDR_{1max} from 131.3 to 125 kW/m^2 is observed.

It should be noted that the time of reaching the maximum of the second peak, τ_{2max}, from the beginning of the test of modern spruce and artificially aged specimens is 195 and 245 s. The total heat release for 2 minutes of the test, THD_{2min}, was 218.2 and 190.7 kW·min/m^2, respectively. The tendency to a decrease in the total heat release for this period, THD_{2min}, is also observed for artificially aged oak wood specimens. It can be concluded that the artificial aging of wood leads to an acceleration of the beginning of charring of the material. However, the increase in thickness of the charred layer and its burnout is extended for a longer period.

Smoke-forming ability of artificially aged coniferous wood depends on the intensity of external heat flow and by values of D_{max} is lower than the smoke-forming ability of modern wood (Fig. 3.28) [21, 50, 51].

Smoke formation factor of artificially aged up to 150 years old spruce wood in flameless combustion mode and external thermal radiation flux of 20 kW/m^2 decreased from $D_{max} = 950$ m^2/kg in the original sample to $D_{max} = 790$ m^2/kg. On the other hand, the toxicity of products of flameless combustion of wood artificially aged up to 150 years was found to decrease under these conditions. The toxicity index of combustion products estimated by the carbon oxide yield, HCL_{50}, changed from 28 g/m^3 for the modern wood sample to 37 g/m^3 for aged spruce wood.

Artificial aging partially removes extractives, hemicelluloses and cellulose from wood, which make a significant contribution to the formation of smoke when the material burns. Increased lignin content also plays an important role in reducing smoke formation during combustion of artificially aged wood. Due to its aromatic polymeric nature, lignin exhibits a high ability to carbonize when heated and exposed to fire. Therefore, less carbon-containing volatile products of lignin decomposition—sources of soot smoke particles—enter the smoke formation zone in the flame.

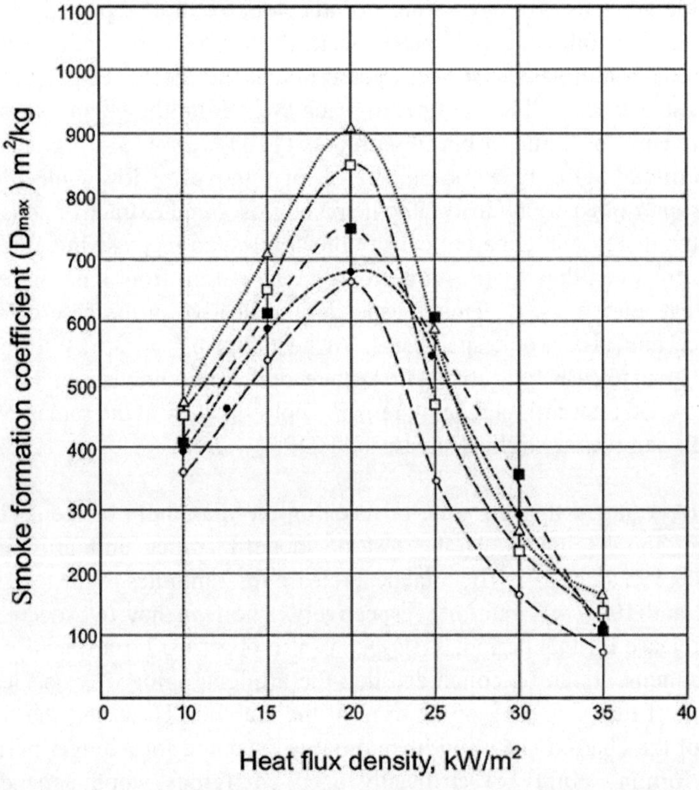

Fig. 3.28 Dependence of smoke formation coefficient (D_{max}) of pine, natural and artificially aged pine timber on heat flux density: 1—pine; 2—artificially aged pine (100–150 years old); 3—pine (1784), a monument of wooden architecture; 4—pine (1759 years old); 5—pine, 1830 (House of the Tolstoys)

References

1. Kisternaya, M. V., & Kozlov, V. A. (2007). *Drevesinovedcheskie aspects of the preservation of historical buildings* (p. 133). Karelian Research Center RAS.
2. Potutkin, G. F., & Shapovalova, L. G. (1990). Changing the properties of wood during long-term operation (by the example of wooden architecture monuments of the Arkhangelsk region). *Woodworking Industry, 10*, 28–30.
3. Isaeva, L. N., & Brukhanova, E. B. (1969). Physico-mechanical properties and anatomy features of Siberian larch wood from the structure 230 years old. In *Proceedings of the Institute of forest and wood of Siberian Branch of Academy of Sciences of USSR*, Krasnoyarsk.
4. Pischik, I. I., Fefilov, V. V., & Burkovskaya, Y. I. (1971). On chemical composition and physical properties of fresh and aged wood. *Izvestiya Vuzov, Lesnoy Zhurnal, 6*, 89–93.
5. Pokrovskaya, E. N. (2003). *Chemical and physical bases of increase in durability of wood. Preservation of wooden architecture monuments with the help of organoelement compounds. Monograph* (p. 104). ASV Publishing House.

6. Tychino, N. A., Fedosenko, I. G., & Baranov, A. V. (2007). Features of the structure and firebioprotection of archaeological wood. *Fire and Explosion Safety, 1*, 19–25.
7. Pishchik, I. I. (2000). Nondestructive assessment of internal stresses in wooden structures by non-traditional methods. In *Structure, properties and quality of wood. Materials of the 3rd international symposium, Petrozavodsk*.
8. Emmanuel, N. M., & Buchachenko, A. L. (1082). *Chemical physics of aging and stabilization of polymers* (p. 360). Nauka.
9. Popov, A. A., Rapoport, N. Y., & Zaikov, G. E. (1987). *Oxidation of oriented and stressed polymers* (p. 232). Chemistry.
10. Pokrovskaya, E. N. (2001). Prediction of wood durability by thermal analysis method. *Fire and Explosion Safety, 6*, 31–32.
11. Mingaleeva, G. R. (1990). *Chemical Encyclopedia* (Vol. 2, pp. 23–24). Sovetskaya Encyclopedia.
12. Mingaleeva, G. R. (1995). *Chemical Encyclopedia* (Vol. 4, p. 822). Sovetskaya Encyclopedia.
13. Nikitin, N. I. (1962). *Chemistry of wood and cellulose* (p. 794). USSR Academy of Sciences, Institute of High-Molecular Compounds.
14. Pavlov, B. A., & Terentyev, A. P. (1972). *Course of organic chemistry* (p. 647). Publishing House.
15. Artemenko, A. I. Organic chemistry. In *Textbook for construction specialties* (p. 560). High School.
16. Medvedeva, E. N., Babkin, V. A., & Ostroukhova, L. A. (2003). Arabinogalactan of larch - properties and prospects for use (review). *Chemistry of Plant Raw Materials, 1*, 27–27.
17. Pischik, I. I. (2004). *Dating long-aged wood by non-destructive methods. Dissertation for the degree of Doctor of Technical Sciences*. Moscow State University of Civil Engineering.
18. Varfolomeev, Y. A., Potutkin, G. F., & Shapovalova, L. G. (1990). Variation of wood properties during long-term operation (by example of Arkhangelsk oblast wooden architecture monuments). *The Woodworking Industry, 10*, 28–30.
19. Kazanskaya, S. Y., Vikhrov, Y. V., & Golman, L. P. (1975). About some features of a chemical composition of archeological wood. *Chemistry of Wood, 2*, 41–44.
20. Sivenkov, A. B., Tarasov, N. I., Kruglov, E. Y., & Lebedev, A. Y. (2011). Influence of duration of exploitation of wood on its fire hazard. *Russian-Serbian Scientific and Analytical Journal, 1*(1), 55–60.
21. Aseeva, R. M., Barbotko, S. L., Serkov, B. B., Sivenkov, A. B., Suleikin, E. V., & Tarasov, N. I. (2009). Effect of operating time of wood on its fire hazardous properties. In *Collection of X international conference "Oligomers-2009"*, Volgograd, VSTU, pp. 270–295.
22. Sivenkov, A. B. (2012). Influence of age of wooden building constructions on values of indices of their fire hazard. *Industrial and Civil Engineering, 3*, 66–67.
23. Alekseeva, T. S., Sivenkov, A. B., & Tarasov, N. I. (2011). *Influence of service life of residential and non-residential wooden structures on the fire hazard properties of wood* (Vol. 3, pp. 7–15). Bulletin of St. Petersburg University of State Fire Service of Russia.
24. Nikitin, V. M., Obolenskaya, A. V., & Shchegolev, V. P. (1978). *Chemistry of wood and cellulose* (p. 368). Forest Industry.
25. Solovjev, V. A. (2004). Kinetic method of assessing the wood destruction ability and activity of fungi. In *Proceedings of the IV International Symposium "Structure, Properties and Quality of Wood-2004"* (Vol. 2, pp. 458–460). St. Petersburg State Forestry Academy.
26. Nevzorov, A. L., & Aksenov, S. E. (2004). Experimental studies of physical and mechanical properties of pile foundations wood. In *Proceedings of IV International Symposium "Structure, Properties and Quality of Wood – 2004"* (Vol. 1, pp. 194–196). St. Petersburg State Forestry Academy.
27. Streltsov, D. Y. (2003). *Investigation of load-carrying capacity of long-lasting wooden structures* (p. 168). Dissertation for the degree of Candidate of Technical Sciences. State Unitary Enterprise Central Research Institute of Building Materials named after V.A. Kucherenko.

28. Roshchina, S. I. (2009). *Strength and deformability of laminated reinforced wooden structures under prolonged load action*. Dissertation for the degree of Doctor of Technical Sciences. Moscow State Academy of Municipal Economy and Construction, p. 330.
29. Varfolomeev, A. Y. (2010). *Damage accumulation in wooden structures during long-term operation in conditions of biological aggression*. Dissertation for the degree of Doctor of Technical Sciences. State Unitary Enterprise Central Research Institute of Building Research named after V.A. Kucherenko, p. 157.
30. Shapovalova, L. G. (1994). *Durability of wooden structures and buildings without the use of chemical means of protection (on the example of wooden architecture of the Arkhangelsk region)* (p. 167). Arkhangelsk State Technical University.
31. Kisternaya, M. V. (2000). *Estimation of wood condition of architectural monuments. Dissertation for the degree of Candidate of Technical Sciences* (p. 124). Institute of Forest of Karelian Scientific Center of RAS.
32. Serkov, B. B., Sivenkov, A. B., Thanh, B. D., & Aseeva, R. M. (2003). Heat emission during wood burning. *Forest Bulletin, 5*(30), 74–79.
33. Serkov, B. B., Sivenkov, A. B., Thanh, B. D., & Aseeva, R. M. (2004). Smoke release during thermal decomposition and combustion of wood. *Forest Bulletin, 2*(30), 99–103.
34. Serkov, B. B., Sivenkov, A. B., Thanh, B. D., & Aseeva, R. M. (2004). Research of toxicity of products of wood burning of different species. *Forest Bulletin, 5*(36), 145–150.
35. Serkov, B. B., Sivenkov, A. B., Thanh, B. D., & Aseeva, R. M. (2005). Thermal decomposition of wood of tropical species. *Forest Bulletin, 2*(38), 70–76.
36. Sivenkov, A. B., Serkov, B. B., Aseeva, R. M., Thanh, D. B., & Ilyin, A. V. (2004). *Research of mass burn rate of different wood species* (Vol. 12). Collection of scientific works of graduate and doctoral students of Moscow State Forest University. Moscow State University of Forestry, pp. 122–132.
37. Pokrovskaya, E. N. (2000). Thermal stability of wood of different duration of exploitation. *Stroitelnye materialy, 9*, 34–35.
38. Sivenkov, A. B. (2015). *Influence of physical and chemical characteristics of wood on its fire danger and efficiency of fire protection: Dissertation for PhD and Doctor of Technical Sciences: 02.00.06 high-molecular compounds* (p. 28). N.N. Semenov Institute of Chemical Physics, Russian Academy of Sciences.
39. Sivenkov, A. B., Tarasov, N. I., Kruglov, E. Y., & Lebedev, A. Y. (2011). Influence of the duration of wood exploitation on its fire hazard. *Russian-Serbian Scientific and Analytical Journal, 1*(1), 55–60.
40. Sivenkov, A. B. (2012). Influence of age of wooden structures of buildings on the values of indicators of their fire hazard. *Industrial and Civil Engineering, 3*, 66–67. p.
41. Pjadichev, E. V., Sivenkov, A. B., & Tarasov, N. I. (2011). Thermal oxidative decomposition of wood of different operational age. *Problems of Risk Management in the Technosphere, 4*(20). Accessed from http://comobzor.ru/registers/media/pi-fs-77-36404-problemy-upravleniya-riskami-v-tehnosfere
42. Grigoryev, G. N., Ani, E. V., Leonovich, A. A., Kulev, D. H., & Zakharov, S. C. (1985). Effect of fire protection on the smoke-forming ability of wood materials. In *Collection of Articles "Theoretical and practical aspects of fire protection of wooden materials"* (pp. 20–23). Latvian SSR Academy of Sciences Institute of Wood Chemistry, Riga, Zinatne.
43. GOST *9.057-75 Unified system of protection against corrosion and ageing (USSPC) materials polymeric, wood, fabric, paper, cardboard. Methods of accelerated climatic ageing tests.*
44. GOST *9.707-81 ЕСЗКС materials polymeric. Methods of accelerated climatic ageing tests.*
45. GOST *20850 Laminated wooden structures. Technical specification (design).*
46. Aseeva, R. M., Barbotko, S. L., Degtyarev, R. V., Serkov, B. B., Sivenkov, A. B., & Tarasov, N. I. (2010). Effect of operating time of wood on its fire properties. *Encyclopedia of Chemical Engineer, 3*, 27–34.
47. *GOST R 53292-2009 (Russia) Fire retardant compositions and substances for wood and wood-based materials. General requirements. Test methods.*

48. Keltsev, N. V. (1984). *Fundamentals of adsorption technology* (p. 592). Mir.
49. Aseeva, R. M., Degtyarev, R. V., Kruglov, E. Y., Serkov, B. B., Sivenkov, A. B., & Tarasov, N. I. (2010). Heat release of wood of different exploitation age. *Bulletin of Moscow State Forest University - Forest Bulletin, 5*(74), 139–143. p.
50. Degtyarev, R. V., Kruglov, E. Y., Suleikin, E. V., Serkov, B. B., Sivenkov, A. B., & Tarasov, N. I. (2010). Effect of wood age on its fire hazard. In *Materials of the IV All-Russian Scientific-Practical Conference: Actual problems of security in the Russian Federation devoted to the 20th anniversary of the Russian Ministry of Emergency Situations*. Ural Institute of State Fire Service of EMERCOM of Russia, pp. 122–124.
51. Degtyarev, R. V., Kruglov, E. Y., & Sivenkov, A. B. (2010). Influence of the term of exploitation of residential and non-residential wooden buildings on fire hazardous properties of wood (in Russian). *Fire and Emergencies: Prevention, Elimination, 2*, 27–35.

Study of the Charring Process of Long-Life Wooden Structures

4

4.1 Studying the Process of Charring of Long-Lived Wood of Natural Ageing by Methods of Thermal Analysis

Studies of the patterns of thermal and thermo-oxidative decomposition of substances and materials are important for creating a fundamental basis for the fire safety of these objects. Decomposition of organic wood substance is one of the main stages of initiation and development of its combustion process.

Decomposition of wood when exposed to heat is a complex process. Wood is a class of materials that can be charred by decomposition. The quantity and quality of the resulting charred layer have a significant impact on the course of flame and smoldering combustion of such materials.

A large number of works are devoted to the study of thermal (pyrolysis) and thermo-oxidative decomposition of wood and its components [1–7]. Because of the anisotropy and structural diversity of wood of different species, differences in their chemical composition, as well as experimental conditions for studying wood decomposition, there is still great uncertainty in the macrokinetic characteristics of this process.

Changing the thermal stability of wood as a result of its long-term use is proven. In world and domestic practice, studies have been repeatedly conducted to establish the dependence of the kinetic parameters of wood thermal decomposition on the time of its operation. So, in [8], it was found that the aging processes occurring in wood largely determine its thermal stability. The data obtained show that the least thermal stability of wood is observed in the first hundred years and after 300 years of its operation. In this work, the author suggested that during these time periods wood is the most fire-hazardous. It was found that in time periods of 150–200 years the activation energy and activation entropy significantly increase, i.e., any chemical processes, including thermal destruction, are hindered [8].

However, at present, the question of the influence of the duration of operation of wooden structures on the features of wood carbon formation, as well as the oxidation process of coke residue, which determines the degree of thermal stability of materials

S. Andrey Borisovich et al., *Fire Hazard and Fire Resistance of Wooden Structures*,
https://doi.org/10.1007/978-3-031-24074-4_4

Table 4.1 Characteristics of samples of wooden structures

Sample number	Name of wooden structures	The place of sampling of wooden structures	Service life, years	[a]ρ, kg/m^3
0	–	Pine timber modern, Arkhangelsk region	–	452
1	Elements of the rafter part of the house	Pine wood, uninhabited house, Republic of Kazakhstan, v. Zhanazhol (North Kazakhstan region)	63	477
2	Internal elements of the structures of the intermediate floor (between the 1st and basement floors)	Pine wood, elements of the wooden ceiling, puppet theater "Gulliver" located at: Kurgan, ul. Sovetskaya, 104	125	588
3	Wooden supporting structures (4 beams and 2 uprights) (4 beams and 2 uprights)	Pine wood, Church of St. Nicholas the Wonderworker, Bryansk region	150	643
4	Elements of the rafter part of the house	Pine wood, uninhabited house, Republic of Kazakhstan, v. Kayindy (Kostanay region)	87	546

[a]Moisture content of wood samples before the fire tests was 12%

and structures made of wood in a fire remains little-studied. It is obvious that numerous destructive influences accompanying the exploitation of wood determine the physical and chemical transformations taking place in wood material and determine the nature and mechanism of thermo-oxidative decomposition of wood.

For the study in this dissertation work, structural elements and load-bearing structures (length from 1.5 to 3.5 m) of solid coniferous wood from different regions at facilities with a service life of 63 to 150 years were taken. The characteristics of the investigated wooden structures are presented in Table 4.1.

For the study, samples of wood modern (Arkhangelsk region, Russia) and wood on the site with wooden load-bearing structures (beams and poles of pine wood) of long service life were taken: the church of Nicholas the Wonderworker in Bryansk region, Russia (1865 built), which went under demolition because of decay (Fig. 4.1).

A set of modern methods of research was involved: methods of thermal analysis (thermogravimetry (TG), differential thermogravimetry (DTG), differential scanning calorimetry (DSC) widely used in the international practice of studying the fire hazard of polymeric materials) [9–13]. The ability to simultaneously measure heat flux (DSC) and change in mass (TGA) over a wide temperature range simplifies the interpretation of results and increases analysis productivity.

The Q600SDT utilizes a rugged, horizontal scale design that supports both TGA and DSC measurements. The sample arm is used to measure the real mass of the sample, while the reference arm is used to correct the TGA measurement for arm elongation using an appropriate algorithm. The SDT Q600 enables simultaneous

Fig. 4.1 General view of the load-bearing wooden beam and support column of the Church of St. Nicholas the Wonderworker (Bryansk region), built in 1865

recording of changes in sample mass (thermogravimetric analysis) and processes that generate or absorb heat (differential scanning calorimetry/differential thermal analysis). The instrument uses two-cup scales (test sample and reference sample) which can be operated at temperatures ranging from room temperature to 1500 °C. The sensitivity of the scales is 0.1 micrograms. The furnace unit allows the composition of the sample atmosphere to be monitored during measurements and automatically changed during the experiment, using both inert and reactive gases. The analyzer incorporates automated furnace movement, automated air cooling, and a touch screen display for easy experiment control.

Material samples in the form of regular shaped fragments were placed in the crucible of the Q600 SDT thermal analyzer. The materials were heated at a rate of 20 °C/min in a nitrogen current to 500 °C. Determination of kinetic parameters of coke oxidation of pine wood samples of various service life was carried out according to differential scanning calorimetry (Borchardt and Daniels method). In the present method, the differential method is used for express estimation of kinetic characteristics.

In the DSC instrument, the difference between the heat flux of the test sample and the flux of the reference crucible is measured as a function of temperature and/or time while heating the test sample and the reference material according to a controlled temperature program in a prescribed atmosphere. The test sample and reference crucible or reference sample are subjected to the same temperature

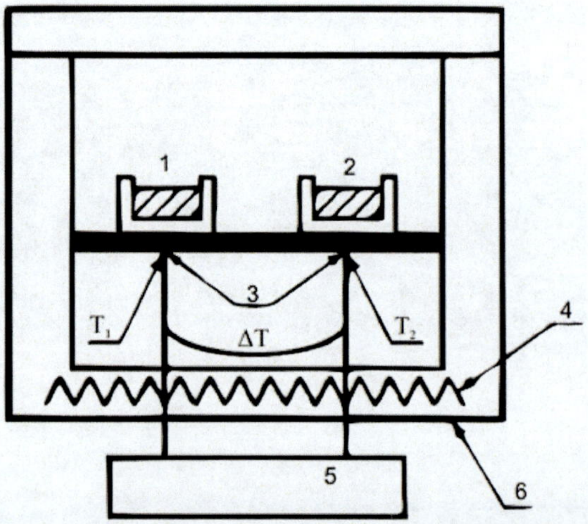

Fig. 4.2 Schematic diagram of DSC device by heat flow: 1—test sample; 2—reference crucible or standard sample; 3—thermocouples; 4—general heater; 5—measuring circuit for; 6—furnace; *T*1—temperature of the test sample; *T*2—temperature of the reference crucible or standard sample; Δ*T*—temperature difference between the test sample and the reference crucible or standard sample

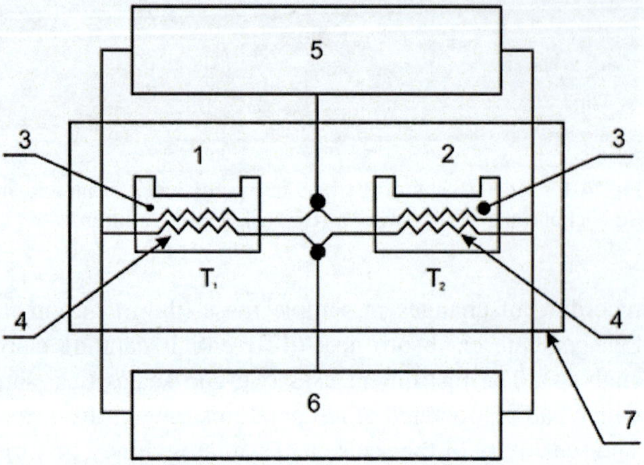

Fig. 4.3 Schematic diagram of DSC device with power compensation: 1—position of testing sample; 2—position of reference crucible or standard sample; 3—thermometers; 4—individual heaters; 5—measuring circuit; 6—power compensation device; 7—heat sink; *T*1—temperature near sample; *T*2—temperature near reference crucible or standard sample

program using a common heater. The temperature difference between test sample and reference crucible or standard sample results from their different heat capacities. From this temperature difference, the difference in heat fluxes between the test item and a reference crucible or standard sample are determined and recorded as a function of the temperature of the reference crucible or standard sample or as a function of time. The diagram of the DSC heat flux meter is shown in Fig. 4.2.

In a power-compensated DSC, individual heaters are used for the test item and the reference crucible or reference sample. The power difference required to maintain the same temperature between the test sample and the reference crucible or standard sample is recorded as a function of time, with the test sample and the reference crucible or standard sample exposed to the same temperature program. The schematic of the DSC instrument with power compensation is shown in Fig. 4.3.

Universal Analysis 2000 software was used in the processing of thermoanalytical curves and kinetic calculation software (Borchardt and Daniels method), which

Table 4.2 Thermal analysis conditions

Sample number	Surveys conducted (conditions)			
	TGA			
	20 °C/min	20 °C/ min	5 °C/ min	DSC
«0»	Nitrogen—air	Air	Air	Air, 5 °C/мин
«1»	Nitrogen—air	Air	Air	Air, 5 °C/мин
«2»	Nitrogen—air	Air	Air	Air, 5 °C/мин
«3»	Nitrogen—air	Air	Air	Air, 5 °C/мин

increased efficiency and simplified curve processing. Universal Analysis 2000 includes an extensive multimedia help system for working with the software and data processing algorithms, containing both a traditional hypertext reference book and a set of videos to help you understand the basics of the different approaches in working with TA Instruments thermal analysis equipment.

We used TA Instruments (USA) Q-600SDT thermal analysis instrumentation, which combines TDG, DTG, and DSC (by heat flux) methods [14], to study the characteristics of thermal (pyrolysis) and thermal oxidative decomposition processes, as well as the oxidative capacity of coke. The ability to simultaneously measure heat flux (DSC) and change in mass (TGA) over a wide temperature range simplifies interpretation of results and increases analysis productivity. The thermoanalyzer allows for simultaneous recording of mass changes (thermogravimetric analysis) and processes accompanied by heat release or absorption (differential scanning calorimetry/differential thermal analysis).

Samples of materials in the form of fragments of regular shape were placed in the crucible of a Q600 SDT thermal analyzer. The sample weights varied between 1.0 and 8.5 mg. The conditions of thermal analysis are presented in Table 4.2.

Figure 4.4 shows TG and DTG—curves in the current of air (heating rate 20 °C/min) of the studied wood samples of different life span.

In the oxidizing environment, there is an earlier stage of the beginning of wood weight loss of long-term natural aging, which leads to a decrease in the rate of weight loss of the samples at the main stage of thermo-oxidative decomposition.

The values of the main thermal indices of wood depending on the service life of wooden structures are summarized in Table 4.3.

In the temperature range, the values of mass loss for samples of different service life have very close values and are characterized by moisture loss in this range. However, in the temperature range of 150–375 °C, with increasing service life of the wooden structure, there is a significant decrease in weight loss of the sample (actually 1.5 times for wood with a service life of 150 years). At the same time, the temperature of maximum mass loss rate decreases (DTG curves).

The greatest changes are undergone by the oxidation stage of the coke residue of long-lived wood. This is especially noticeable for wood samples with a service life of 150 years. Oxidation rate of coke increases by 1.6 times compared to modern wood samples (Table 4.3, Fig. 4.5).

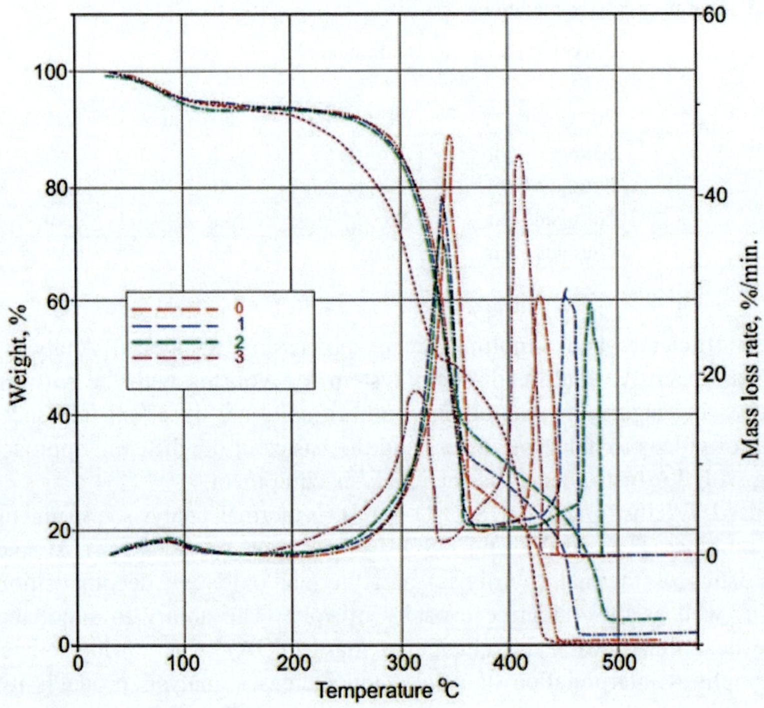

Fig. 4.4 TG and DTG curves of wood samples of different life span in air (20 °C/min) (numbering of samples in Table 4.1)

Table 4.3 Results of thermal analysis of wood of different life cycles in air, heating rate 20 °C/min

Sample number				
	«0»	«1»	«2»	«3»
Temperature range 30–150 °C				
Weight loss, %	6.1	5.8	6.3	6.2
Temperature range 150–375 °C				
Weight loss, %	66.5	63.2	58.3	44.2 (150–340 °C)
Maximum DTG, °C/A, %·min^{-1}	343/47.5	338/40.5	338/36.0	310/18.4
Temperature range 375–575 °C (coke oxidation in air, Fig. 4.5)				
Oxidation rate coke, %·min^{-1}	28.7	29.8	28.0	45.4

To study the features of the thermal decomposition process (pyrolysis) of wood of different life cycles, the materials were heated at a rate of 20 °C/min in a nitrogen current up to 500 °C. Then, the carrier gas was changed from nitrogen to air. The thermodestruction and thermal oxidation process was recorded by both TGA signal and DSC. The combined TGA, DTG, and DSC curves are shown in Fig. 4.6.

TG and DTG curves show that after moisture removal (when heated up to 150 °C), there is a noticeable weight loss of long-lived wood samples compared to modern wood samples. During this period, apparently (up to 250–270 °C) the removal of

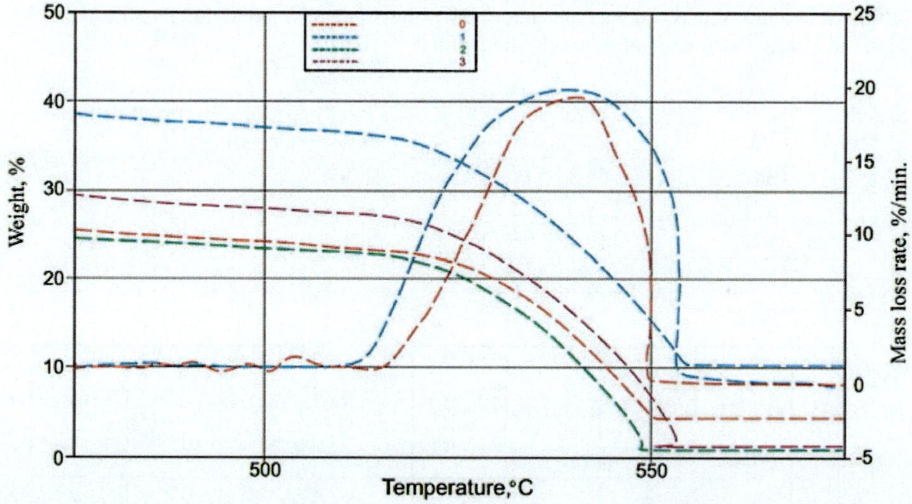

Fig. 4.5 TG and DTG curves of wood samples of different life cycle in the interval of 500–575 °C in air (20 °C/min) (numbering of samples in Table 4.1)

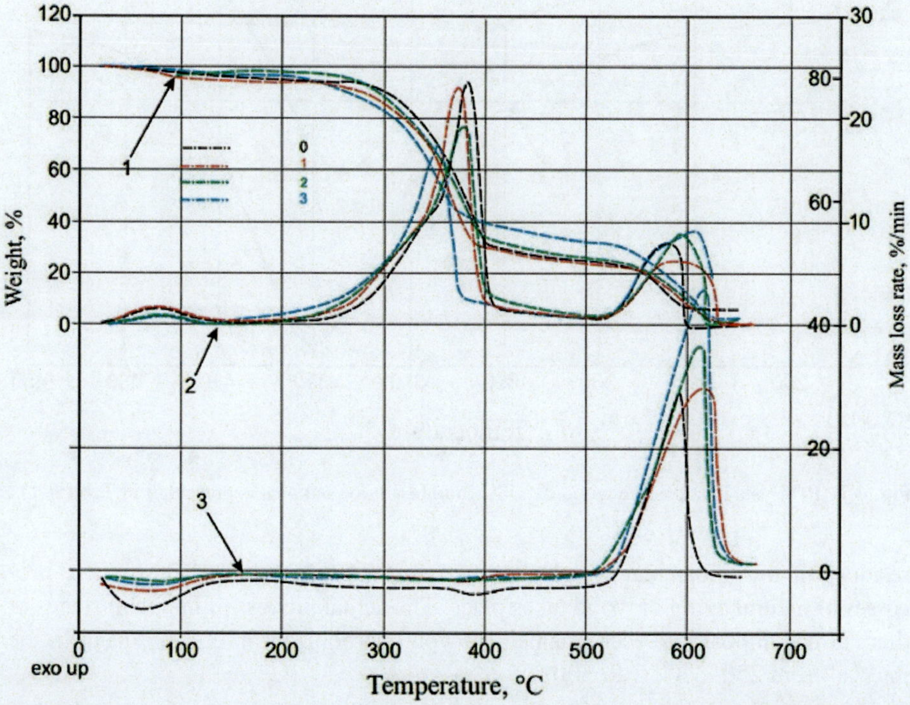

Fig. 4.6 The results of thermal analysis (1—TG, 2—DTG, 3—DSC) of the studied wood samples of different life cycle in nitrogen-air environment (20 °C/min) (numbering of samples is presented in Table 4.1)

Table 4.4 Values of maximum temperature and pyrolysis rate of wood of different life cycles by DTG curve at the main stage of decomposition (300–400 °C)

Sample number	Peak temperature, °C	Pyrolysis rate by DTG (%/min)	Pyrolysis heat (J/g) (based on total mass of the sample)
0	382	23.8	174.4
1	372	23.0	132.4
2	379	19.4	125.2
3	356	17.6	38.2

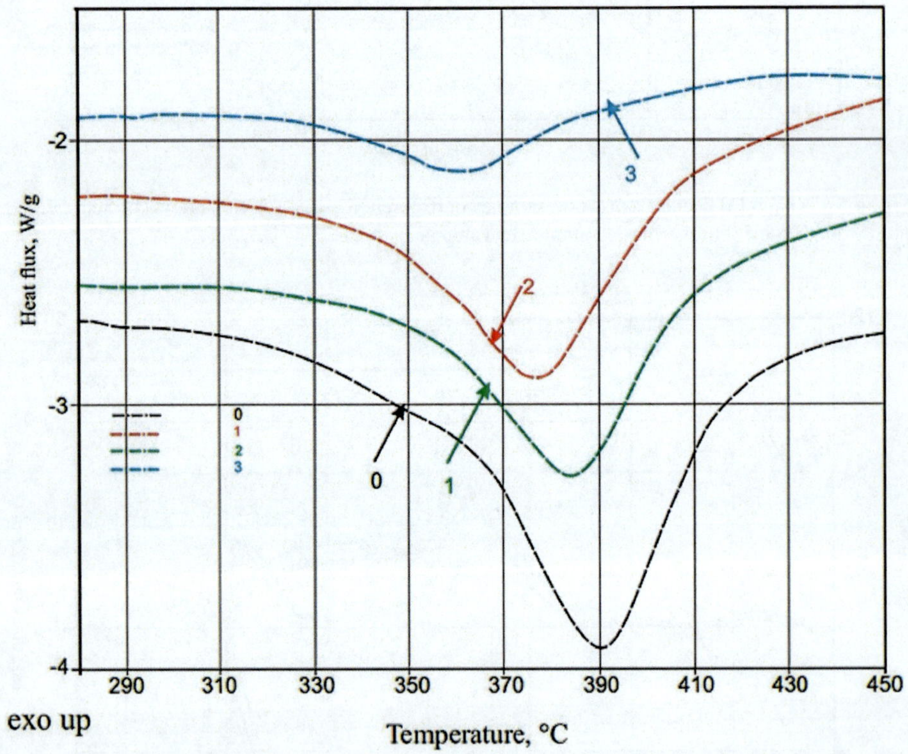

Fig. 4.7 Pyrolysis heat of pine wood by DSC (numbering of samples is presented in Table 4.1).

residues of low molecular weight products formed as a result of destructive processes of natural aging of wood takes place. The actual process of the main stages of thermal decomposition, accompanied by wood carbonization, is carried out in the interval from 250–270 °C to 500 °C.

Decrease of weight loss rate of long-lived wood samples at the main stage is confirmed by acceleration of charring and decrease of pyrolysis heat values. So, when wood lifetime increases up to 150 years, pyrolysis heat is 38.2 J/g, which is 4.5 times less than for modern wood samples (Table 4.4).

Graphical change of pyrolysis heat values during the main stage of wood thermal decomposition is shown in Fig. 4.7.

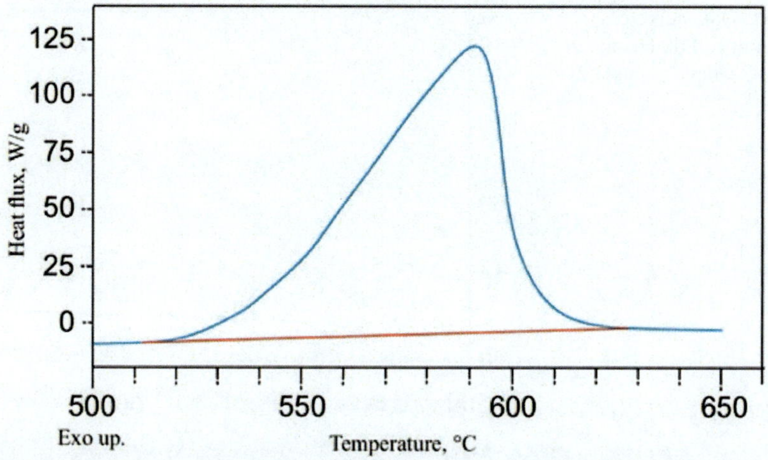

Fig. 4.8 DSC curve of oxidation of carbon residue of wood sample (sample 0)

Fig. 4.9 DSC curve of oxidation of carbon layer of wood residue (sample 1)

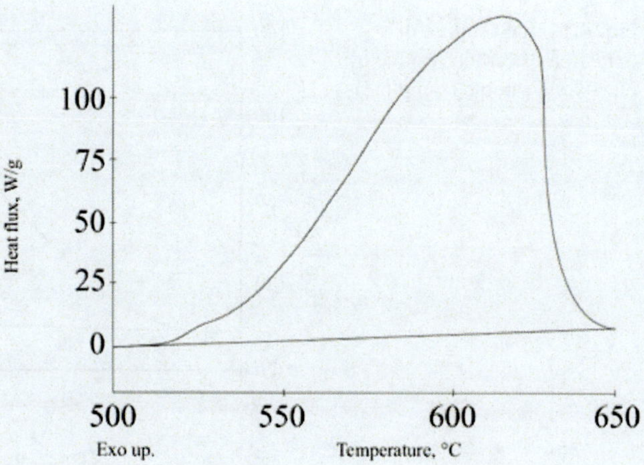

For a more detailed study of the oxidation process of coke residue depending on the duration of operation of wooden structures, we calculated the kinetic parameters of thermal decomposition of the studied wood samples in the temperature range of 500 °C and above. Kinetic parameters were calculated by the Borchardt-Daniels method using DSC curves.

To calculate the kinetic parameters of the thermal decomposition process of the studied samples, DSC curves (Figs. 4.8, 4.9, 4.10, and 4.11), dependences of the reaction rate constant on temperature (Figs. 4.12, 4.13, 4.14, and 4.15), and

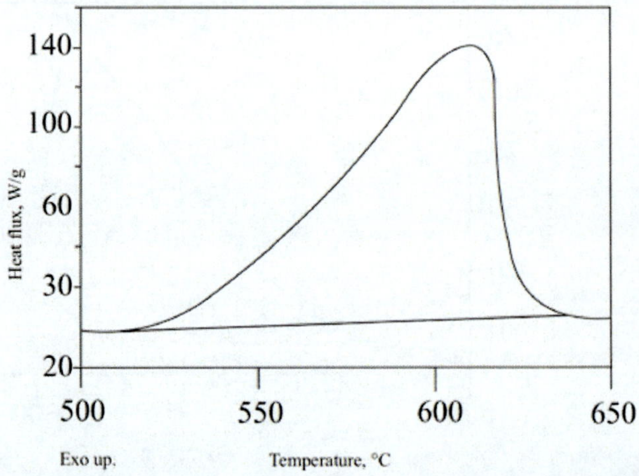

Fig. 4.10 DSC curve of oxidation of carbon residue of pine wood sample (sample 2)

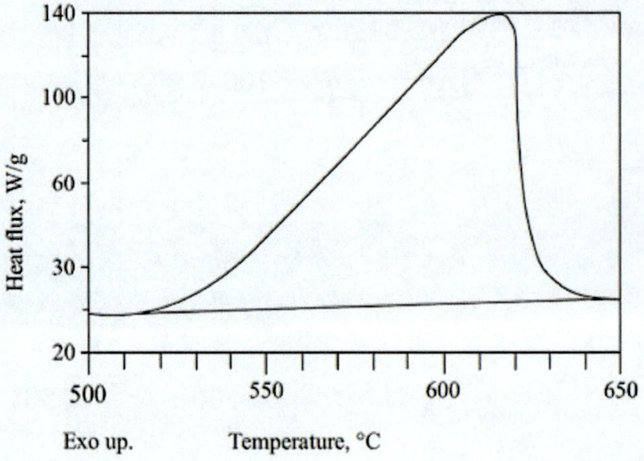

Fig. 4.11 DSC curve of oxidation of carbon residue of pine wood sample (sample 3)

logarithm of the reaction rate constant on temperature (Figs. 4.16, 4.17, 4.18, and 4.19) were built for wood samples with varying age of operation.

The calculated data for estimation of kinetic parameters are presented in Tables 4.5, 4.6, 4.7, and 4.8.

The Borchardt and Daniels kinetics (*B*/*D*) calculates the activation energy (E_a), the pre-exponential multiplier (Z), and the reaction heat effect (ΔH), reaction order (n), and rate constant (K) from one DSC scan.

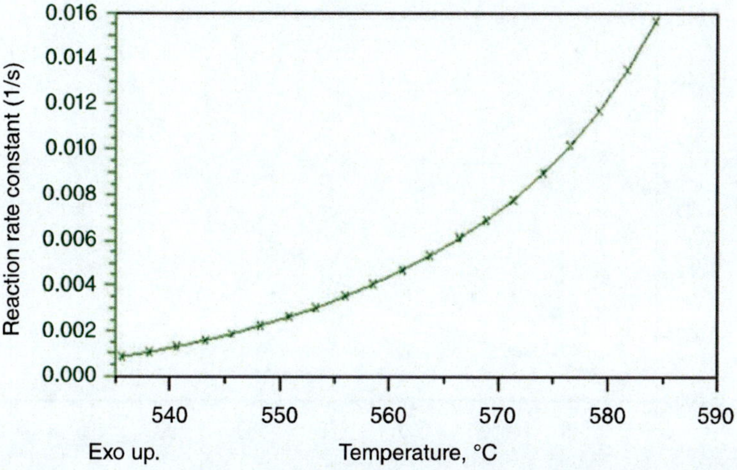

Fig. 4.12 Dependence of the reaction rate constant on temperature in the study of modern wood (sample 0)

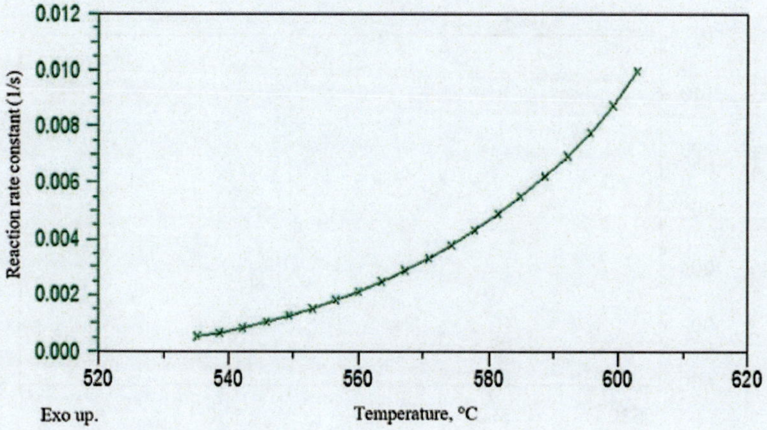

Fig. 4.13 Dependence of the reaction rate constant on temperature in the study of modern wood (sample 1)

The method assumes that the reaction follows nth order kinetics and obeys a general kinetic equation:

$$\frac{d\alpha}{dt} = k(T)[1-\alpha]^n, \tag{4.1}$$

or

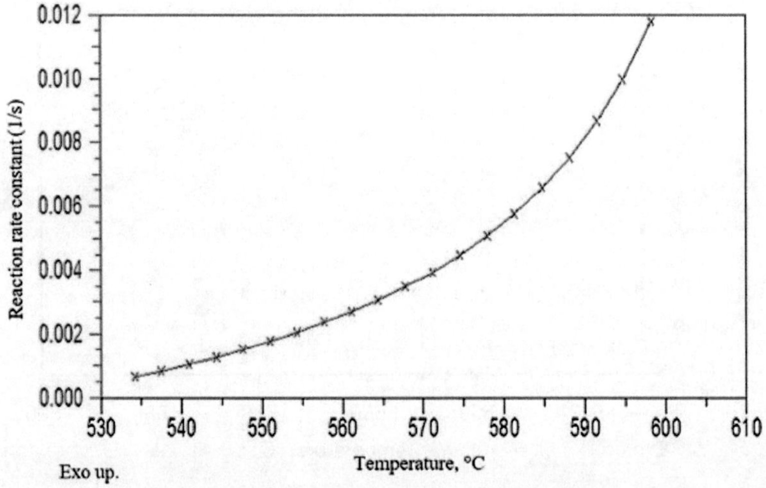

Fig. 4.14 Dependence of the reaction rate constant on temperature (sample 2)

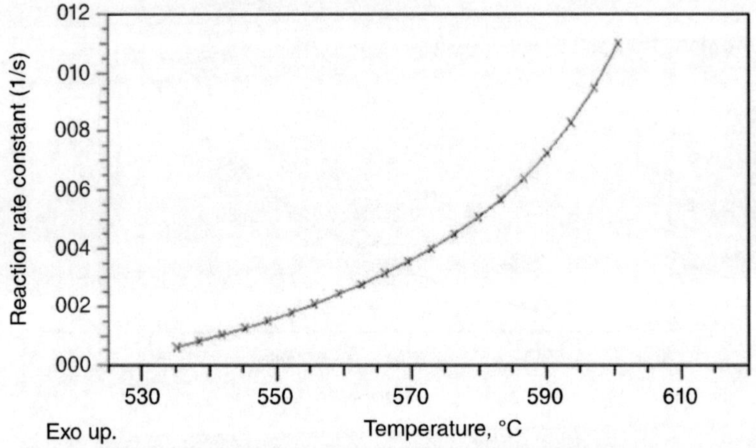

Fig. 4.15 Dependence of the reaction rate constant on temperature (sample 3)

$$\frac{dH}{dt} * \frac{1}{\Delta H} = k(t)\left(1 - \frac{\Delta H_T}{\Delta H_0}\right)^n, \tag{4.2}$$

where $d\alpha/dt$—the reaction rate (1/s); α—the degree of transformation (conversion)—$\Delta H_T/\Delta H_0$; ΔH_T—the heat effect at temperature T; ΔH_0—the total heat flux; $k(T)$—the reaction rate constant; and n—the reaction order.

The developers of the method assumed that the reaction obeys Arrhenius' law:

$$k(T) = Ze^{-E_a/RT}, \tag{4.3}$$

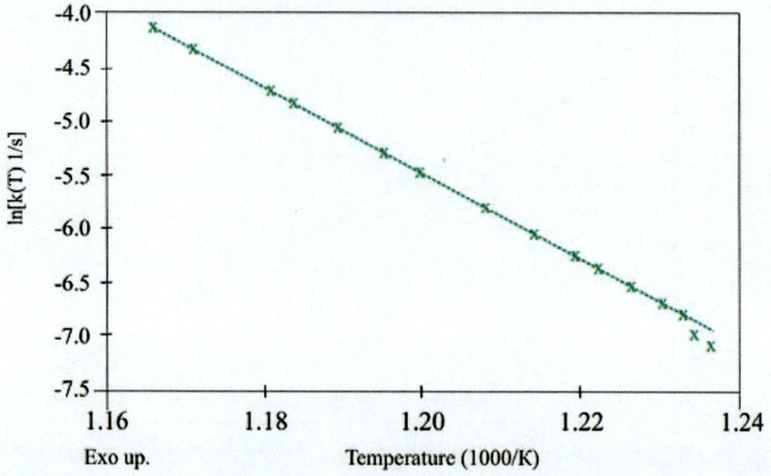

Fig. 4.16 Dependence of the logarithm of the reaction rate constant on temperature (sample 0)

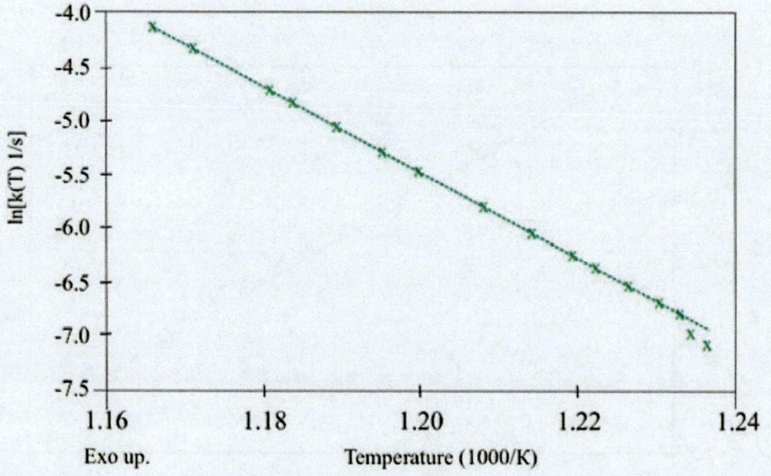

Fig. 4.17 Dependence of the logarithm of the reaction rate constant on temperature (sample 1)

where E_a is the activation energy (J/mol); *Z*—the pre-exponential or frequency factor (1/s); *R*—the gas constant (8.314 J/mol·K).

Подставляя уравнение (4.2) в уравнение (4.1), и взяв логарифмы, получим:

$$d\alpha/dt = Z\, e^{-Ea/RT}\, (1-\alpha)^n. \quad (4.4)$$

$$\ln[d\alpha/dt] = \ln(Z) - E_a/RT + n\, \ln[1-\alpha]. \quad (4.5)$$

или

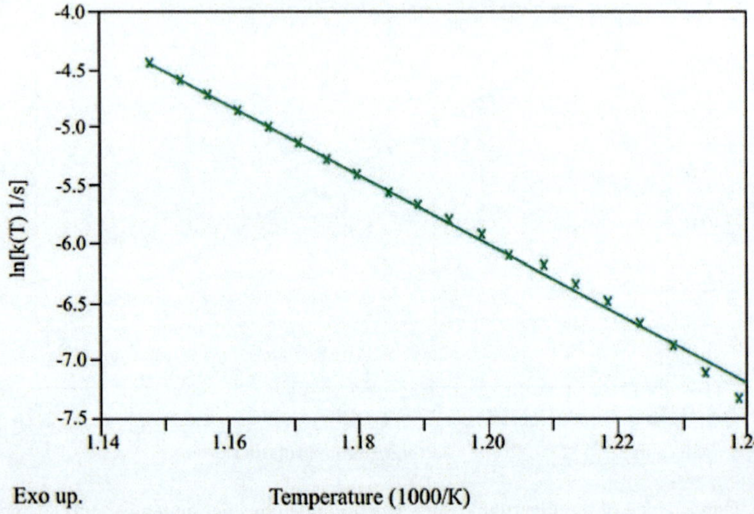

Fig. 4.18 Dependence of the logarithm of the reaction rate constant on temperature (sample 2)

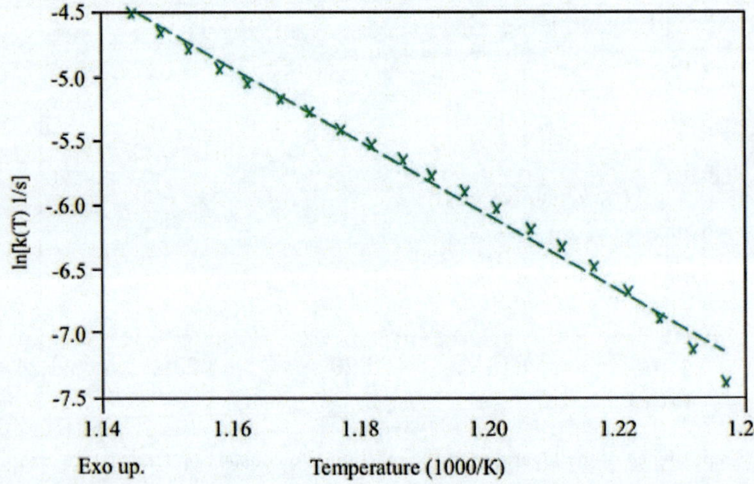

Fig. 4.19 Dependence of the logarithm of the reaction rate constant on temperature (sample 3)

$$\ln\left(\frac{dH}{dt} * \frac{1}{\Delta H_0}\right) = \ln(Z) - \frac{E_a}{RT} + n \cdot \ln\left(1 - \frac{\Delta H_T}{\Delta H_0}\right).$$

Equation (4.4) can be solved using multiple linear regression (Eq. 4.5):

$$z = a + \text{в} \cdot x + \text{c} \cdot y, \tag{4.6}$$

where two main parameters ($d\alpha/dt$ and α) are determined from the DSC experiment. α—the degree of transformation (conversion)—$\Delta H_T/\Delta H_0$; ΔH_T—the heat effect at temperature (T); ΔH_0—the total heat flux.

Table 4.5 Calculated rate constant and half-life data for a modern wood sample

Weight: 1.2000 g					
Temperature, °C	Temperature, K	Temperature, 1000/K	$k(T)$, 1/s	ln $k(T)$, 1/s	Half-life, min
520.00	793.15	1.26	0.0003769	−7.88	30.6
530.00	803.15	1.25	0.0007031	−7.26	16.4
540.00	813.15	1.23	0.001292	−6.65	8.94
550.00	823.15	1.21	0.002338	−6.06	4.94
560.00	833.15	1.20	0.004172	−5.48	2.77
570.00	843.15	1.19	0.007342	−4.91	1.57
580.00	853.15	1.17	0.01275	−4.36	0.906
590.00	863.15	1.16	0.02187	−3.82	0.528
600.00	873.15	1.15	0.03704	−3.30	0.312
610.00	883.15	1.13	0.06199	−2.78	0.186
620.00	893.15	1.12	0.1026	−2.28	0.113

Table 4.6 Calculated rate constant and half-life for sample 1

Weight: 1.4000 g					
Temperature, °C	Temperature, K	Temperature, 1000/K	$k(T)$, 1/s	ln $k(T)$, 1/s	Half-life, min
520.00	793.15	1.26	0.0003174	−8.06	36.4
530.00	803.15	1.25	0.0005067	−7.59	22.8
540.00	813.15	1.23	0.0007995	−7.13	14.4
550.00	823.15	1.21	0.001248	−6.69	9.26
560.00	833.15	1.20	0.001927	−6.25	6.00
570.00	843.15	1.19	0. 002944	−5.83	3.92
580.00	853.15	1.17	0.004455	−5.41	2.59
590.00	863.15	1.16	0.006677	−5.01	1.73
600.00	873.15	1.15	0.009915	−4.61	1.17
610.00	883.15	1.13	0.01459	−4.23	0.792
620.00	893.15	1.12	0.02129	−3.85	0.543
630.00	903.15	1.11	0.03080	−3.48	0.375
640.00	913.15	1.10	0.04420	−3.12	0.261
650.00	923.15	1.08	0.06294	−2.77	0.184

The stage of exothermic process of oxidation of the charcoal layer of wood of long-term natural aging on the DTG and DSC curves recorded a significant increase in the mass loss rate, the calculated thermal effect of the coke oxidation reaction, and the heat release rate of the thermal oxidation process increases (Table 4.9).

The values of the effective parameters of the coke oxidation process (activation energy and the logarithm of the pre-exponent) indicate a decrease in the energy input needed to start this process. The results obtained by methods of thermal analysis indicate a shift of the carbon formation process to a lower-temperature area for

Table 4.7 Calculated rate constant and half-life for sample 2

Weight: 1.3980 g.					
Temperature, °C	Temperature, K	Temperature, 1000/K	$k(T)$, 1/s	ln $k(T)$, 1/s	Half-life, min.
520.00	793.15	1.26	0.0003978	−7.83	29.0
530.00	803.15	1.25	0.0006364	−7.36	18.2
540.00	813.15	1.23	0.001006	−6.90	11.5
550.00	823.15	1.21	0.001573	−6.45	7.34
560.00	833.15	1.20	0.002434	−6.02	4.75
570.00	843.15	1.19	0.003727	−5.59	3.10
580.00	853.15	1.17	0.005650	−5.18	2.04
590.00	863.15	1.16	0.008482	−4.77	1.36
600.00	873.15	1.15	0.01262	−4.37	0.916

Table 4.8 Calculated rate constant and half-life for sample 3

Weight: 1.6680 g.					
Temperature, °C	Temperature, K	Temperature, 1000/K	$k(T)$, 1/s	ln $k(T)$, 1/s	Half-life, min
510.00	783.15	1.28	0.0002486	−8.30	46.5
520.00	793.15	1.26	0.0003958	−7.83	29.2
530.00	803.15	1.25	0.0006229	−7.38	18.5
540.00	813.15	1.23	0.0009696	−6.94	11.9
550.00	823.15	1.21	0.001493	−6.51	7.74
560.00	833.15	1.20	0.002275	−6.09	5.08
570.00	843.15	1.19	0.003433	−5.67	3.37
580.00	853.15	1.17	0.005130	−5.27	2.25
590.00	863.15	1.16	0.007595	−4.88	1.52
600.00	873.15	1.15	0.01114	−4.50	1.04
610.00	883.15	1.13	0.01621	−4.12	0.713
620.00	893.15	1.12	0.02338	−3.76	0.494
630.00	903.15	1.11	0.03345	−3.40	0.345
640.00	913.15	1.10	0.04748	−3.05	0.243
650.00	923.15	1.08	0.06689	−2.70	0.173

long-life wood, acceleration of charring, as well as the formation of specific properties of carbon residue and its high oxidative and calorific capacity, which should eventually lead to acceleration of thermal damage of wooden structures.

The establishment of high oxidative and calorific ability of the formed coal residues and the kinetic parameters of the coke oxidation process can be used for the prognostic assessment of fire hazard and fire resistance of wooden constructions of different operation time. So, for example, for wooden structures with a service life of 100–150 years, we can predict a significant heat release under fire exposure and

Table 4.9 Characteristics of the oxidation process of carbon residue for wood of different life cycles

Sample number	Calculated parameter by DSC curves				
	Rate heat release rate[a], W/g	Heat effect[b], J/g	Activation energy, kJ/mol	Logarithm of the pre-exponent, log (1/min)	Given reaction order
0	30.2	15,318	331.0	20.15	1
1	28.9	25,218	247.9	14.60	1
2	37.3	21,580	249.8	14.82	1
3	46.6	22,176	240.7	14.23	1

[a]Heat release rate is referred to the original mass of the sample
[b]Heat effect is referred to the coke mass

Table 4.10 Value of the thermal conductivity coefficient for wood of different service life with fire-retardant compositions (across the fibers)

Sample name	Thermal conductivity coefficient, W/m·K
Modern pine timber	0.096
Long-life pine timber 81 years	0.107

significant thermal damage, which will eventually have a negative impact on the fire resistance of wooden structures.

Analysis of the obtained results shows that at the initial and main stage of thermal decomposition of long-lived wood specimens, due to early carbon formation, there is a decrease in the rate of weight loss of the specimens. However, with further increase in temperature, the negative role of carbon residue is manifested, due to its high oxidation capacity and the release of a significant amount of heat. For a wood sample with a life span of 150 years, the rate of heat release is actually 1.5 times higher compared to samples of modern wood.

Thus, at the main stage (300–400 °C) of thermal decomposition of long-lived samples, there is a marked decrease in the rate of mass loss. This is due to the earlier course of the charring process of long-lived wood of natural aging compared to samples of modern wood, as well as the properties and structure of the resulting charcoal layer.

It has been established that the thermal-oxidative stability of wood, as well as the energy and oxidative capacity of the formed charcoal layer, significantly changes as a result of temperature-oxidizing effects.

Calculation of kinetic parameters, in particular, activation energy, of thermo-oxidative decomposition of wood samples showed that with increasing wood life there is a decrease in the values of activation energy. An important feature of the oxidative process of coke wood of natural aging is the ability at lower energy costs, compared to modern wood and the emergence and course of the oxidative process, which has an exothermic nature.

Fig. 4.20 General view of the device for measuring the thermal conductivity of materials

4.2 Studying the Structure and Thermal-Physical Characteristics of Long-Lived Timber Layers

During the experimental evaluation of the thermal conductivity of the studied samples under steady-state heating conditions, it was found that the value of the thermal conductivity coefficient for wood with a lifetime of 81 years is 1.11 times higher compared to modern samples (Table 4.10).

These results can be explained by the change in the morphological structure of long-term natural aging wood and the increase in its density compared to samples of modern wood in 1.2 times.

The essence of the method consists in determining the thermal conductivity coefficient (λ) of solid composite materials under the stationary thermal regime.

A thermal conductivity meter (Fig. 4.20) is used for investigations, the principle of operation of which is based on creating a stationary heat flow directed perpendicularly to the face of the sample.

The heat flux density and the temperature of the opposite face faces are recorded during the study. The instrument (Fig. 4.20) is designed to measure λ of materials on the samples: parallelepiped with the size of sides 100 × 100 mm and circular disks with the diameter from 90 to 100 mm.

Calculation of thermal conductivity (λ) (effective thermal conductivity) at steady-state thermal conditions is performed by a computing device according to the formula

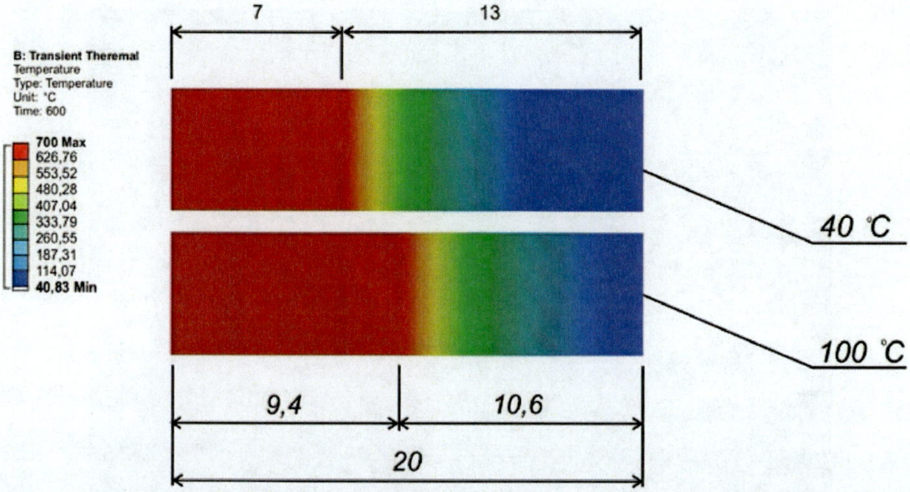

Fig. 4.21 Simulation results of the heating of the 20-mm layer of the structure made of wood of two types at the temperature regime of the fire chamber

$$\lambda = \frac{h \cdot q}{T_{н} - T_{x}}, \tag{4.7}$$

where λ—effective thermal conductivity, W/(m·K)

h—thickness of the measured sample, mm

q—the density of steady-state heat flux passing through the measured sample, W/(m^2)

T_n—temperature of hot edge of the measured sample, K

T_x—temperature of cold face of the measured sample, K

This is due to the higher density values of the wood material. However, for the formed charcoal layer of long-lived wood, the values of wood thermal conductivity coefficients are lower than for samples of modern wood. For the charcoal layer of modern wood, the thermal conductivity coefficient was 0.08 W/m·K, while for wood with a service life of 150 years, it was 0.05 W/m·K. This can be attributed to the structural features of long-lived wood, as well as to the peculiarities of the formation of a fine-cell charcoal layer and its heat-insulating properties.

To compare the thermal conductivity of different types of wood, the duration of simulation time was taken equal to 10 min. According to test results and resulting charring rate, at the time of 10 min, the temperature in the fire chamber was 700 °C; the depth of charring of modern and long-life wood construction was 7 and 9.4 mm, respectively. Thus, the distance from charring front to the thermocouple, located at the depth of 20 mm, was 13 and 10.6 mm, and wood temperature according to thermocouple readings was 40 and 100 °C, respectively. The above unsteady heat transfer conditions were simulated in ANSYS 17 Academic system (Fig. 4.21).

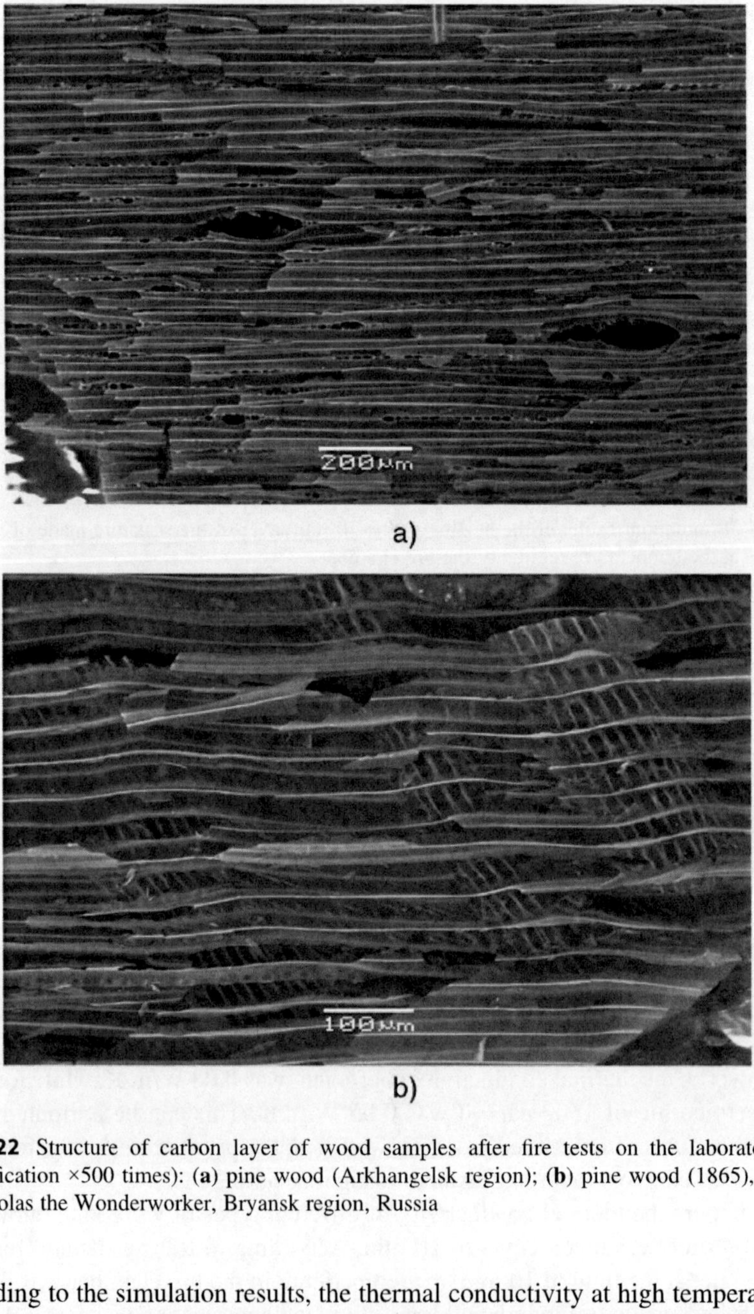

Fig. 4.22 Structure of carbon layer of wood samples after fire tests on the laboratory unit (magnification ×500 times): (**a**) pine wood (Arkhangelsk region); (**b**) pine wood (1865), Church of Nicholas the Wonderworker, Bryansk region, Russia

According to the simulation results, the thermal conductivity at high temperature of long-life wood exceeds the thermal conductivity of modern wood by 30–35%.

The subsequent cycle of research was devoted to the study of the specifics of carbonized layers with increasing duration of operation of wooden structures.

Evaluation of qualitative changes in the structure of the carbonized residue was performed on a scanning digital electron microscope JSM-5610LV (Japan) in

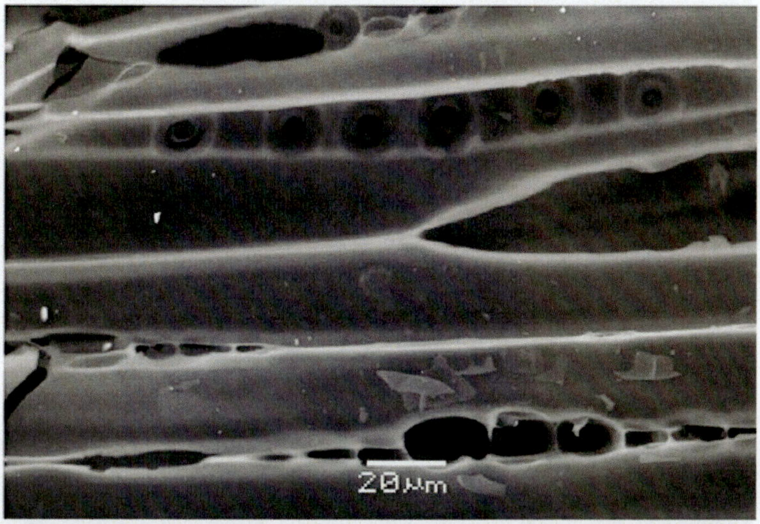

a)

b)

Fig. 4.23 Structure of carbon layer of wood samples after fire tests on the laboratory unit (magnification ×1000 times): (**a**) pine wood (Arkhangelsk region); (**b**) pine wood (1865), Church of Nicholas the Wonderworker, Bryansk region, Russia

reflected electrons (BSE COMPO). A thin film of carbon was pre-sprayed on the wood samples to drain the electrostatic charge arising from the interaction of the sample with the electron beam. Images were captured at a surface magnification of 500 to 1000 times [15].

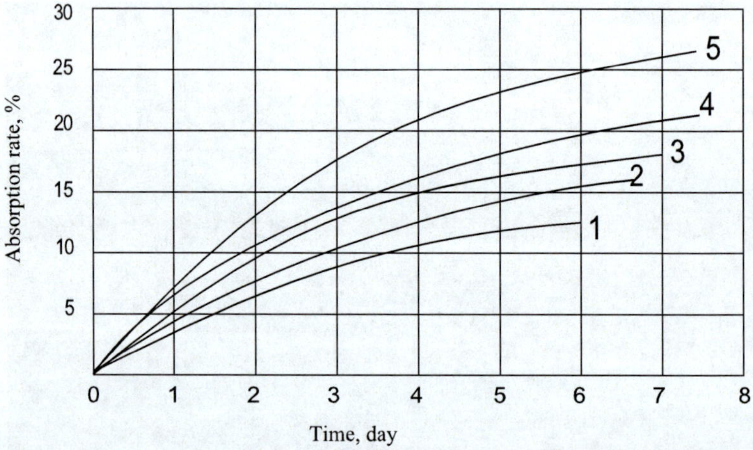

Fig. 4.24 Kinetic curves of benzene vapor sorption for coal residues of pine timber of different lifetimes: 1—modern pine wood; 2—pine wood, non-residential house (1952), Republic of Kazakhstan, Zhanazhol village (North Kazakhstan region); 3—pine wood, non-residential house (1928), Republic of Kazakhstan, v. Kaiyndy (Kostanay region); 4—pine wood (1890), elements of the wooden ceiling, Gulliver puppet theater; 5—pine wood (1865), Church of Nicholas the Wonderworker, Bryansk region

When considering the surface of carbonized layers (magnification of 500 times), we can conclude that, for modern wood samples, formation of lamellar structures is observed, between which pore-like holes of unequal size are located. For the carbonized layer of long-lived wood, there is a transition from a lamellar to a more open structure in which a large number of regular-shaped pores can be distinguished (Fig. 4.22).

When the studied surface is increased by 1000 times for the carbonized residue of modern wood (Fig. 4.23), we can indeed observe a random and asymmetrical arrangement of pores, and the carbonized layer of long-lived wood acquires an ordered "honeycomb-like" structure. This indicates a more developed inner surface of the charcoal residue.

To study the characteristics of the surface coal layer, the method of vapor sorption was used. The adsorbent was coal powder of a fine fraction (0.3–0.5 mm). Samples of charcoal were taken from wood samples tested on a small-scale plant under standard fire temperature conditions. Benzene was chosen as the adsorbate because this substance has a high surface tension and can wet the wood capillary surface. Kinetic curves of adsorption of carbon residues of the studied samples are shown in Fig. 4.24.

Figure 4.24 shows that the maximum values of adsorbate increment are for coal residue of pine wood with a life of 150 years (26.76%). This is actually 2 times higher than the value of adsorbate increment of coal samples for modern wood (12.8%). The coal residue of modern wood is characterized by rapid adsorbent vapor saturation at the initial stage and further stabilization of the curve. Complete saturation of samples with benzene vapor occurred within 6 days.

Table 4.11 Characteristics of the porous structure of coal

№ sample (corresponds to Fig. 4.24)	Name of characteristics of the coal structure			
	a_m, %	V_Σ, sm^3/g	d_m, нм	S_{sp}, m^2/g
1	14.8	0.4726	25.03348	377.5744
2	16.53	0.5234	20.56678	823.7534
3	18.48	0.5968	18.51378	1038.637
4	21.81	0.6327	14.65459	1554.8574
5	26.76	0.6812	12.80586	1645.7461

Slow saturation at the initial stage (2–3 h), further intensification of the process, and uniform saturation are characteristic for the carbon residue of long-term natural aging wood. As the test results show, the period of benzene vapor absorption for naturally aged wood has a longer period compared to the sample of modern wood. Complete saturation of samples with benzene vapor occurred on the 8th day. This confirms the statement about a more developed structure of the charcoal layer for long natural aging wood.

To explain the obtained results, the characteristics of the porous structure of charcoal residues were calculated: total pore volume (V_Σ), average pore radius (d_m), and specific surface area of charcoal layers (S_{sp}). The results of calculations of characteristics of coal bed structure are given in Table 4.11.

Thus, with increasing service life of wooden structures in fire conditions, active intensification of the charring process is observed. Some decrease in mass loss in the main area of wood thermal decomposition is due to the peculiarities of fine-pored structure of coal layer and, apparently, to a decrease in its thermal conductivity. However, further, as the temperature increases, the formed carbon structure with increased specific surface area and specific pore volume contributes to its increased oxidation, burnout, and flameless (smoldering) combustion. Changing the chemical composition of wood and significant transformation of the structure as a result of its long-term use leads to acceleration of charring of wooden structures, lowering the temperature threshold of loss of the original mechanical characteristics of wood, which will affect the fire resistance limit of wooden structures.

4.3 Influence of Operating Time on Charring Intensity and Physical-Mechanical Characteristics of Long-Term Natural Aging Wood

To study the process of charring of long-term natural aging wood, fire tests were conducted under one-sided heating on the installation for determining the flammability of building materials according to GOST 30402-96 (Russia) and on a small-scale laboratory installation under the conditions of standard fire temperature regime. In these studies, it is important to establish the effect of the duration of operation of wooden structures on the parameters of charring of wood in different conditions of fire exposure (isothermal and dynamic heating modes).

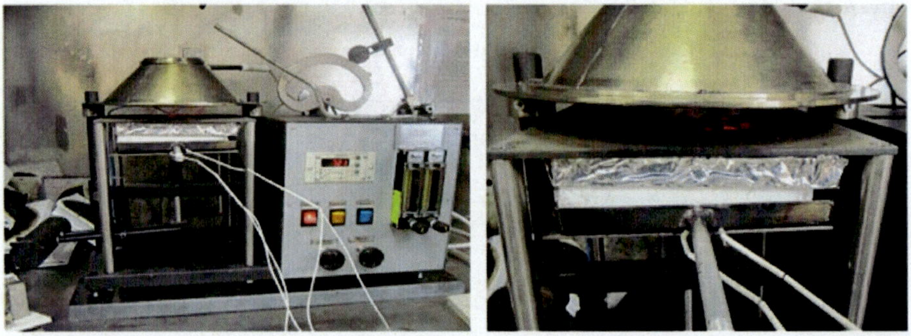

a. b.

Fig. 4.25 General view of the installation to assess the flammability of building materials according to GOST 30402-96 (Russia): (**a**) installation before tests; (**b**) wood sample at the time of testing

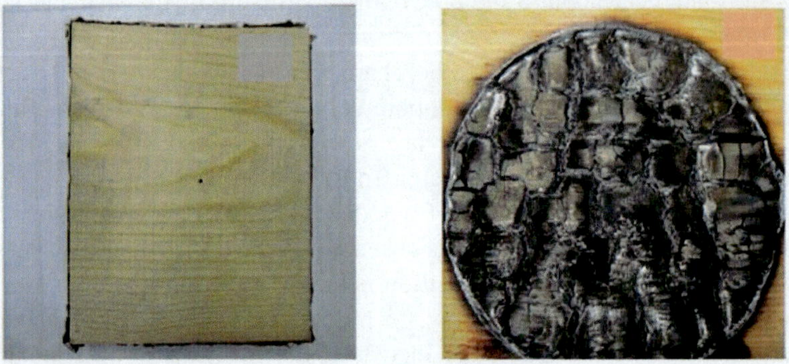

a. b.

Fig. 4.26 Specimen of modern pine before (**a**) and after (**b**) the test

To establish the parameters of charring in the mode of isothermal heating of wood samples, an apparatus for evaluating flammability (GOST 30402–96, Russia) was used.

Figure 4.25 shows the apparatus for assessing the flammability of building materials at the time of fire tests.

Tests were conducted under the influence of external radiation heat flux density of 20 kW/m^2—a boundary regime between pre-ignition mode and flame combustion without the use of an additional gas burner ignition. The dimensions of the wood samples were 165 × 165 mm and 20 mm thick. The direction of heat flow was chosen across the fibers of the wood composite. Before the tests, the wood specimen was wrapped in foil with its surface open in the center in the form of a circle 12 cm in diameter. Figure 4.26 shows a general view of a modern wood specimen before and after the fire tests.

The results show that even in the absence of flame combustion, there was charring of the samples (Table 4.12).

Table 4.12 Results of estimation of charring parameters of tested samples on the installation according to GOST 30402-96 (Russia)

Sample name	Charring depth, mm	Charring rate, mm/min	Time of the beginning of smoldering combustion (charring)
Modern pine (Arkhangelsk region, Russia)	15	0.5	6 min 53 s
Pine timber (1890), elements of wooden ceiling, Gulliver Puppet Theatre, 104 Sovetskaya str.	20	0.7	5 min 40 s
Pine timber (1865), Church of St. Nicholas the Wonderworker, Bryansk region, Russia	24	0.8	2 min 20 s

a. b.

Fig. 4.27 General view of a small-scale laboratory setup: (**a**) with an open outer panel and a fire chamber; (**b**) with a sample holder

a. b.

Fig. 4.28 General view of the chamber (**a**) and the heat flux sensor (**b**) of the small-scale laboratory setup

a. b.

Fig. 4.29 Fixing the sample in the installation holder (**a**) and its placement in the furnace firing chamber (**b**)

For wood with increasing service life, the charring rate increased by 30–40%. It is interesting to note that the visual time of onset of flameless (smoldering) combustion, expressed by characteristic glow, for wood of natural aging began much earlier, actually on the second minute, which indicates an earlier charring compared to the modern wood sample.

Intensification of charring process for long-term natural aging wood is also observed under standard fire temperature conditions.

A small-scale laboratory setup was used to conduct research under standard fire temperature conditions. The general view of small-scale laboratory unit, its main structural elements, and mounting for samples are shown on Figs. 4.27, 4.28, and 4.29.

The main elements of a small-scale installation are as follows:

1. Fire chamber in which a specified temperature regime is maintained.

1. Gas supply system into the fire chamber
2. Control and measuring equipment

Holes are provided in the walls of the fire chamber, intended for installing thermocouples of control and measuring equipment.

The body of the small-scale laboratory setup is made of red brick, size 1030 × 1200 × 905 mm. Inside the furnace, there is a fire chamber, with the dimensions of 770 × 390 × 320 mm. On the front side of the furnace in the brickwork, there is an opening of rectangular shape, measuring 390 × 370 mm, for communication with the fire chamber. In the rear part of the brickwork, there is a gas burner and a pipeline with forced air supply with an air regulating valve to ensure normal burning of the flame. In the upper part of the furnace, there is a 250 mm diameter exhaust pipe with an adjustable damper.

a. b.

Fig. 4.30 Sample of pine wood before (**a**) and after (**b**) the test

a b

Fig. 4.31 Sample of pine wood after fire tests at the time of determination of charring parameters

A gas flare with separate gas drive is used to start the furnace. The gas is supplied from a 50-liter cylinder. The gas pressure reducer is used to reduce the gas pressure. Gas supply to the combustion chamber is carried out through a gas pipeline. The flow of supplied gas into the combustion chamber is regulated by means of a gas cock. The gas pressure in the cylinder is controlled by means of a manometer. After the gas flare is lit and placed in the furnace next to the burner, the main drive of the gas burner is opened, after which the gas supply is shut off and removed, if stable flame burning in the furnace is established.

The test duration was up to 30 minutes. The sample was fixed using a special holder in the form of a frame. Petal thermocouples of chromel-alumel type were used to measure temperature. Tests were conducted for each series on three samples,

Table 4.13 Results of estimation of charring parameters of wooden structures elements on the laboratory unit under standard fire temperature conditions

Sample name	[a]ρ, kg/m^3	δ_c, mm	v, mm/min	ρ_c, кг/м^3	τ_0, s
Pine wood modern	452	16.5	0.62	252	224 c.
Pine wood, uninhabited house (1952), Republic of Kazakhstan, v. Zhanazhol (North Kazakhstan region)	477	24	0.94	246	280 c.
Pine timber, non-residential house (1928), Republic of Kazakhstan, v. Kayindy (Kostanay region)	546	28	1.1	237	295 c.
Pine timber (1890), elements of wooden ceiling, Gulliver Puppet Theatre, 104 Sovetskaya str.	588	31	1.26	224	325 c.
Pine timber (1865), Church of St. Nicholas the Wonderworker, Bryansk region, Russia	643	35	1.43	220	335 c.

Note: [a]Moisture content of the wood samples was 12%

having the form of a square, with a side of 150 mm. The thickness of the samples was 30–40 mm.

For the fire tests, wood specimens were prepared with a size of 150 × 150 mm and a thickness of 20 mm. The most important parameters during tests on a small-scale unit were determination of temperature on the surface and inside the specimen, time of self-ignition of the specimen, and depth of charring for specimens of given size and thickness.

Figures 4.30 and 4.31 show wood specimens in the course of fire tests, as well as after fire action at the moment of determination of charring parameters.

Table 4.13 shows the results of estimation of charring parameters (thickness of carbon layer (δ_c), charring rate (v), density of carbon residue (ρ_c), time of the beginning of flame, or flameless (smoldering) burning (τ_0) of tested samples on laboratory unit under standard temperature regime of fire) [16, 17].

The results, presented in Table 4.13, show that with increasing service life of wooden structures, there is an intensification of charring process. So, under standard temperature conditions for wooden structures with operating life of 150 years (Church of Nicholas the Wonderworker, Bryansk region, Russia), charring rate increases in 2.6 times compared with modern wood. For long-term wood samples, the time of the beginning of flame combustion increases by 1.2–2.3 times during fire tests on the laboratory setting.

The density of the formed charcoal residue with increasing life of wood significantly decreases. This may indicate that the formed charcoal layer has a developed porous structure and lower thermophysical characteristics compared to samples of modern wood. The formation of such charcoal structures, along with a decrease in the amount of volatile components in the wood of long-term natural aging, leads to increased resistance of the material to flammability. For all examined samples of long-lived wood, the time of the beginning of flame combustion increases.

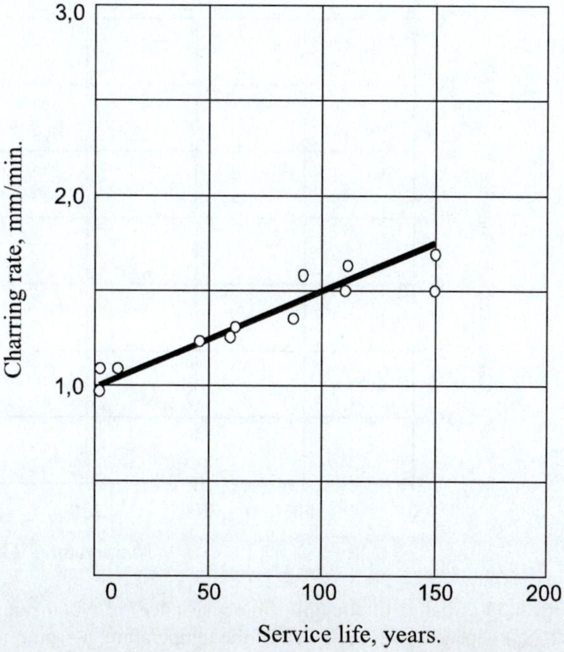

Fig. 4.32 Effect of natural aging of wooden structures on the charring rate of wood (spruce, pine) under standard fire temperature conditions

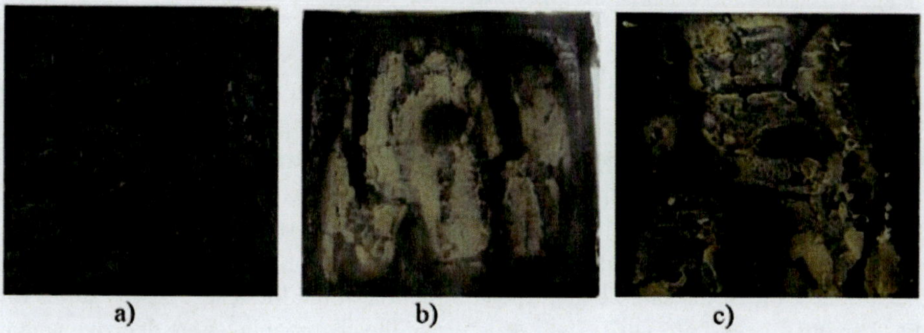

Fig. 4.33 Photographs of the surface of wood samples after fire tests on a small-scale unit: (**a**) pine wood (Arkhangelsk region); (**b**) pine wood (1890), elements of the wooden ceiling, Gulliver Puppet Theater, Kurgan, Sovetskaya str. 104; (**c**) pine wood (1865), Church of St. Nicholas the Wonderworker, Bryansk region, Russia

The conducted analysis of results on estimation of intensity of charring processes of long-term natural aging wood (service life up to 150 years) in conditions of exposure to standard fire temperature regime can be represented in the form of a linear relationship (Fig. 4.32).

The results show an increase of charring rate up to the value of 1.72 mm/min for wooden structures with service life up to 150 years.

Figure 4.33 shows photos of the surface of the examined wood samples after fire tests on a small-scale fire stove under the conditions of standard fire temperature regime.

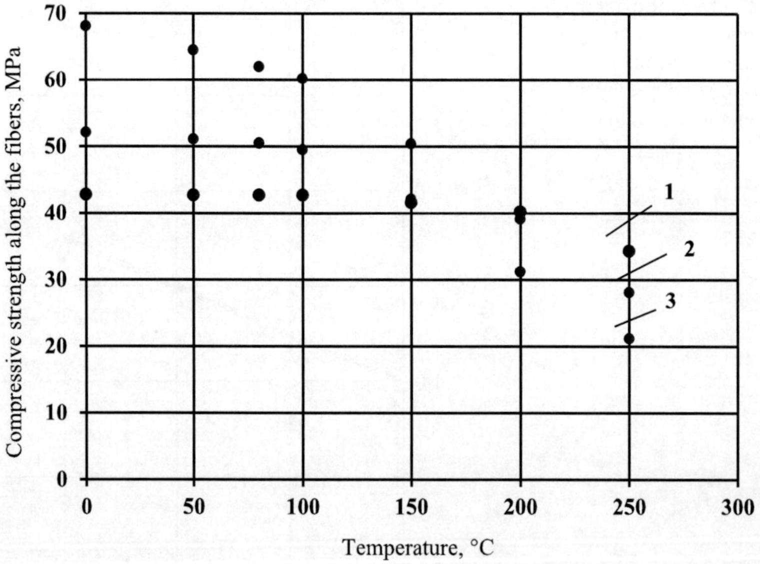

Fig. 4.34 Change of strength (time resistance) of wood samples in compression along the fibers (R_{comp}) depending on the value of the temperature: 1—pine wood (Arkhangelsk region); 2—pine wood (1890), elements of the wooden ceiling, puppet theater "Gulliver"; 3—pine wood (1865), Church of Nicholas the Wonderworker, Bryansk region, Russia

From the presented photos, we can see that modern pine wood is characterized by significant charring with the formation of deep shrinkage cracks. For long-term samples, along with a high degree of thermal damage, "traces" of over charring of the upper carbonized structures due to their high oxidative activity, intensive heat release, and creation of a high-temperature field in the pyrolysis zone of the surface layer are characteristic. Visually, it is accompanied by a longer characteristic glow in the surface zone of the sample after the completion of the fire test. These features of the charring process, properties, and structure of the charcoal layer are decisive when considering the peculiarities of behavior of long-life wooden structures under fire conditions and their fire-resistance.

The results of estimation of wood density values testify to the fact that with the increase of wood structures service life, the increase of this physical indicator value is observed due to the removal during operation of extractable components and volatile fractions from wood, shrinkage, and transformation of the morphological structure of wood substrate. However, when analyzing the density of carbonized residues formed when the samples were exposed to standard fire temperatures, there is a decrease in the density of carbonized residues. This may indicate a significant change in the charcoal layer structure for long-term natural aging wood.

The intensity of charring has a direct impact on the intensity of loss of bearing capacity of wooden structures during exposure to fire.

The most important indicators when considering the fire resistance of wooden structures are the physical and mechanical characteristics of wood. In this work, special attention was paid to the change in the strength of the examined wood samples when exposed to different temperatures. Temperature effect on the samples was carried out in a muffle furnace lasting from 30 min to 2 h. The change in compression strength of timber structures along fibers was assessed using a TIME WDW-50E electromechanical universal testing machine. More than 200 examined samples were analyzed.

The results show that when the temperature increases for wood specimens of different operating life, there is a decrease in their density and the value of compressive strength (tensile strength) along the fibers (Fig. 4.34).

It is important to note that for long-term natural aging wood, the decrease in compressive strength occurs at a lower temperature compared to modern wood. For long-life wooden structures already at 80 °C, there is a significant loss of the initial mechanical properties and at 250 °C by 60%. For the investigated dry samples of modern pine wood, the initial decrease in mechanical properties is observed only at the temperature of 150 °C and above [18].

It can be assumed that the strength limits of long-term wood samples under the influence of temperatures up to 250 °C are affected by changes as a result of natural wood aging in its chemical composition and the intermolecular interaction of the main components of wood. Partial degradation of cellulose fibers, the latter together with lignin are responsible for the mechanical strength of the material, has a particularly negative effect during wood aging.

The activity of the process of carbonization and charring with increasing service life of long-term natural aging wood shifts to the area of lower temperatures compared to modern wood, which also contributes to a more intensive loss of bearing capacity of wooden structures.

References

1. Thanh, B. D. (2006, May 26). *Influence of chemical components on fire properties of tropical wood in Vietnam*. PhD Candidate of Technical Sciences. Academy of State Fire Service, p. 184.
2. Mamleev, V., Bourbigot, S., Le Bras, M., & Yvon, J. (2009). The facts and hypothesis relating to the phenomenological model of cellulose pyrolysis. Independence of the steps. *Journal of Analytical and Applied Pyrolysis, 84*(1), 1–17.
3. Cordero, T., Rodriguez-Maroto, J. M., Garcia, F., & Rodriguez, J. J. (1991). Thermal decomposition of wood in oxidizing atmosphere. A kinetic study from non-isothermal TGA experiments. *Thermochimica Acta, 191*, 161–178.
4. Fumo, E. (1999). Non-isothermal kinetics of wood degradation and combustion. In *Proceedings of the 8th International Conference on Fire Science and Engineering, Interflam'99*, Edinburgh, Scotland, Vol. 2, pp. 873–884.
5. Di Blasi, C., & Branca, C. (2004). Global intrinsic kinetics of wood oxidation. *Fuel, 83*(1), 81–87.
6. Fang, M. X., et al. (2006). Kinetic study on pyrolysis and combustion of wood under different oxygen concentrations by using TG-FTIR analysis. *Journal of Analytical and Applied Pyrolysis, 77*(1), 22–27.
7. Kashiwagi, T., & Nambu, H. (1992). Global kinetic constants for thermal oxidative degradation of cellulosic paper. *Combustion and Flame, 88*, 345–368.

8. Pokrovskaya, E. N., Pischik, I. I., Smirnov, N. V., & Naganovsky, Y. K. (2000). Thermal stability of wood of different duration of operation. *Building Materials, 9*, 34–35.
9. ASTM E 2041-2013. *Standard test method for estimating kinetic parameters by differential scanning Calorimeter using the Borchardt and Daniels method.*
10. Khemminger, V. (1990). Calorimetry. *Theory and Practice, 1990*, 176.
11. Shestak, J. (1987). *Theory of thermal analysis: Physical and chemical properties of solid inorganic substances* (p. 456). Mir.
12. Berstein, V. A., & Egorov, V. M. (1990). Differential-scanning calorimetry in physical chemistry of polymers. *Chemistry, 1990*, 256.
13. GOST R 55134-2012 (ISO 11357-1:2009). *Plastics. Differential scanning calorimetry (DSC). Part 1. General principles.*
14. Makishev, J. K., & Sivenkov, A. B. (2016). *Study of charring process of wooden structures of different service life by methods of thermal analysis*. Technologies of technosphere safety. Accessed from http://ipb.mos.ru/ttb/2016-1
15. Anokhin, E. A., Makishev, J. K., & Sivenkov, A. B. (2016). *Influence of the structure of the carbonized layer on the fire resistance of wooden structures with a long service life.* Tekhnologii tehnosfernoy sfery. Accessed from http://ipb.mos.ru/ttb/2016-1
16. Almenbaev, M. M., Artsybasheva, O. V., & Sivenkov, A. B. (2014). The study of the charring rate of wooden structures long term operation. *Proceedings of Southern Federal University. Technical Sciences, 7*, 246–254.
17. Makishev, J. K., & Sivenkov, A. B. (2016). Fire resistance of wooden structures of long life. *Fire and Explosion Safety, 3*, 246–254.
18. Kharitonov, V. S. (1992). *Bearing capacity of bendable laminated wooden structures of massive section under heat impact: Thesis for the degree of Candidate of Technical Sciences: 05.26.01*. All-Russian Research Institute of Fire Protection, p. 190.

Investigation of the Effect of the Long Service Life of Timber Structures on their Fire Hazard and Fire Resistance

5

5.1 Fire Hazards of Wood and Wood Structures of Different Ages

5.1.1 Fire Hazard Properties of Long-Lived Wood

When considering the effect of long-term use of wooden structures on their fire hazard, the most important issue is to identify the relationship between changes in the chemical and elemental composition, as well as in the morphological structure of wood and its fire hazard characteristics.

When studying the fire hazard of materials, modeling the processes of emergence and development of fire, and solving many practical problems related to ensuring the fire safety of objects, use the so-called lower heat of total combustion, which is an absolute value less than Q_{h} (standard enthalpy of combustion) by the condensation heat of the water formed during combustion. In this case, D.I. Mendeleev's equation, taking into account the moisture contained in the substance, has the form

$$Q_{\mathrm{low}} = 339.4\mathrm{C} + 1257\mathrm{H} - 108.9(\mathrm{O} + \mathrm{N} - \mathrm{S}) - 25.1(9\mathrm{H} + \mathrm{W}), \mathrm{kJ/кg}, \quad (5.1)$$

where C, H, O, N, S, and W are the content of the corresponding elements and moisture, %.

The lowest heat of the total combustion of wood samples is determined by the standard method of GOST 147-74, as well as by the elemental composition. To analyze the elemental composition of absolutely dry samples, quite often an automatic device Carlo Erba 1106 (Italy) is used. It was necessary to find out the effect of natural aging of wood samples on the enthalpy of complete combustion. For this purpose, dry wood samples of different degrees of natural aging were used. Values of the lower heat of combustion were determined experimentally with the help of IKA-calorimeter C 5000 and were also calculated according to the elemental composition according to Mendeleev D.I. equation. Table 5.1 shows the results obtained.

S. Andrey Borisovich et al., *Fire Hazard and Fire Resistance of Wooden Structures*,
https://doi.org/10.1007/978-3-031-24074-4_5

Table 5.1 Effect of operating time of wooden structures on the heat of complete combustion of wood

	Elemental composition, %				
Sample name	C, %	H, %	O, %	$Q_{low}{}^{e}$, kJ/g	$Q_{low}{}^{c}$, kJ/g
Pine wood ($\rho = 459$ kg/m^3)	52.14	5.91	41.95	19.4	19.1
Long-life pine timber (81 years) ($\rho = 554$ kg/m^3)	48.8	8.34	42.86	22.8	22.4
Long-life pine (200 years) ($\rho = 482$ kg/m^3)	50.5	8.47	41.03	23.6	23.7

Note: $Q_{low}{}^{e}$—experimental net calorific value, kJ/g;
$Q_{low}{}^{c}$—calculated net calorific value, kJ/g

With increasing operational age of wooden structures, physical and chemical transformations occur in wood, which lead to a decrease in the carbon component, and the content of hydrogen and oxygen increases proportionally to each other. The calculated and experimental evaluation of the lower heat of total combustion of samples shows that with the increasing service life of wooden structures, there is a tendency to increase the values of the lower heat of total combustion of wood. This is due to a decrease in the elemental composition of carbon and an increase in hydrogen, which is four times greater than carbon in terms of its contribution to the heat of complete combustion. The observed effects are in good agreement with the results of the analysis of changes in the chemical composition of wood as a result of its aging.

With increasing operational age of wooden structures, physical and chemical transformations occur in wood, which lead to a decrease in the carbon component, and the content of hydrogen and oxygen increases proportionally to each other. The calculated and experimental evaluation of the lower heat of total combustion of samples shows that with the increasing service life of wooden structures, there is a tendency to increase the values of the lower heat of total combustion of wood. This is due to a decrease in the elemental composition of carbon and an increase in hydrogen, which is four times greater than carbon in terms of its contribution to the heat of complete combustion. The observed effects are in good agreement with the results of the analysis of changes in the chemical composition of wood as a result of its aging.

The contribution of some individual wood components to the value of its lower calorific value is obvious. The content of the main chemical constituents of wood samples was determined according to the usual accepted methods. To determine the contribution of lignin, which is of aromatic nature and has a high calorific potential, we analyzed the experimental data on combustion heat values obtained using the example of pine wood of long-term natural aging from different regions of the Russian Federation. A direct correlation of the values of the lowest heat of complete combustion of coniferous pine wood of different ages with the content of lignin in its chemical composition was established (in the range of 25–40%).

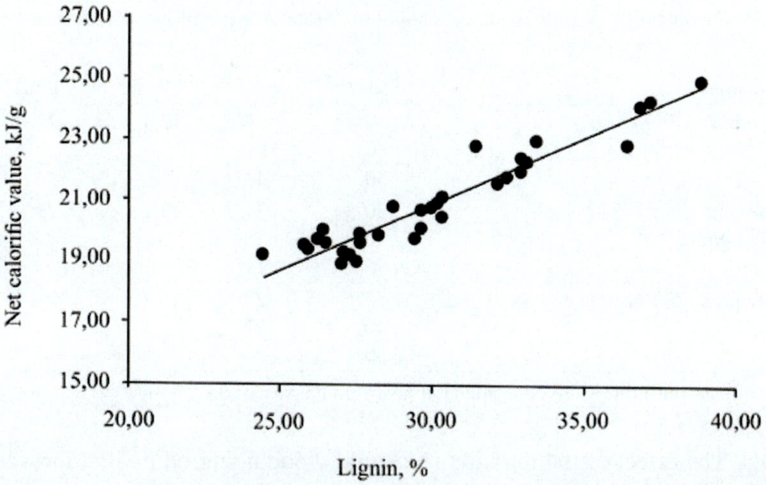

Fig. 5.1 Influence of lignin content in pine wood on the value of its lower heat of combustion (humidity of wood 12%)

Using linear approximation, a regression equation was obtained using the least squares method:

$$Q_{\mathrm{low}} = 8.02 + 0.429 \cdot X_{\mathrm{L}}, \mathrm{кJ/g}. \quad (5.2)$$

The sample size $n = 33$; correlation coefficient $r = 0.96$: coefficient of determination $r^2 = 0.92$; Fisher's test $F = 371$; explained variance $\upsilon = 0.92$.

Visualization of the equation is shown in Fig. 5.1.

The dependence demonstrates the possibility of constructing a direct correlation of the values of the lower calorific value of complete combustion of softwood pine of different lifetimes with the content of lignin in its chemical composition. The values of the lower heat of combustion of wood include the contribution to the calorific value of various components, including the carbohydrate part, aromatic components, and other compounds having a different chemical nature. The established dependence demonstrates the increasing contribution of lignin to the values of the lower calorific value when its content in the wood complex increases.

Of great interest in the aspect of fire hazard and fire resistance of wooden structures is the question of the effect of natural wood aging on such fire hazard characteristics as flammability, mass burn rate, smoke-forming capacity, the intensity of the charring process taking into account the analysis of the chemical composition of wood, and its morphological structure.

Determination of flammability of wood samples was carried out on the device according to GOST 30402 (Russia). The unit was equipped with an additional device for recording the mass loss during the test. Using this method, it was found that with increasing service life of wooden structures at the same intensity of external heat flux with increasing volume weight (density) of wood increases the time delay to ignition (τ_{ign}) and the critical heat flux of ignition (q_{cr}) samples. These experimental results are in good agreement with the thermal theory of ignition of different substances and

Table 5.2 Flammability of pine wood depending on the service life of wooden structures

Sample name	q_e, kW/m^2	τ_{ign}, s.	q_{cr}, kW/m^2	$^a\Psi_{max}$, g/m^2·s
Pine wood ($\rho = 459$ kg/m^3)	30	21	12.5	27.7
	40	9		
	50	4		
Long-life pine timber (81 years) ($\rho = 554$ kg/m^3)	30	31	14.1	21.1
	40	20		
	50	11		
Long-life pine (200 years) ($\rho = 482$ kg/m^3)	30	27	13.8	24.4
	40	17		
	50	9		

[a]Values of mass burn rate were determined at heat flux density of 35 kW/m^2

materials. The effect of the duration of natural wood aging on its ignition values is demonstrated in Table 5.2.

Table 5.2 shows that the ignition delay time of each wood specimen depends on the value of external radiation heat flux density. Under the same test conditions, 81- and 200-year-old pine wood shows higher values of τ_{ign} compared to a modern wood specimen. The same trend is observed for the critical heat flux density of ignition q_{cr}.

We can note a stable correlation between some physical properties of wood and its fire hazard. When considering the fire hazard of wood of different types and species, the most important indicator is the density of the wood composite, which is a reflection of the features of its morphological structure. For example, a useful pattern from the scientific and practical point of view was found when studying the effect of wood density on its resistance to combustion (Fig. 5.2). Density values for the examined samples of coniferous and deciduous wood of different degrees of natural aging are summarized in Table 5.3.

If we analyze wood samples of materials or structures at the initial stage of their operation period, the question arises about the possibility of the influence, in some cases, of the chemical composition of the wood composite on the induction period of its ignition. Therefore, the obtained dependence may not always be fair, since in some cases the chemical composition rather than the wood density is predominant during the ignition process of materials. This situation, for example, is typical for tropical wood species of Vietnam, which have in their chemical composition a fairly large number of extractive components that have a predominant influence on the ignition time of the material [1]. However, as a result of long-term aging of wood samples, its chemical composition undergoes significant changes that affect thermostable extractive components and part of the carbohydrate components. It can be assumed that the contribution of such components to the ignition of wood as a result of its long-term aging will be leveled. One way or another, the validity of the obtained dependencies seems to be more reliable for a specific sample of wood samples, taking into account the climatic factor. As a result of changes in the macrostructure and chemical composition of wood materials, all their properties, including fire hazard characteristics, change. With a considerable service life of

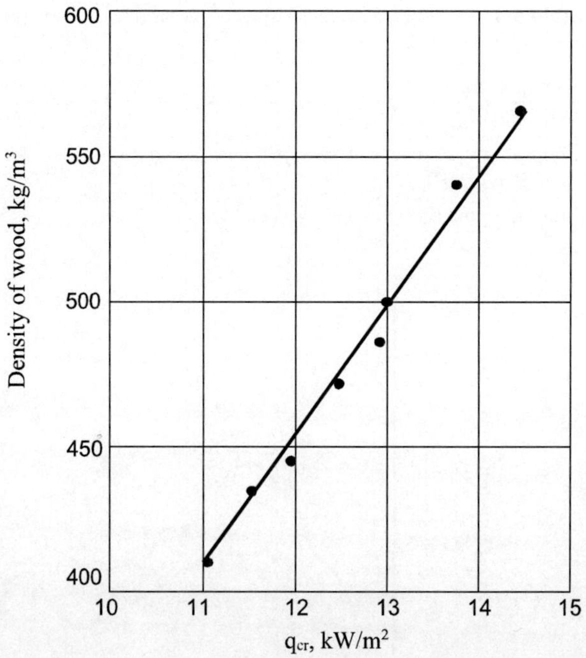

Fig. 5.2 Dependence of the critical surface heat flux density of wood (q_{cr}) on its volume mass (ρ_{12})

Table 5.3 Values of wood density of different degrees of natural aging

Wood sample, year of operation	Wood density (ρ_{12}), kg/m^3
Pine (central strip of Russia)	440
Birch (central strip of Russia)	565
Oak (central strip of Russia)	635
Spruce (central strip of Russia)	390
Oak, Sergiev Lavra, 1650	739
Pine, Tolstoy House, 1830	410
Spruce (artificially aged, (the central strip of Russia)	405
Pine (artificially aged, the central strip of Russia)	482
Birch (artificially aged, the central strip of Russia)	705
Oak (artificially aged, central strip of Russia)	603
Spruce, (dwelling house, Vologda district, v. Semigory), 150 years old	420
Pine, Volkostrov, Kizhi, Chapel of Peter and Paul, 300 years old	587

wooden buildings, the observed transformations of wood substance have a non-linear nature in time, which affects the fire hazard of materials.

The linear regression equation for the samples in question in this case can be represented as follows:

Table 5.4 Fire hazard characteristics of wood of different life cycles

Sample name	q_{low}, kJ/g	D_{max}, m^2/kg	Ψ, kg/m^2·c	q_{cr}, kW/m^2
Pine wood ($\rho = 459$ kg/m^3)	19.4	430	0.0277	12.5
Long-life pine timber (81 years) ($\rho = 554$ kg/m^3)	23.6	340	0.0211	14.1
Long-life pine (200 years) ($\rho = 482$ kg/m^3)	22.8	375	0.0244	13.8

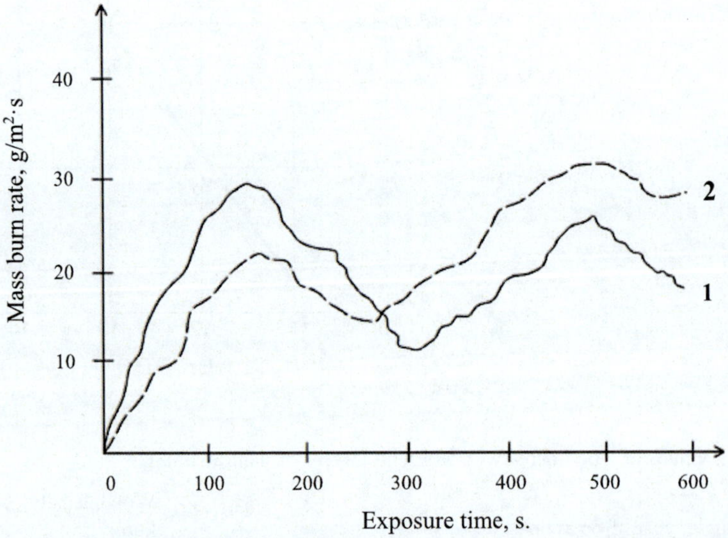

Fig. 5.3 Mass burn-up rate for burning samples of modern pine (1) and pine with a lifetime of 81 years (2) at a heat flux density of 35 kW/m^2

$$q_{cr} = (\rho_{12} + 62.8)/43.3, \tag{5.3}$$

where q_{cr}—the critical surface heat flux density, kW/m^2; ρ_{12}—the density of wood with a moisture content of 12%.

Despite the increase in the calorific value of wood of long-term natural aging, a decrease in the smoke-forming capacity can be observed, as well as an increase in the flammability resistance of the wood material (Table 5.4).

At the same time, the maximum mass rate of wood burnout Ψ_{max} at external heat flux of 35 kW/m^2, which determines the maximum rate of material heat release under fire action, decreases at the main stage of the burning process (Fig. 5.3) for long-lived wood. However, later there is an increase in mass rate of burnout of wood of long natural aging up to 1.3 times, which is directly reflected in charring characteristics of wooden structures.

Burn rate of wood samples is accompanied by formation of carbon layer in the surface zone, the value of which depends significantly on the degree of aging of wood material and the density of external radiation heat flux. Figure 5.4 shows the

Fig. 5.4 Effect of heat flux density on the thickness of the coal layer: 1—birch; 3—oak; 3—birch (50 years); 4—oak (50 years); 5—spruce; 6—pine; 7—spruce (150 years); 8—pine (150 years)

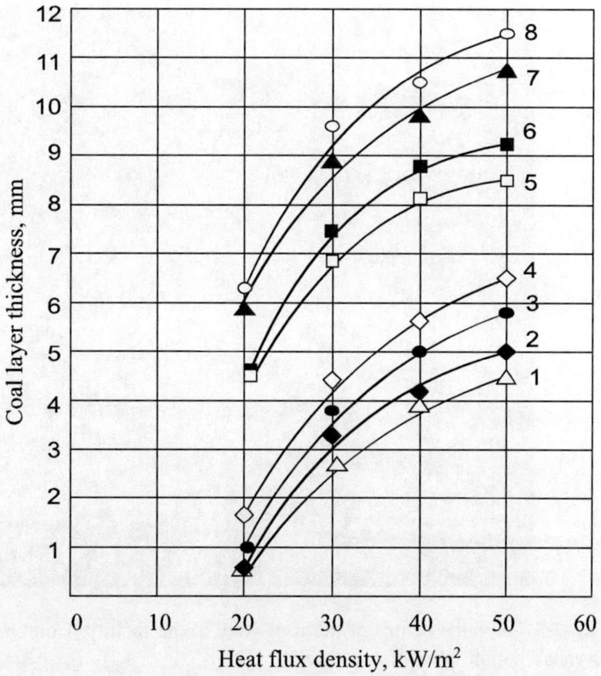

results of fire tests to estimate the depth of charring for wood of different species, taking into account the artificial aging up to the age of 150 years.

For the wood samples under consideration, there is a natural tendency: with increasing density of external radiation heat flux, there is an increase in thickness of charring layer. At the same time, for all wood species after the procedure of artificial aging, there is an increase in the depth of the carbon layer, which is obviously directly related to the charring rate, which determines the fire hazard and fire resistance of wooden structures. By increasing the thickness of the carbon layer, the effect of reducing the combustibility of wood is usually achieved, which is the basis for reducing the fire hazard of many polymeric materials.

With regard to the fire resistance of wooden structures, the acceleration of charring processes plays a more negative role due to the reduction of the working section of the structure involved in the retention of the mechanical load. However, in this case, it is not only the thickness of the resulting charred layer that is important, but also its properties and structure.

Changes in the chemical composition of wood during long-term natural aging will be reflected in those indicators of fire hazard of materials that are directly related to the chemistry of this process. Such indicators may include the smoke-forming ability coefficient. Smoke-forming ability was determined by a standard method according to GOST 12.1.044 (Russia) in the mode of decomposition and smoldering in a wide range of heat flux densities. Smoke-forming ability of elements of wooden structures changes nonlinearly with the growth of their service life. Pine samples have the lowest smoke-forming coefficient at the service life of wooden structures of

Fig. 5.5 Wooden beams of the roof vault at the facility: Church of Nicholas the Wonderworker, Bryansk region (year of construction 1865)

Table 5.5 Results of tests of elements of wooden structures under standard fire temperature conditions

Sample name	[a]ρ, kg/m^3	δ_c, mm	v, mm/min	ρ_c, kg/m^3	τ_0, s
Pine timber, load-bearing wooden beam (wooden beam, section 150 × 150 mm, 1.5 m long)	452	13	0.5	258	310
Pine timber (1865), Church of St. Nicholas the wonderworker, Bryansk region (wooden beam, section 150 × 150 mm, length 1.5 m)	643	33	1.3	215	420

[a]Moisture content of wood samples before the fire tests was 12%

20 years. This result generally agrees with the dynamics of changes in other wood properties in the process of natural aging (Table 5.4).

To study the charring process of elements of solid wooden structures, fire tests were carried out on a small-scale installation under the conditions of standard fire temperature regime. Fire tests were carried out on samples of wooden beams with a cross-section of 150 × 150 mm and a length of 1.5 m. Sampling of wooden beams was carried out at the site: the Church of St. Nicholas the Wonderworker, Bryansk region (year of construction 1865), which was demolished because of its dilapidated state. The wooden beams of the vaulting were in satisfactory condition, not subject to biodegradation (rotting) (Fig. 5.5).

The results of tests of wooden structure elements under standard fire temperature conditions are presented in Table 5.5.

Fig. 5.6 Photo of wood beam specimen after fire tests

The experimental data obtained are a confirmation of the results of fire tests in the conditions of isothermal heating during the tests on the flammability assessment unit. There is resistance to ignition of solid wood element by 1.35 times, as well as an increase in the intensity of charring of wood by 2.6 times. At the same time, there is a tendency to decrease the density of the formed charred layer, as in the case of testing samples under the conditions of one-sided radiant heating. A photo of a wooden beam specimen with the service life of more than 150 years after fire tests is presented in Fig. 5.6.

The presented photo (Fig. 5.6) shows significant thermal damage to the surface of the wooden beam with rather deep shrinkage cracks. The analysis of test results under different conditions of fire exposure testifies to an abnormal degree of thermal damage of wooden structures of long-term operation. The values of charring rate for wood in conditions of long-lasting operation can increase by 1.5 and more times in conditions of fire tests (fire). Considering the results obtained, it is of interest to conduct fire tests to assess the fire hazard and fire resistance of full load-bearing and enclosing wooden structures of different service life in standard conditions of fire tests.

5.1.2 Fire Risk of Long-Term Wooden Structures

While carrying out studies, the main attention was paid to changes in temperature indices, the degree, and the character of thermal damage of the examined samples of wooden structures because wood thermal-oxidative decomposition and combustion is a process proceeding with considerable heat emission at all its stages.

Fig. 5.7 Fragment of the enclosing wooden wall from the outside

In this connection, the central task is to determine characteristic distinctions in the value of temperature growth—a heat effect index, the degree of thermal damage, and the intensity of charring (depth and rate of charring) of the examined samples under fire test conditions according to GOST 30403 (Russia) for fencing wooden structures of different service lives.

Samples of wooden structures of a non-residential house are located at the address: Russia, Yaroslavl region, Borisoglebsky district, Kedskoe village. The building was built in 1935 (age—81 years), built of wooden pine log elements with a diameter of 220 to 240 mm. The elements of wooden structures are preserved in satisfactory condition without mechanical and biological damage. The general view of the enclosing wooden structures at the object under consideration is presented in Figs. 5.7 and 5.8.

Selected samples of wooden structures were subjected to tests in a fire stove under the conditions of standard fire temperature conditions according to the procedure established by GOST 30403. The fire test time of the investigated enclosing wooden structures was 15 and 45 min.

Fig. 5.8 Fragment of the enclosing wooden wall from the inside

Similar tests were carried out on modern wooden structures with pine logs of 140 × 140 mm cross-section used as elements. The structural elements were naturally dried to a moisture content of 13.8–15.1%.

The average density value of the long-term wood samples was 554 kg/m^3, and for the modern wood samples 459 kg/m^3. The structures measuring 1300 mm in width and 2400 mm in height were mounted in the fire stove. General view of the mounted samples of fencing wood structures is shown in Figs. 5.9 and 5.10.

To register the temperature during the tests, thermoelectric transducers (TETs) were installed in the heat and fire chambers, as well as inside along the thickness of wooden structure specimens, with a spacing along the depth of the structure of 20 mm.

The scheme of FET installation is presented in Fig. 5.11; the places of thermocouple installation are shown in Table 5.6.

Thermoelectric transducers were installed in the heat and fire chambers, as well as inside along the thickness of the wooden structure specimens, to record the temperature during the tests with a spacing of 20 mm along the depth of the structure.

Fig. 5.9 Specimen of a long-life wooden structure before the fire test

The temperature changes inside the fire chamber and the heat chamber for the specimen of a long-life timber structure and the specimen of a modern timber structure at a test time of 15 min are shown in Figs. 5.12, 5.13, 5.14, and 5.15, respectively.

The results shown in Figs. 5.12 and 5.13 show that at the initial stage (5–10 min of fire tests), there is a decrease in the average temperature value in the fire chamber for the enclosing wooden structure (service life of 81 years) compared to the sample of modern wooden structure. A similar effect was also found for wood structures with a service life of 150 years in fire tests for fire resistance according to GOST 30247.0 and GOST 30247.1 (Russia). The decrease in temperature for the long-term natural aging structure in comparison with the samples of modern wooden structure at the initial stage of tests is apparently caused by the formation of a characteristic carbonized residue with a more developed specific surface and a fine-porous structure. The presence of such a structure was previously established in [2]. Subsequently, this effect, due to an increase in temperature and intense thermal burnup of the structure, is leveled. During further testing, a noticeable increase in the average temperature in the fire chamber is observed.

In the heat chamber (Figs. 5.14 and 5.15), the detected effect appears at a later stage and persists until the end of the experiment. In spite of this, at a time interval of

Fig. 5.10 A sample of a modern wooden building envelope before the fire tests

5 to 10 min for a sample of wooden structure (81 years of service life), a higher temperature by several tens of degrees is observed in the fire chamber and heat chamber compared to the sample of modern wooden structure. The detected effect should be taken into account in general when assessing the fire and environmental situation at the site, in particular, the possible increase in the thermal regime and the toxicological effect of combustion products under fire conditions.

It is a well-known fact that elevated room temperature has a significant effect on the development of human body intoxication and contributes to the accelerated development of poisoning signs. It has been shown that with an increase in ambient temperature to 45 °C and a decrease in oxygen concentration to 11%, the toxic effect of flue gases significantly increases [3].

The results of estimating the value of temperature increase along the thickness of wooden structures are also a reflection of physical and chemical processes of structure formation and oxidation of the carbon layer of the studied samples. In fact, during the whole time of the experiment inside the sample of wooden structure of the long-term operation both in the geometrical center of the heated surface (fire chamber) and in the control zone (heat chamber), the temperature is kept by several tens of degrees lower than for samples of modern structures made of wood. The structure and thickness of the charcoal layer formed play a paramount role in this

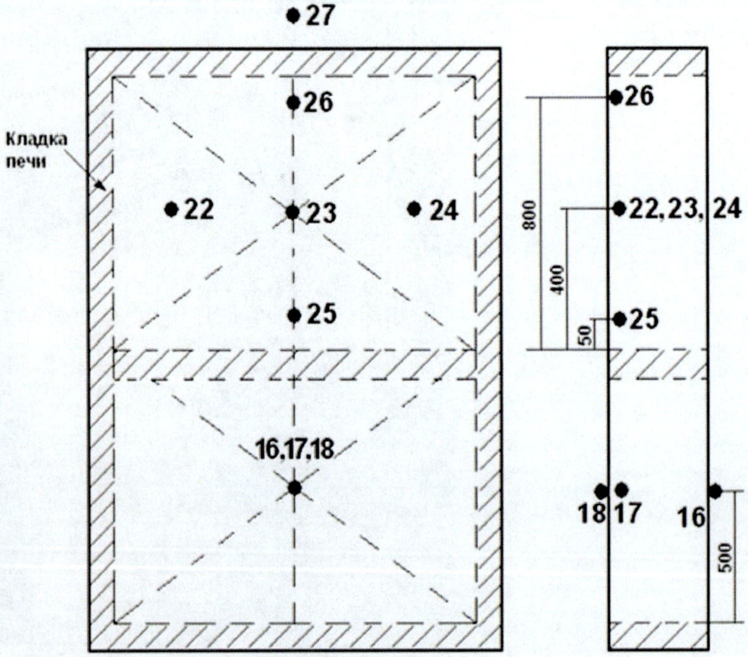

Fig. 5.11 Scheme of thermocouple installation during the fire tests of the enclosing wooden structures

Table 5.6 Installation locations of thermoelectric transducers in the furnace and inside the sample

Location of thermoelectric converters	Thermoelectric transducer numbers
Along the axis of the opening for the exit of gases from the heat chamber	27
At the junction of the specimen	26
In the geometric center of the heated surface of the sample	16, 17, 18
In the interface planes of the material layers of multilayer structures	22, 23, 24, 25

Note: The distance of the thermoelectric transducers to the end or joint of the specimen is not less than ½ of the specimen thickness

case. According to the results of fire tests, the values of depth and rate of charring of the examined samples of fencing wooden structures were determined (Table 5.7).

The values of charring parameters presented in Table 5.7 show that long-term natural aging of wooden structures leads to intensification of wood charring process more than 1.2 times in comparison with the sample of modern wooden structure. As a result of prolonged natural aging of wood, there is a significant change in its thermal stability, as well as structural features and properties of charring residue. The most important stages of wood thermal-oxidative decomposition that determine the features of its combustion are the main decomposition stage and the stage of coke

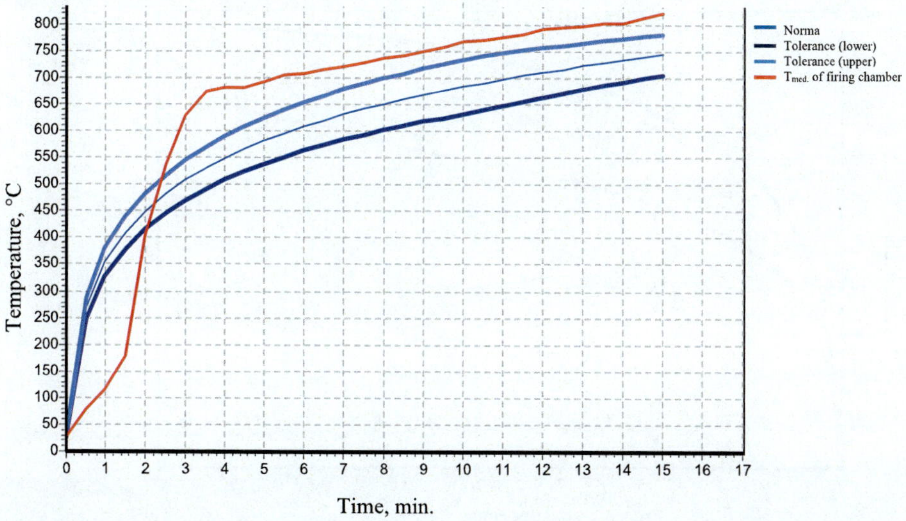

Fig. 5.12 Change of temperature in the fire chamber during the test of the long-life timber envelope

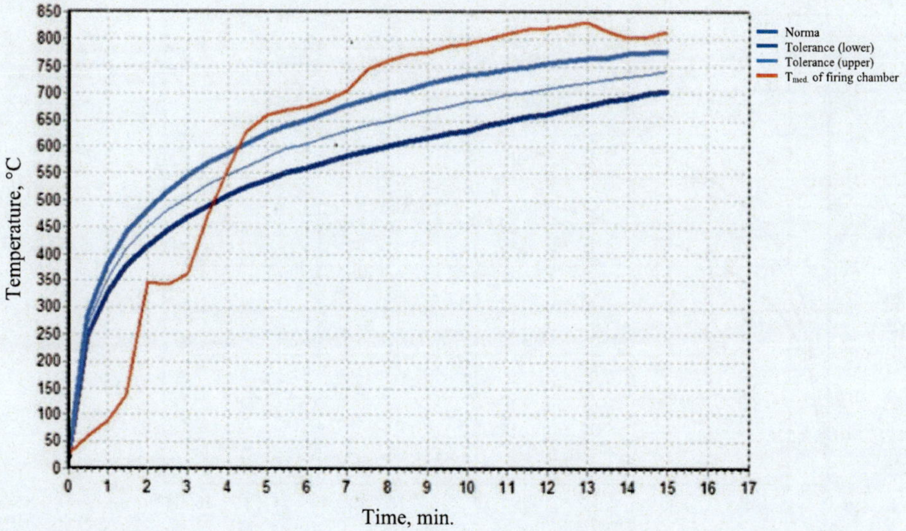

Fig. 5.13 Change of temperature in the fire chamber when testing a modern wooden building envelope

residue formation and oxidation, which determines the energy of the wood substance combustion process [1]. Given that the greatest heat release during thermo-oxidative decomposition is associated with oxidation of the carbonized layer at high temperature, it was decided to increase the duration of fire tests to 45 min under standard temperature fire conditions.

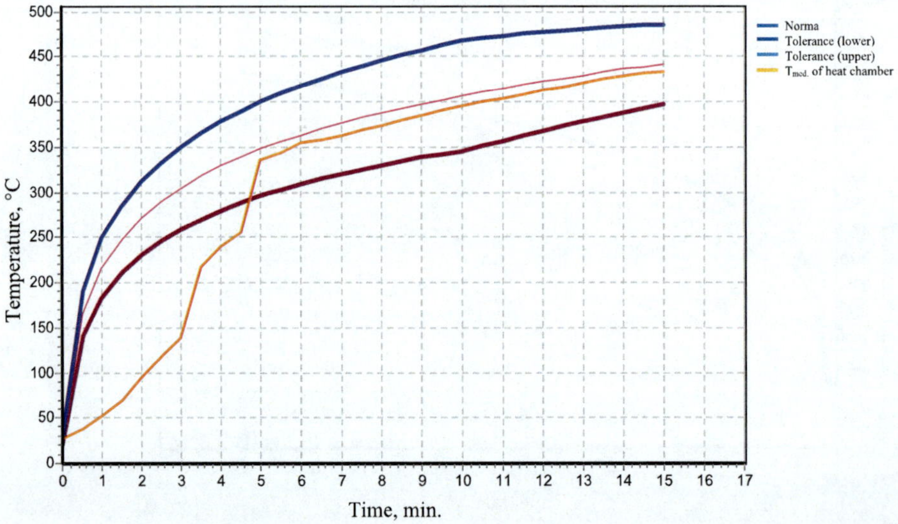

Fig. 5.14 Change of temperature in the heat chamber during the test of a long-life wooden building envelope

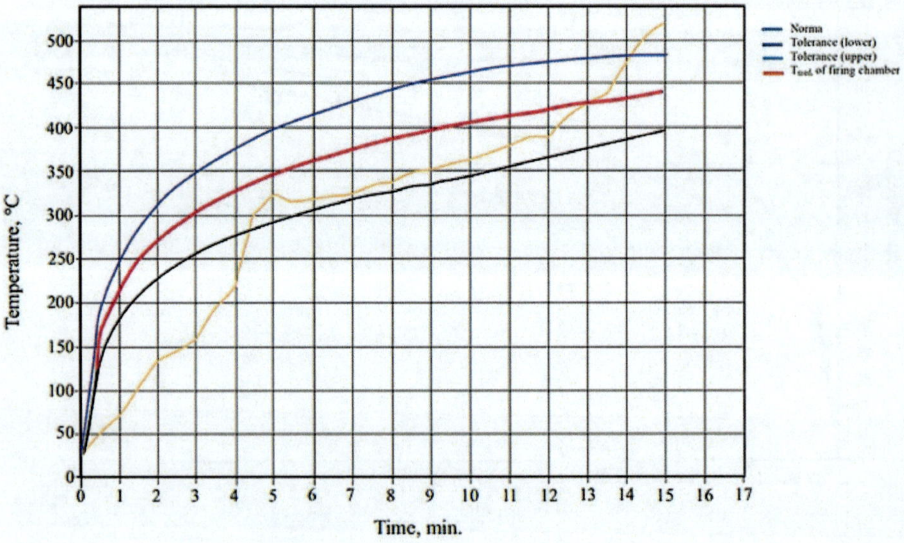

Fig. 5.15 Change of temperature in the heat chamber when testing a modern wooden building envelope

For the purpose of comparative analysis, during the experiment, visual observations were made through the observation window of the fire chamber, temperature registration inside the fire chamber and heat chamber, as well as inside the samples of wooden structures. The temperature values in the fire chamber and heat chamber are shown in Figs. 5.16, 5.17, 5.18, and 5.19, respectively.

Analysis of experimental data shows that the dynamics of temperature change in the fire chamber during combustion of a long-life wooden fencing structure and a

Table 5.7 Charring characteristics of samples of wooden structures according to the results of fire tests according to GOST 30403 (test time 15 min)

Measurement location (from the bottom of the sample)	An example of a wooden structure with a long service life		An example of a modern wooden structure	
	Charring depth, mm	Charring speed, mm/min	Charring depth, mm	Charring speed, mm/min
Bottom beam (0.10 m from the bottom level of the unit)	19	0.95	13	0.76
Thermocouple installation location in the firing chamber (0.75 m from the bottom of the unit)	15	0.88	12	0.70
Thermocouple installation location in the heat chamber (1.8 m from the bottom of the unit)	8	0.4	3	0.17

Note: the sample of long-life wooden structure has thermal damage along the entire height, and the sample of modern wooden structure has thermal damage at a height of up to 1.7 m, and the depth of charring in this place is 0.3 mm

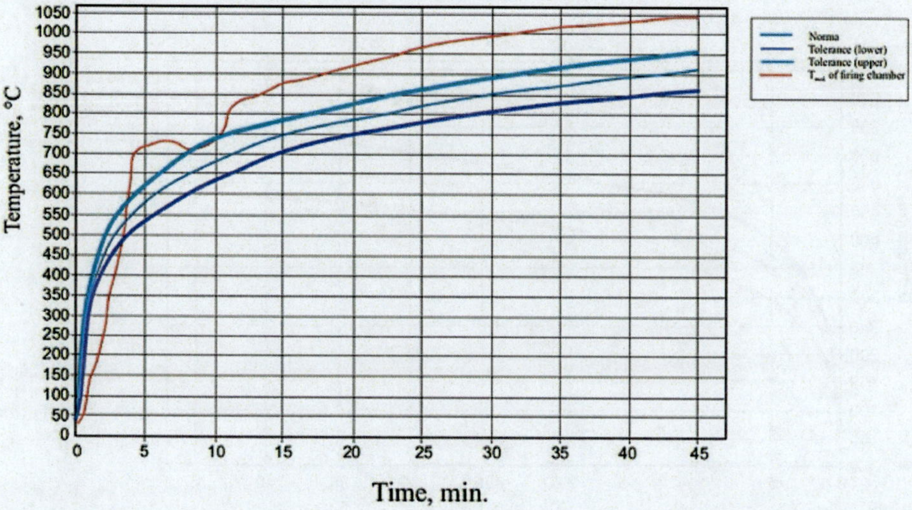

Fig. 5.16 Change of temperature in the fire chamber during the test of the long-life timber envelope

modern wooden structure has characteristic differences (Figs. 5.16 and 5.17). So, in the time interval from the beginning of the test to 5 min, the temperature increase is most intensive in the sample of modern wooden structure, reaching its peak at 740 ° C. Then, there is a decrease in temperature due to the beginning of the process of active carbon formation followed by reaching the stationary mode of heating of the structural specimen.

Similar temperature change at the initial stage is also observed for the sample of the wooden structure of a long-life structure. Active heat generation, intensive

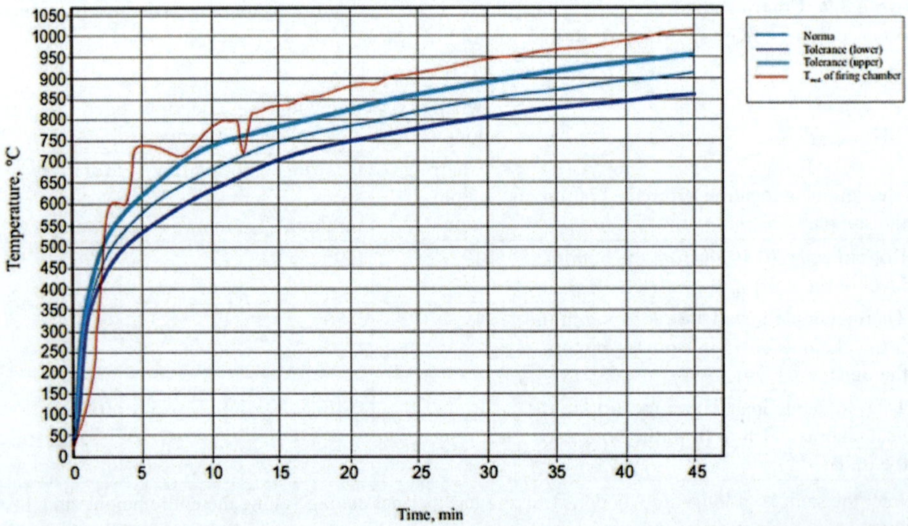

Fig. 5.17 Change of temperature in the fire chamber when testing a modern wooden building envelope

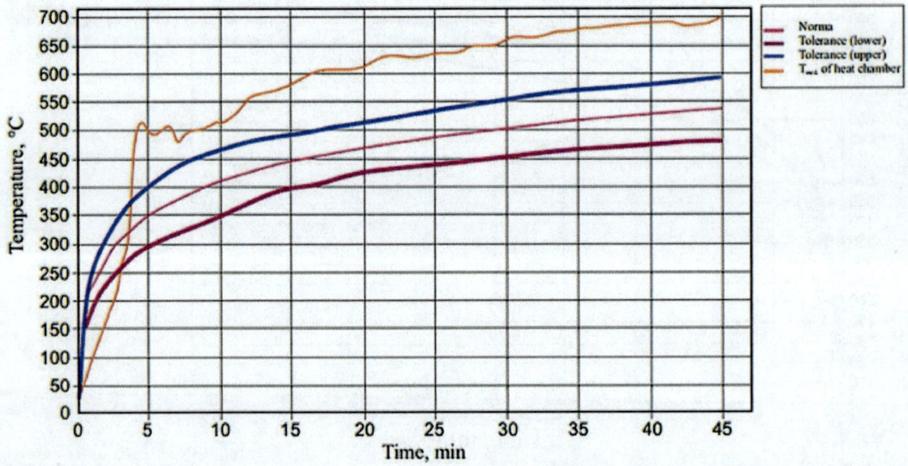

Fig. 5.18 Change of temperature in the heat chamber during the test of a long-life wooden building envelope

decomposition, and charring of such wooden structures at temperatures above 300–350 °C may be due to an increase in the content of aromatic component—lignin, which has high energy during combustion as a result of natural aging of wood [1].

In the time interval from 7 to 10 min, the appearance of a local extreme area associated, apparently, with the active formation of the carbon layer on the surface of the structure is observed. The contribution of lignin, which has an aromatic reticular

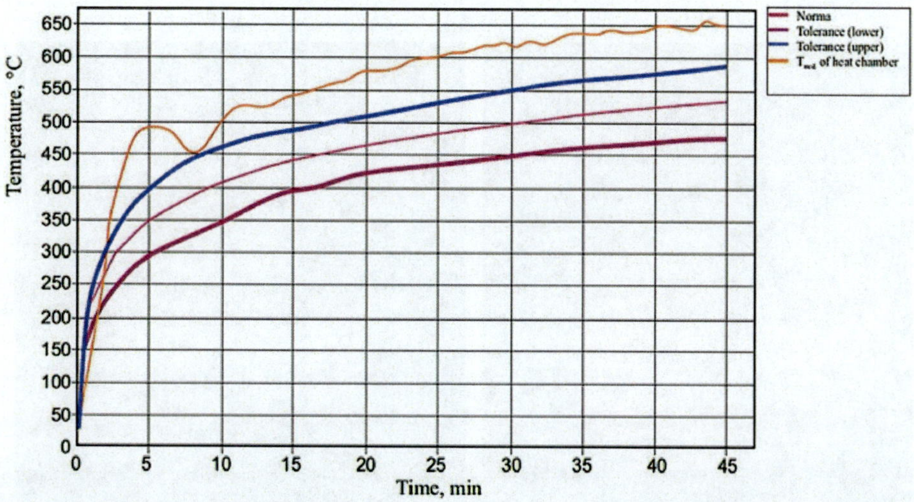

Fig. 5.19 Change of temperature in the heat chamber when testing a modern wooden building envelope

structure, is also more significant in the processes of carbon formation compared to the carbohydrate part of the wood.

Subsequently, in the time interval from 10 to 15 min, intense temperature growth is observed due to the active course of oxidative processes of the carbon residue. After the tenth minute of testing, the average temperature in the fire chamber until the end of the experiment for the long-term wooden structure was actually higher by 50 °C compared with the sample of modern wooden structure. A similar picture is observed in the heat chamber (Figs. 5.18 and 5.19).

At the end of the experiment, the average temperature in the fire chamber was 1050 °C. As a result of visual observation at the end of the test, an abnormal glow of the coke surface layer for the wooden structure of long-term natural aging (Fig. 5.20) associated with an active oxidative process of the coal residue was observed.

It can be assumed that the oxidation rate of wood carbonization products is affected by the nature of the formed porous charcoal structure associated with the thermophysical properties of the carbonized layer. Thus, in the work of Japanese scientists [4], a significant role of the thermal conductivity of the charcoal layer in the intensity of oxidation and the maximum rate of charcoal burnout was shown. The observed effect is a consequence of the fact that the latter parameter reflects the character of the coal bed porosity and its reactivity.

In the heat chamber of the test facility, as well as in the fire chamber, a visible temperature increase with a significant thermal effect is observed for the specimen of the long-life wooden structure (Figs. 5.16 and 5.18). The temperature value in the heat chamber after the tenth minute of the test throughout the experiment for the specimen of the long-life wooden structure is higher than that of the modern wood structure by 50 °C.

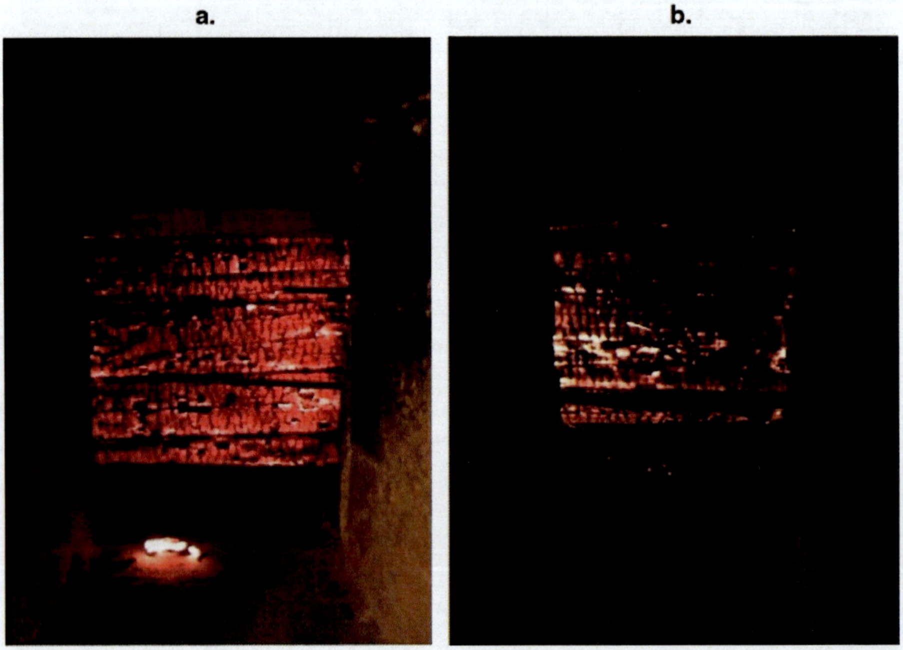

Fig. 5.20 Visual glow of structure surface during oxidation of coke residue (after turning off the gas burner): (**a**) fencing wooden structure of long service life; (**b**) fencing modern wooden structure

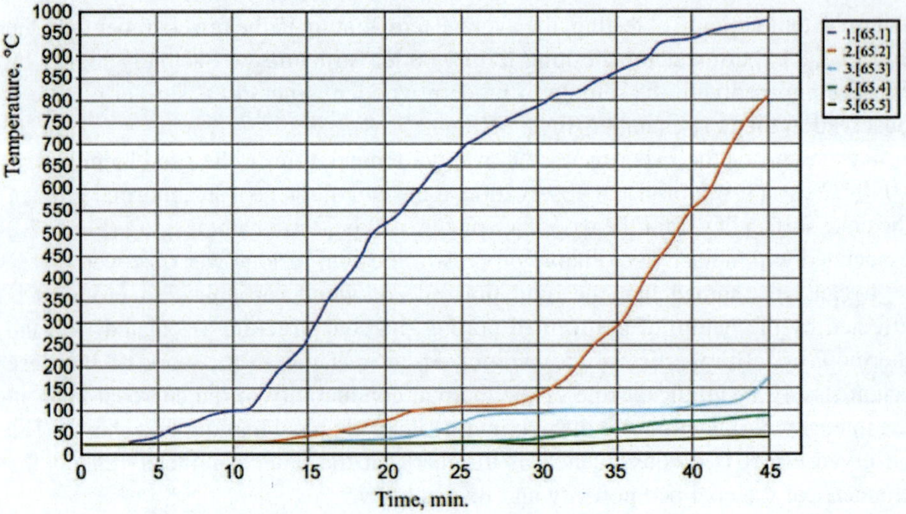

Fig. 5.21 Temperature change in the geometric center of the heated surface of the sample of a long-life wooden structure (fire chamber). Depth of thermocouple placement from the exposed surface of the sample: 1–20 mm; 2–40 mm; 3–60 mm; 4–80 mm; 5–100 mm

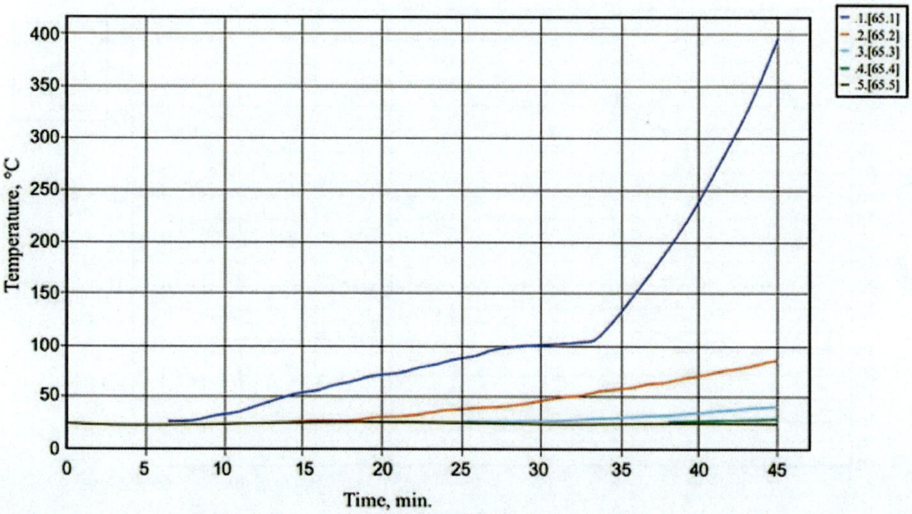

Fig. 5.22 Temperature change in the geometric center of the heated surface of the sample of modern wooden structure (fire chamber). Depth of thermocouple placement from the exposed surface of the sample: 1–20 mm; 2–40 mm; 3–60 mm; 4–80 mm; 5–100 mm

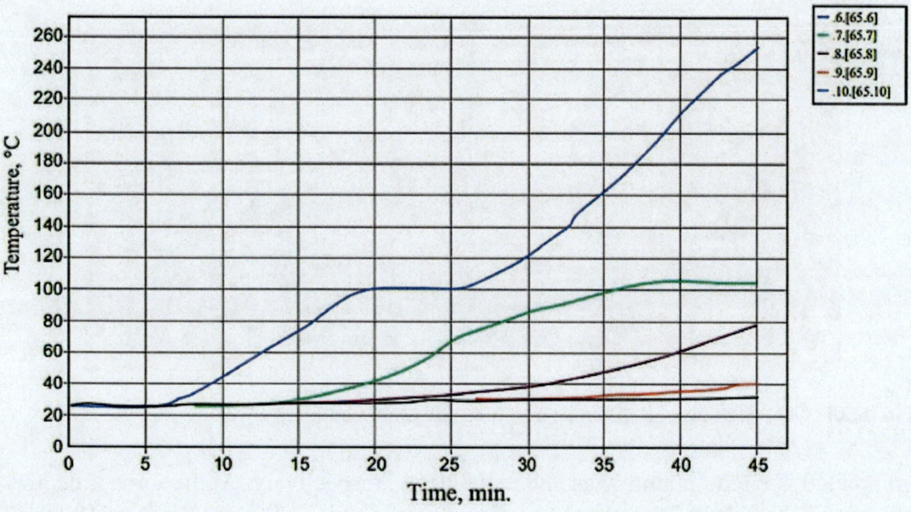

Fig. 5.23 Temperature change in the control zone for a sample of long-life wooden structure (heat chamber). Depth of thermocouple placement from the exposed surface of the sample: 6–20 mm; 7–40 mm; 8–60 mm; 9–80 mm; 10–100 mm

Figures 5.21, 5.22, 5.23, and 5.24 show the results of temperature measurements inside the samples of wooden structures for the fire chamber and heat chamber of the test oven (test duration was 45 min).

The heating of long-term wooden structures in thickness is almost twice as high as that of a modern wooden structure specimen. Thus, for thermocouples installed at a depth of 20 mm from the surface of the samples by the time the experiment was

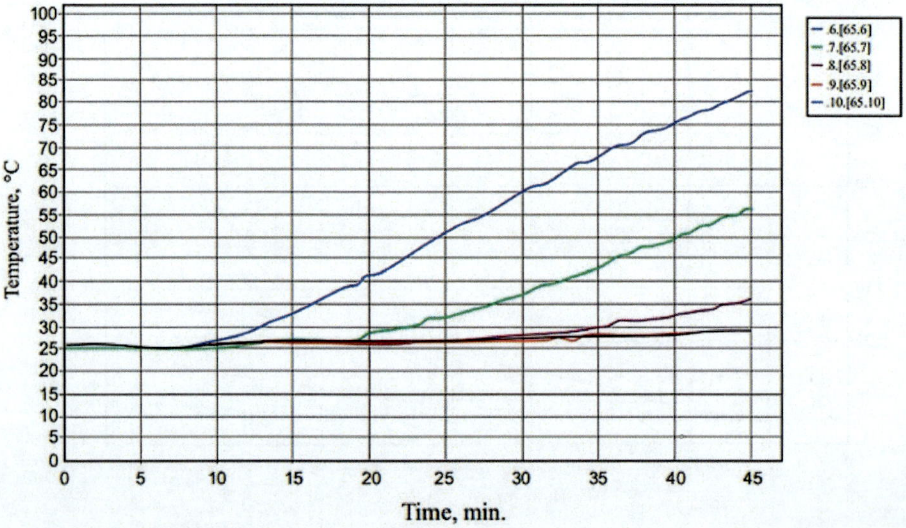

Fig. 5.24 Temperature change in the control zone for a sample of modern wooden structure (heat chamber). Depth of thermocouple placement from the exposed surface of the sample: 6–20 mm; 7–40 mm; 8–60 mm; 9–80 mm; 10–100 mm

Fig. 5.25 General view of a specimen of a long-life timber building envelope after the test

completed, the temperature was 980 and 390 °C, respectively. At the same time, the temperature inside the sample of wooden structure of long life at a depth of 100 mm for 45 min of testing was 15 ° C and for the sample of modern wooden structure only 3 °C.

It is obvious that the structures made of long-lived wood are characterized by a higher rate of advancement of the temperature front along the thickness of the sample. This is largely due to the high degree of thermal burnout of the structure and intensification of charring process for long-term natural aging wood. According to the results of fire tests in accordance with GOST 30403, charring rate of modern wood was 0.7 mm/min and that of long-term wood 0.94 mm/min.

General view of samples of fencing wooden structures after fire tests is presented in Figs. 5.25 and 5.26.

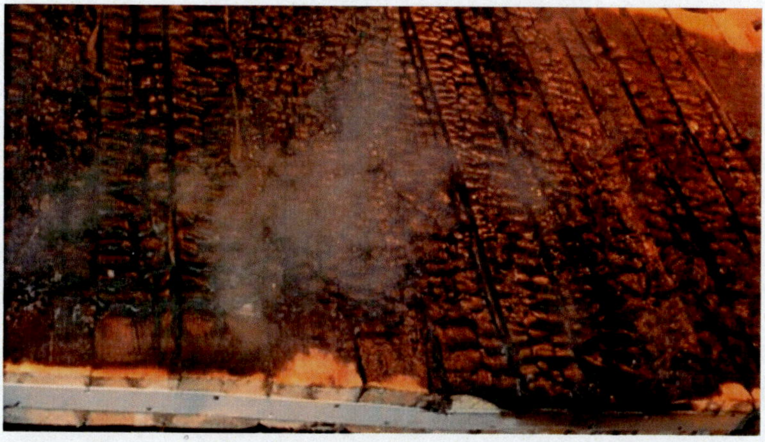

Fig. 5.26 General view of the specimen of the fencing modern wooden structure after the test

Table 5.8 Charring characteristics of samples of wooden structures according to the results of fire tests according to GOST 30403 (test time 45 min)

Measuring location (from the bottom of the sample)	An example of a wooden structure with a long service life		An example of a modern wooden structure	
	Charring depth, mm	Charring speed, mm/min	Charring depth, mm	Charring speed, mm/min
Bottom beam (0.10 m from the bottom level of the unit)	60	1.03	30	0.6
Thermocouple installation location in the firing chamber (0.75 m from the bottom of the unit)	55	0.94	35	0.7
Thermocouple installation location in the heat chamber (1.8 m from the bottom level of the unit)	30	0.51	13	0.26
Upper beam (2.3 m from the bottom level of the unit)	20	0.34	15	0.3

According to the results of fire tests, measurements of charring depth were made with estimation of charring rate of wooden structures, which are presented in Table 5.8.

The established parameters of charring of long-term wooden structure sample, as in the case of 15-min fire exposure, significantly exceed in their values the thickness and rate of charring for modern wooden structure. The highest charring rate values are characteristic of the fire chamber of the test oven, especially in the lower part of the envelope structures. The value of charring rate in this place of the sample for the fencing wooden structure (81 years of service life) is 1.7 times higher compared with the sample of modern wooden structure. At the same time, the thickness of charring of the sample of naturally aged wooden structure increased by 2 times. The effect of intensification of the process of charring of wooden structures of long-term natural aging is also observed in a heat chamber. At that, charring rate for a wooden structure of long natural aging in the place of installation of thermocouples is 1.96 times higher than the value of charring rate for a sample of modern structure.

According to the results of the fire tests, the following features of the fire behavior of the long-life wooden structures have been revealed:

- According to the results of the study, it was found that during the fire action of 45 min on the sample of wooden structures (81 years of service life), throughout the experiment the temperature was actually 50 °C higher compared to the sample of modern wooden structure both in the fire chamber and in the heat chamber. This circumstance can lead to deterioration of fire and environmental conditions, including increased thermal regime and toxicological hazard in buildings and structures.
- It was found that intensive heat generation for a sample of wooden structure of long natural aging is accompanied by a higher rate of burnout and intensive charring. The long service life of the timber structure has a significant effect on the degree of thermal damage to the sample and the values of the charring rate (the charring rate increases by a factor of 1.7 in the fire chamber) as well as the heating intensity over the whole thickness of the sample. This testifies to the fact that prolonged natural aging of wooden structures leads to easier conditions of wood carbonization and acceleration of oxidation of carbon residue.
- The results of recording the temperature increase inside the structures along the thickness showed that the temperature for long-term wooden structures at a distance of 20 mm from the surface of the sample was more than 2 times higher in comparison with the modern structure, which indicates a significant degree of thermal burnout and charring of the sample. According to simulation results, thermal conductivity at high temperature of long-life wood exceeds thermal conductivity of modern wood by 30–35%.
- The identified features of fire hazard of long-life wooden structures are necessary in the development of ways to reduce their fire hazard, including the development and application of fire protection means; the selection of substances and compounds, allowing to reduce the intensity of heat loss and affecting the structure and oxidative capacity of carbon layer; as well as the intensity of charring process.

5.2 Fire Resistance of Long-Life Wooden Structures

5.2.1 Method of Evaluation of Fire Resistance of Building Structures in Accordance with GOST 30247.0 and GOST 30247.1

The tests were conducted in accordance with the requirements of GOST 30247.0 "Building structures. Test methods for fire resistance. General Requirements" and GOST 30247.1 "Building Structures. Test methods for fire resistance. Bearing and enclosing structures" (Russia).

The general view of the installation for assessment of fire resistance of building structures in accordance with GOST 30247.0 and GOST 30247.1 is presented in Fig. 5.27.

The fire resistance tests of the presented samples considered the limiting condition—loss of bearing capacity (R). Loss of bearing capacity for bendable structures is characterized by the collapse of the structure or the occurrence of limit deformations.

These standards apply to load-bearing, self-supporting, and hinged walls and partitions without openings, coverings and ceilings without openings with suspended ceilings (if used to increase the fire-resistance limit of the structure) or without them, columns and pillars, beams, transoms, arches, trusses and frames, as well as other supporting and enclosing structures.

When establishing the fire-resistance limits of structures in order to determine the possibility of their use in accordance with the fire protection requirements of regulatory documents, the methods established by this standard should be applied.

Fig. 5.27 General view of the fire oven according to GOST 30247

To standardize the fire-resistance limits of load-bearing and enclosing structures, the following limit states shall be used:

- For columns, beams, trusses, arches, and frames—only loss of bearing capacity of the structure and assemblies—R
- For external load-bearing walls and coatings—loss of bearing capacity and integrity—R, E, for external non-load-bearing walls—E
- For non-load-bearing internal walls and partitions—loss of insulating capacity and integrity—E, I
- For load-bearing internal walls and fire barriers—loss of bearing capacity, integrity, and insulating capacity—R, E, I

Temperature conditions in the fire chamber were in accordance with GOST 30247.0. The hot junction of furnace thermocouples was set at a distance of 900 mm from the edge of the wall of the fire chamber and at a distance of 100 mm from the surface of the sample structure. Excess pressure in the firing space of the furnace at the height of 2/3 of the furnace opening was (10 ± 2) Pa after 5 min after the beginning of the test.

The specimens in the center were loaded with a test load equal to 100 and 150 kg. When the load was applied, a condition was provided that in case of possible deformation of the structure, the weights would not shift and would not affect the value of the fire-resistance limit due to changes in heat exchange conditions with the environment. The load was set immediately prior to the start of the test. Temperature values in the fire chamber during the tests did not exceed permissible values defined by GOST 30247.0. Excess pressure in the fire chamber at a distance of 100 mm from the surface of the sample was maintained in the range (8–12) Pa after 5 min from the start of testing.

5.2.2 Study of Heating Intensity of Long-Life Wooden Structures under Standard Fire Temperature Conditions

In this work, an experimental assessment of fire resistance of long-life wooden structures under conditions of large-scale fire tests was carried out. Fire tests in accordance with GOST 30247 were conducted to determine the time from the beginning of a one-sided thermal effect to the limiting state of the structure by the fire-resistance parameter R, the depth and rate of charring, and the heating time of the wooden beam by depth.

A sample of the pine timber bearing beam from the Arkhangelsk region (timber humidity 12.0%, density of 452 kg/m^3) and a sample of the pine timber bearing beam from the building (Church of St. Nicholas the Wonderworker, Bryansk region, 1865 (timber humidity 12.0%, density of 643 kg/m^3)) were submitted for testing. The cross-section of the wooden structures was 200x200 mm, length 3500 mm.

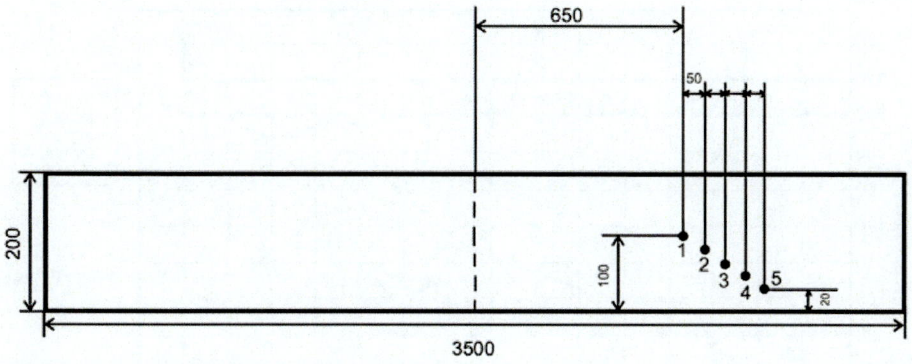

Fig. 5.28 Schematic diagram of thermoelectric transducers installation inside the wooden beam sample

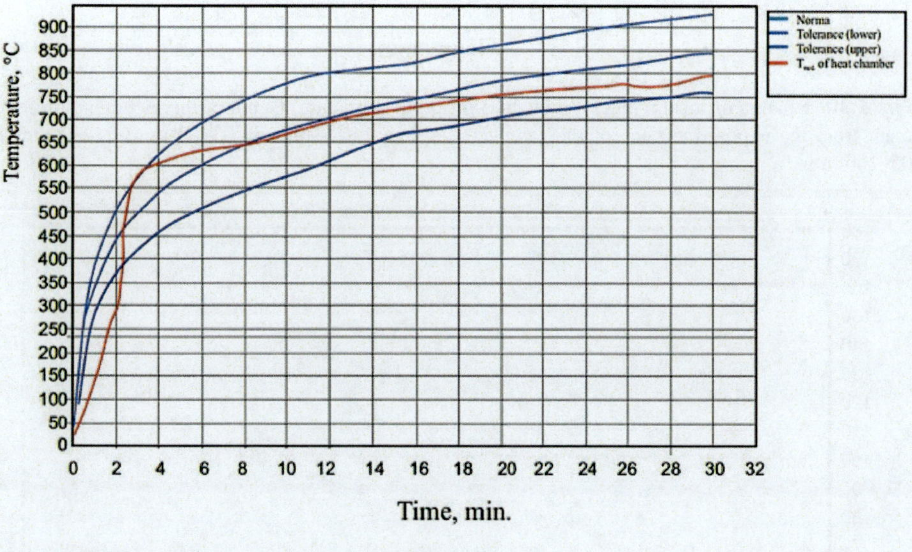

Fig. 5.29 Temperature mode in the fire chamber

According to the results of fire tests, charring rate, char thickness, ignition time, and temperature values along the thickness of the beam were evaluated (placement of thermoelectric transducers inside the sample is shown in Fig. 5.28).

The sample in the center was loaded with a test load of 150 kg. The test time for all specimens was 30 min. When the load was applied, a condition was ensured that in case of possible deformation of the structure, the weights would not shift and would not affect the fire-resistance limit due to changes in the heat exchange conditions with the environment. The load was set immediately prior to the start of the test.

The results of measurements of the temperature regime in the fire chamber and the temperature heating of the modern wooden beam by thickness are shown in Figs. 5.29, 5.30, and 5.31.

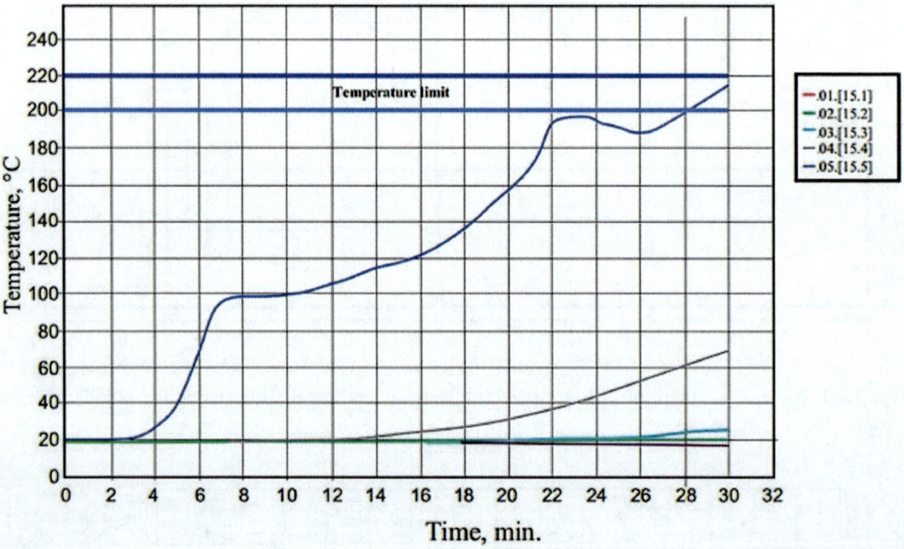

Fig. 5.30 Change of local temperature values inside the structure. Depth of thermocouple placement from the exposed surface of the sample: 01–20 mm; 02–40 mm; 03–60 mm; 04–80 mm; 05–100 mm

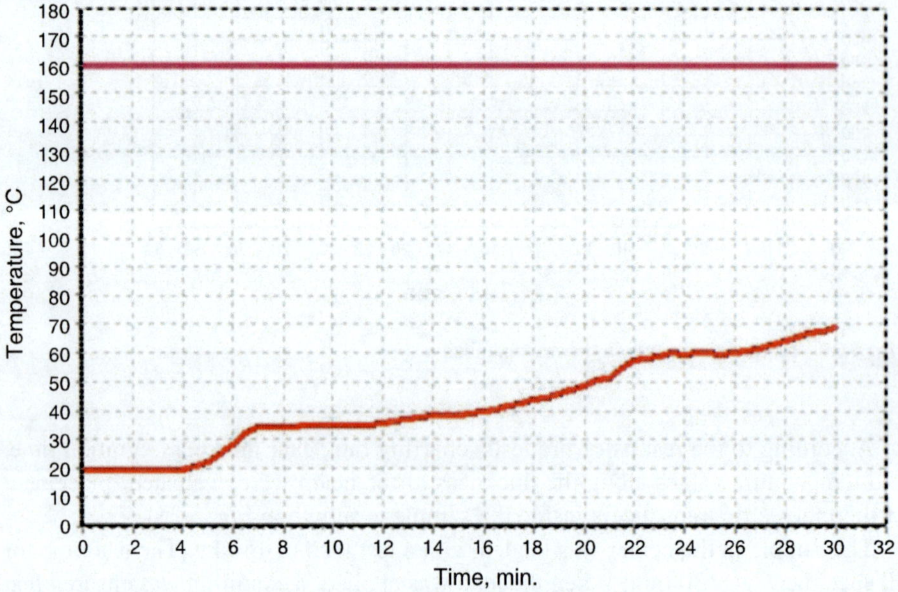

Fig. 5.31 Change of average temperature inside the wooden structure

Values of temperature in the fire chamber during tests did not exceed allowable values defined in GOST 30247.0 Excess pressure in the fire chamber "Setting (Furnace)..." at a distance of 100 mm from the surface of the sample in 5 min from the start of testing was maintained in the range (8–12) Pa.

Fig. 5.32 Appearance of the specimen during the test (outside the furnace)

Fig. 5.33 Appearance of the sample during the test (inside the furnace)

The appearance of the wooden structure specimen during the test outside and inside the oven is shown in Figs. 5.32 and 5.33.

The values of depth and charring rate are presented in Table 5.9.

The ignition time of the beam was 5 min and 20 s.

The results of measurements of the temperature regime in the fire chamber and the temperature heating of the samples are shown in Figs. 5.34, 5.35, and 5.36.

The values of temperature in the fire chamber during the tests did not exceed the allowable values defined according to GOST 30247.0. Excess pressure in the fire chamber "Installations (Furnaces)..." at a distance of 100 from the surface of the sample after 5 min from the start of testing was maintained in the range (8–12) Pa.

Table 5.9 Values of the depth and charring rate of a modern wooden beam

Indicator	Distance from the end part along the length of the beam, m.								
	In the end section	0.5	1.0	1.5	1.75	1.5	1.0	0.5	In the end section
Charring depth, mm	17	17	16	15	15	15	16	16	17
Charring rate, mm/min	0.69	0.69	0.64	0.6	0.6	0.6	0.64	0.64	0.69
Average charring speed—0.6 mm/min									
Average charring depth—15 mm									

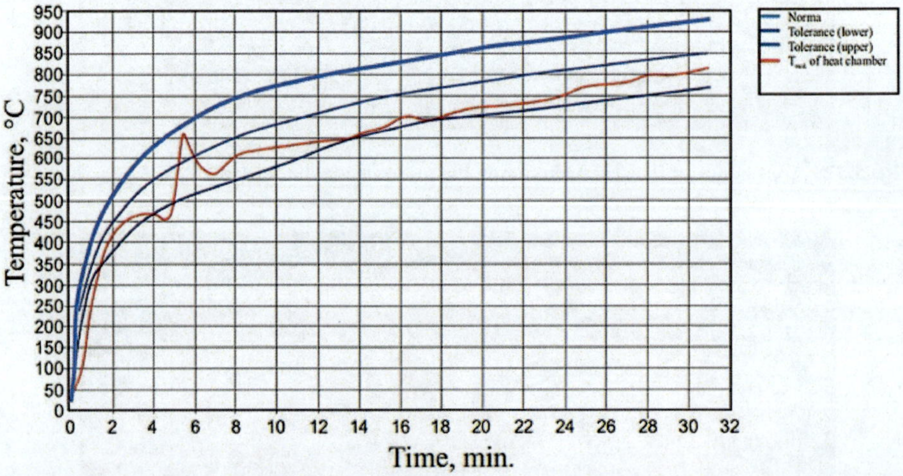

Fig. 5.34 Temperature mode in the fire chamber

The appearance of the wooden structure specimen during the test outside and inside the oven is shown in Figs. 5.37 and 5.38.

Values of depth and charring rate of wooden beam (150 years of operation) are presented in Table 5.10.

The ignition time was 6 min and 40 s.

The most important thing in this series of tests seemed to be fixation of charring parameters of wooden structures tested under standard fire temperature conditions. The results presented in Tables 5.9 and 5.10 show that for long-life wooden structures during fire tests, the intensity of charring process increases actually 2.4 times in comparison with modern beams. Values of charring rate for modern wood are comparable to the values of this indicator presented in normative and scientific literature [5–7].

The important thing is that the discrepancy between the results of charring rate measurements on a laboratory fire oven and a large-scale installation is not more than 15%, which allows us to predict the values of charring parameters when testing on small samples under conditions of one-sided heating.

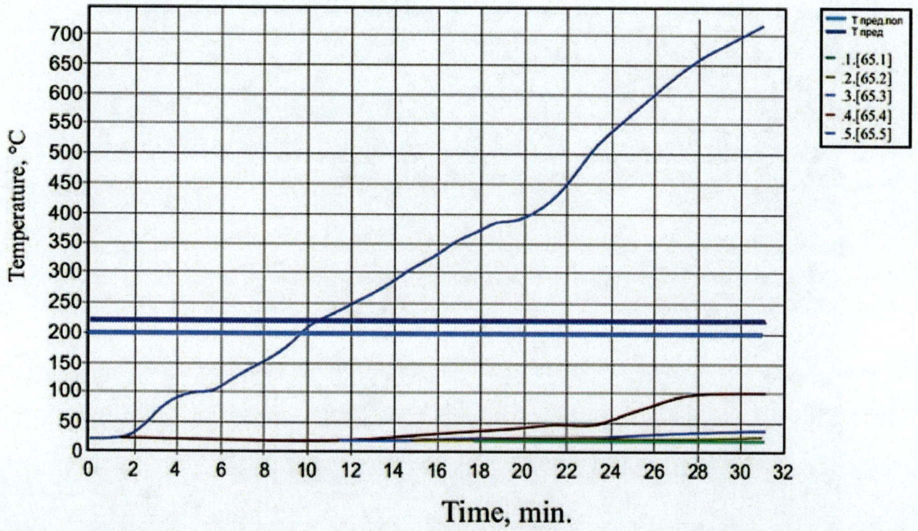

Fig. 5.35 Diagram of local temperature values inside the structure. Depth of thermocouple placement from the exposed surface of the sample: 01–20 mm; 02–40 mm; 03–60 mm; 04–80 mm; 05–100 mm

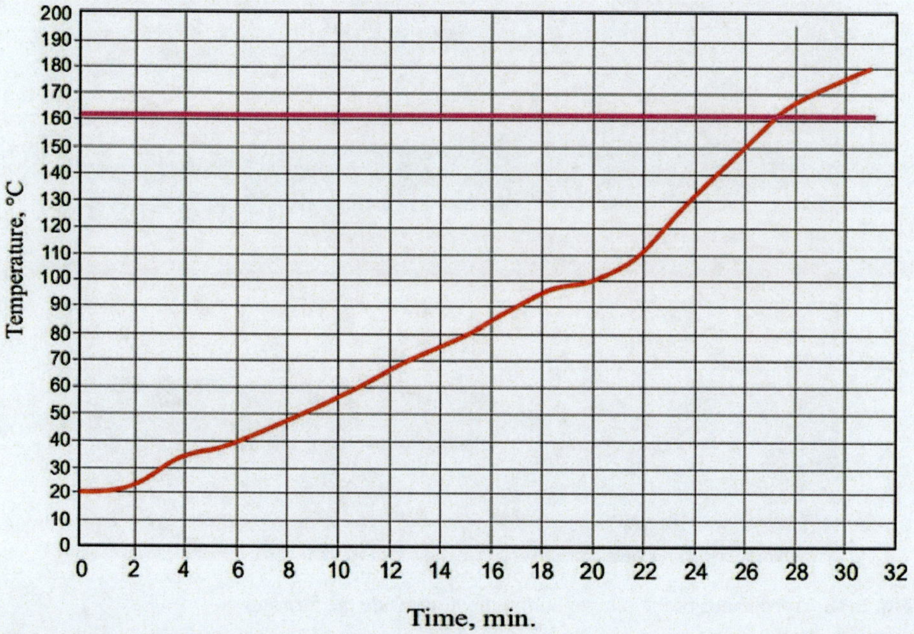

Fig. 5.36 Graph of the average temperature inside the structure

Fig. 5.37 Appearance of the specimen during the test (outside the furnace)

Fig. 5.38 Appearance of the sample during the test (inside the furnace)

Table 5.10 Values of depth and rate of charring of wooden beam (150 years of operation)

Indicator	Distance from the end part along the length of the beam, m.								
	In the end section	0.5	1.0	1.5	1.75	1.5	1.0	0.5	In the end section
Charring depth, mm	27	27	25	26	26	26	26	27	28
Charring rate, mm/min	1.11	1.11	1.03	1.1	1.1	1.1	1.1	1.11	1.2
Average charring speed—1.1 mm/min									
Average charring depth—26 mm									

5.2.3 Investigation of the Effect of Operating Life on the Fire-Resistance Limits of Wooden Structures

In order to determine the effect of structural features of the charcoal layer formed under fire conditions on the intensity of heating and fire resistance of wooden structures, fire tests were conducted in accordance with the requirements of GOST 30247 (Russia) in conditions of standard fire temperature regime.

The following samples were tested: a pine timber-bearing beam from Arkhangelsk region (humidity of wood—12.0%, density of wood—452 kg/m^3) and a pine timber-bearing beam from the building: the Church of St. Nicholas the Wonderworker, Bryansk region, 1865 (humidity of wood—12.0%, density of wood—643 kg/m^3). The cross-section of the wooden structures was 180 × 180 mm and length 2500 mm. Samples in the center were loaded with a test load of 100 kg and simultaneously exposed to standard fire temperature conditions.

For the next series of tests, the main task was to determine the fire-resistance limit of wooden structures with a section of 180x180 mm and length 2.5 m. The specimens were loaded with a test load of 100 kg. Two samples of wooden beams (1—a modern beam; 2—a beam with a service life of 150 years) were simultaneously exposed to standard fire temperature conditions. Figures 5.39 and 5.40 show pictures of the wooden beams with load before and during the fire tests.

According to the results of fire tests, the temperature values were determined along the thickness of the beam (the placement of thermoelectric transducers inside the sample is shown in Fig. 5.41).

As a result of observations, it was found that for a long-life wooden beam, the deflection of the structure in the central part was 10% after 25 min of fire tests and 15% after 32 min. The collapse of the beam (fire-resistance limit of the structure in terms of loss of load-bearing capacity) occurred at 37 min.

For a modern beam, the deflection of the structure in the central part was 15% only at the 74th minute. The fire-resistance limit of this structure in terms of loss of load-bearing capacity occurred at 81 min.

The values of temperatures inside the wooden beam samples confirm the previously stated assumption about the formation of a coal layer, which has a low value of the thermal conductivity coefficient, for long-life wooden structures.

Fig. 5.39 General view of wooden beam specimens under load before fire tests

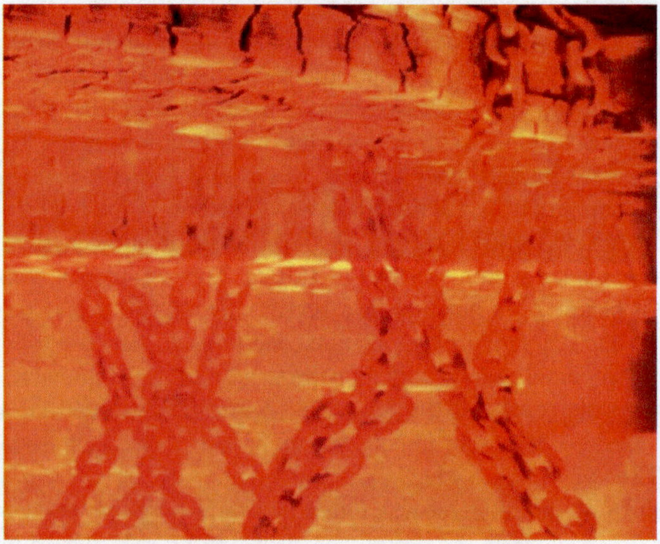

Fig. 5.40 General view of wooden beam specimens under load during fire tests

At the time of the collapse of a wooden beam with a service life of 150 years, the temperature at a distance of 20 mm from the outer surface of the structure was 320 °C. In contrast, this temperature at 37 min for modern wood was 400 °C.

The presented results for modern wooden structures (Fig. 5.42a) reflect the classical behavior of wooden building structures under the influence of standard fire temperature regime. The processes of loss of initial mechanical characteristics

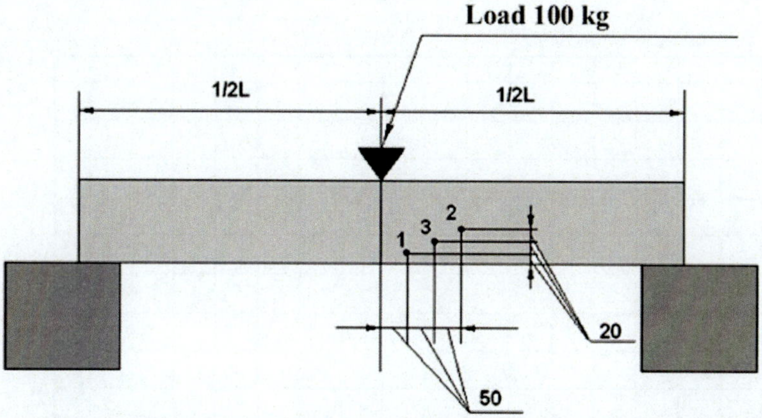

Fig. 5.41 Schematic diagram of thermoelectric transducers installation inside the wooden beam sample

and charring of wood are observed under the temperature values for four-sided heating of a structure established in normative and reference documents. It is possible to predict with sufficient accuracy the value of average charring rate of 0.7 mm/min for modern wooden structures. In this case, the thickness of the layer of wood, which has lost its original mechanical properties, is 7 mm. A different picture is observed for wooden structures of long-term natural aging (150 years of operation) (Fig. 5.42b). The charring rate for these wooden structures should be 1.2 mm/mm, and the temperature of the beginning of charring should be 250 °C.

Heating along the depth of the sample for a wooden structure of a long service life in general is more intensive compared to the sample of a modern wooden beam, which is due to the intensification of the charring process with the formation of a specific charcoal layer structure for wood of a long natural aging.

On the whole, a significant reduction of fire-resistance limit of wooden structures with the increase of their service life up to 150 years (by 2.2 times) is caused by the increase of rate of charring of structures, decrease of temperature indices of the carbonization process and loss of initial mechanical characteristics of wood, and increase of oxidative and calorific capacity of the formed carbonized layer.

In general, it is possible to note the following general features of behavior of wooden structures of long service life (based on the results of research of wood samples from the central and northern strip of Russia, as well as the northern and southwestern part of Kazakhstan):

- The influence of operating time on the fire resistance of wooden structures was established. A new hazard, previously not taken into account in fire-resistance standards and assessments, was found—the reduction of fire resistance of wooden structures of long service life. A significant effect of the time factor of operation is due to the increase in the charring rate of wooden structures. Thus, for a load-

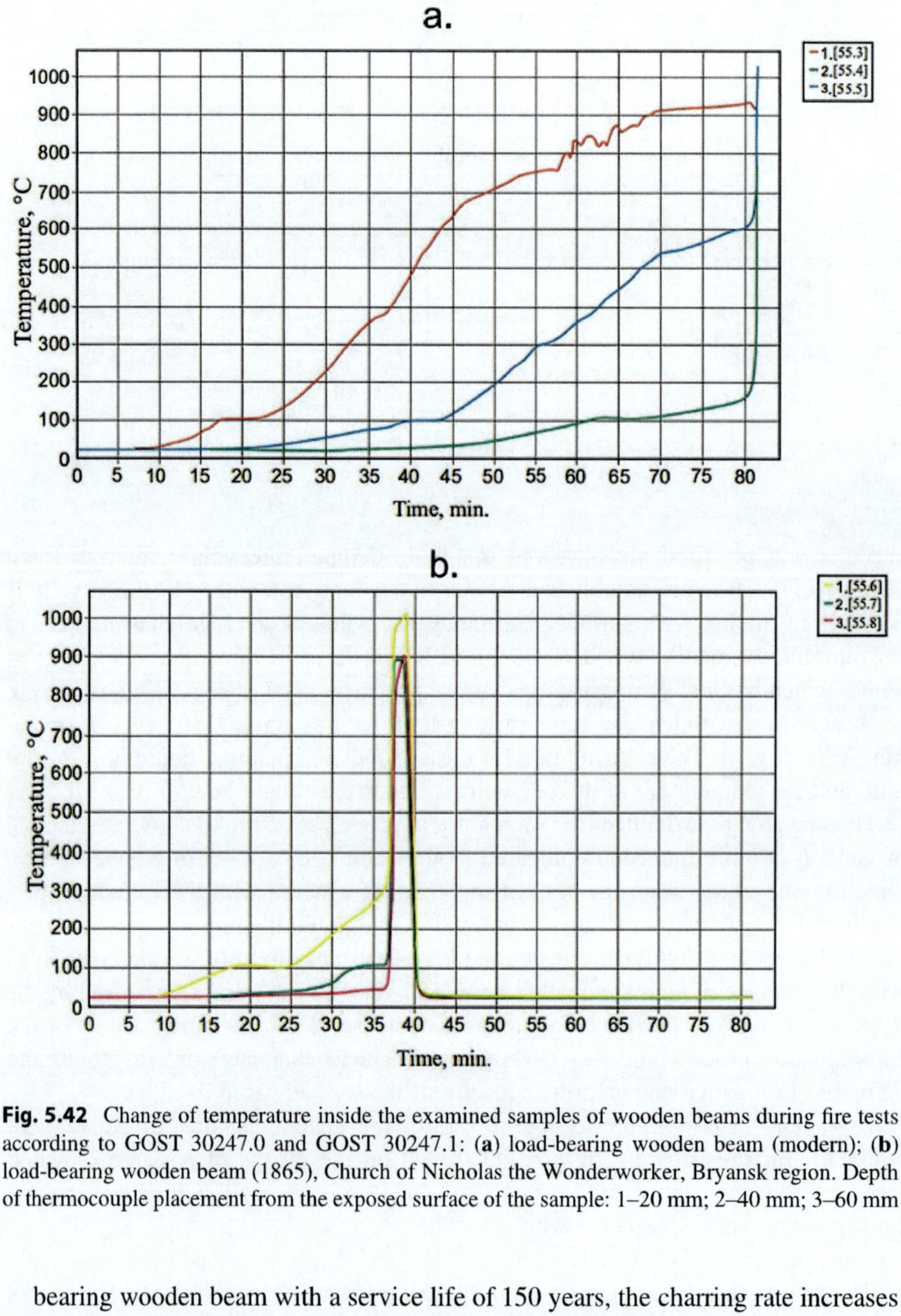

Fig. 5.42 Change of temperature inside the examined samples of wooden beams during fire tests according to GOST 30247.0 and GOST 30247.1: (**a**) load-bearing wooden beam (modern); (**b**) load-bearing wooden beam (1865), Church of Nicholas the Wonderworker, Bryansk region. Depth of thermocouple placement from the exposed surface of the sample: 1–20 mm; 2–40 mm; 3–60 mm

bearing wooden beam with a service life of 150 years, the charring rate increases on average by 1.8 times compared with modern softwood structures.

– The thermal influence on the change of mechanical, physicochemical, and thermophysical properties of wooden structures with different duration of operation is shown. The formed charcoal layers are characterized by a higher oxidation capacity and higher values of heat release rate; it contributes to significant thermal damage of wooden structures.

- Pyrolysis intensity and the beginning of charring process of wooden structures with increasing service life shifts to the area of lower temperatures (from 270 °C to 250 °C).
- It has been experimentally established that the changes taking place in physical, chemical, and mechanical properties of long-life wood under the influence of high temperatures or fire lead to a general decrease in the fire-resistance limit of wooden structures in terms of loss of bearing capacity by 2.2 times. The paper shows the possibility of prognostic qualitative estimation of intensity of charring process of wooden constructions and their behavior in conditions of fire with the use of laboratory and physicochemical methods of examination of small samples of wooden constructions elements.
- Considering a large variability of conditions and terms of operation of wooden structures in buildings and constructions, varieties of constructions and peculiarities of their application, loading schemes, and types of stressed states, it is necessary to continue research in the scientific direction of studying the effect of duration of operation on fire hazard and fire resistance of wooden structures.

To carry out a calculated assessment of the fire-resistance limits of long-life wooden structures, it is necessary to take into account the following:

- Time resistance when tested under loading conditions A [7] with a moisture content of wood 12% according to GOST for the corresponding types of tests for pine wood with a service life of 100–150 years can be set to 50–70 MPa for compression along the fibers, depending on the type of structure and the features of its loading and operating conditions. At the temperature of 250 °C, time resistance to compression along the fibers of the wood of long service life (up to 150 years) decreases by 60% in comparison with the initial strength values.
- Density of softwood pine (service life of 100–150 years) is taken 600–650 kg/m^3 (at moisture content of 12%).
- The temperature of the beginning of charring of long-life wood (service life of 150 years) is 250 °C.
- Conditional rate of charring (speed of movement of charring front), including the effect of angular curvatures, for softwood structures must be set depending on the service life: 50 years—0.8 mm/min, 80 years—0.9 mm/min, 125 years—1.1 mm/min, 150 years—1.2 mm/min.
- Due to uneven temperature distribution over the cross-section of wood behind the charring front, when determining the geometric dimensions of the cross-section at any point in time of fire exposure, a layer of overheated above 80 °C long-life wood should be excluded, taking into account the 12 mm thick corner roundings.
- Under the conditions of the standard fire temperature regime, the ignition time for long-life wooden structures (100–150 years) is 1.2 times higher than for modern timber structures and can be assumed to be 6.5 min.

References

1. Sivenkov, A. B. (2015). *Influence of physical and chemical characteristics of wood on its fire danger and fire protection efficiency: Dissertation . . . D. in Technical Sciences: 02.00.06*, p. 289.
2. Makishev, J. K. (2016). *Fire resistance of wooden structures of long service life: Dissertation. ... candidate of technical sciences: 05.26.03*, p. 163.
3. Ilichkin, V. S. (1993). *Toxicity of combustion products of polymeric materials. Principles and methods of determination* (p. 136). Chemistry.
4. Himoto, K., & Tanaka, T. (2004). A Burning Model for charring materials and its application to the compartment fire development. *Fire Science and Technology, 23*(3), 170–190.
5. Nagruzova, L. P., Romanenkov, I. G., & Rykov, R. I. (1996). *Bearing capacity of wooden structures under fire* (p. 305). Scientific.
6. EN 1995-1-2. (2004). *Eurocode 5: Design of timber structures - Part 1-2. General rules - Structural fire design. EN 1995-1-2. Eurocode 5: Design of Timber Structures - Part 1-2. General Rules - Structural Fire Design.*
7. *Code of Rules of the Russian Federation 64.13330.2017. Wood structures (Timber structures).*

Effectiveness of Fire Protection to Reduce Fire Hazard and Increase Fire Resistance of Wooden Structures

6

6.1 Rationale for the Choice of Fire Protection Directions for Long-Life Wooden Structures

Fire protection of wooden structures is carried out by means of structural and nonstructural methods. Depending on the method of application to the surface of the protected material, in the case of non-constructive fire protection, a distinction is made:

- Application of fire-retardant paints
- The application of plasters and coatings
- Impregnation of wood with flame retardants

Conceptually, types of fire protection for long-term wooden structures can be divided into the main areas: physical (structural) and chemical methods of fire protection, as well as various variants of their combination (Fig. 6.1).

Among the abovementioned methods of fire-retardant treatment of wooden structures, the use of impregnating compositions is currently the most widespread, which is explained by the high manufacturability of their use, low cost, minimum weighting of structural elements, and, in addition, the ability to preserve the wood texture.

Development and study of fire-retardant properties of various compositions devoted a lot of research, some of which were reviewed in [1–3]. Despite the fact that, in fact, high flame-retardant effectiveness of blowing flame retardants and paints has been proved [4]; currently, among the means of passive fire protection, the "classic" flame-retardant impregnation compositions based on aqueous solutions of various low molecular weight compounds are still widely used. The popularity of such compositions, with their low fire protection ability, is due to the relative cheapness and ease of production.

Such compositions are now being developed in many foreign countries [5–14]. For example, in the USA in 2008, a patent [9] was published for a

S. Andrey Borisovich et al., *Fire Hazard and Fire Resistance of Wooden Structures*,
https://doi.org/10.1007/978-3-031-24074-4_6

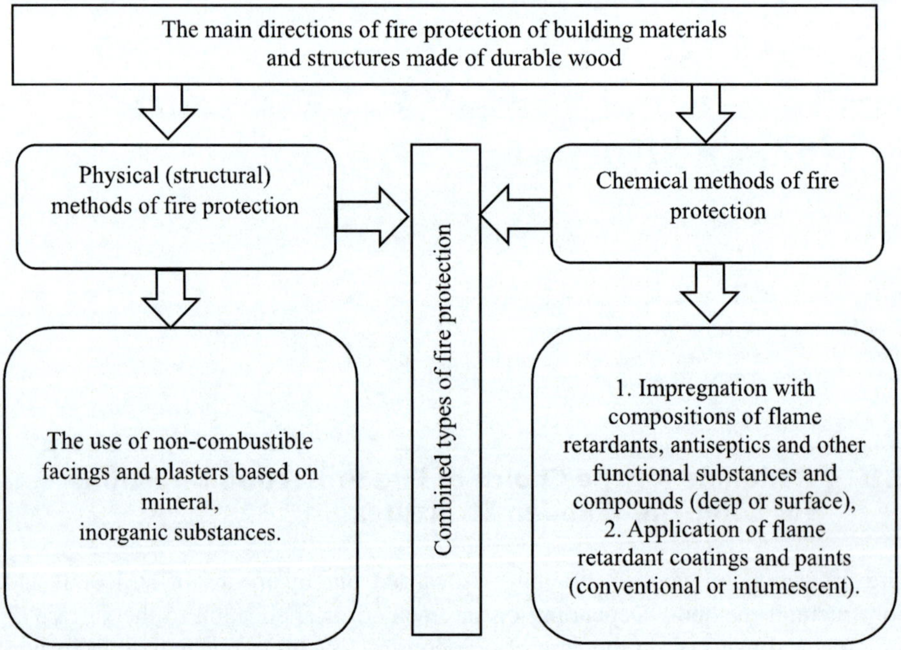

Fig. 6.1 The main directions of fire protection of wooden structures

fire-retardant composition including ammonium chloride and camphor oil dissolved in water; in China in 2015, several patents [10–14] were published for fire retardant compositions for wood based on aqueous solutions of organic and inorganic compounds.

More effective in terms of providing fire protection, as mentioned above, are bloating flame retardants and paints of combined action. The mechanism of action of most such coatings includes several simultaneous processes: mechanical protection of the surface through the formation of a thermal insulating foamy coke layer, a change in the charring mechanism of wood, and the formation of non-combustible or phlegmatizing gases. In addition, there are also compositions that provide only the creation of a foamed thermal insulating layer.

Typical chemical compositions of intumescent-type flame-retardant coatings are as follows:

- Substances that form a porous coke layer on the surface to be protected (usually serve as film-forming components)
- Catalysts for formation of carbon skeleton (accelerate dehydration, cyclization, and crosslinking reactions of polymeric molecules)
- Foaming agents which give off non-inflammable gases during thermal decomposition (H_2O, CO_2, NH_3, HCl), bloating components (thermoexpandable graphite)

From the chemical point of view, there are several directions of obtaining fire retardant compositions and paints for wood.

The first and the simplest in terms of technology is modification of film-forming systems on the basis of water- or organosoluble oligomeric and polymeric products (urea-formaldehyde, alkyd resins, etc.) by introducing flame retardants and gas forming agents.

The resulting paint and lacquer coatings mainly have a flame-retardant effect due to the formation of a carbonaceous foam layer on the surface of the protected material.

The most effective is considered to be the use of fire protection systems based on innovative chemical technologies, including chemically active flame retardants with classical phosphorus-, boron- and nitrogen-containing compounds, as well as coke-forming and blowing (intumescent) fire-protective compositions, which include functional chemical complexes consisting of classical flame retardants, gas formers, polyhydrate and aromatic compounds, thermoplastic polymers, and mineral fillers in their composition.

The mechanism of bloating (intumescence) implies the use of flame-retardant bloating coatings up to 1 mm thick which form a dense thermal insulating foamcoke layer (bloating multiplicity up to 100) on the protected surface under fire action, Fig. 6.2.

The use of bloating systems allows for 2–3-layer application (consumption of 650 g/m^2) to provide group I of fire protection efficiency according to GOST R 53292 (Russia). Mass loss during fire tests of wood samples with intumescent flame-retardant paints based on classical bloating system, polyhydrate compounds, and thermoplastic polymers and those based on classical bloating system, polyhydrate compounds, and mineral additives is 8.4 and 3.2%, respectively. In addition to the relative simplicity of such coatings, their advantage is the high fire protection efficiency. A subjective disadvantage of a number of paint and varnish coatings can be called a concealment of the wood structure, which in some cases may limit the use of coatings for aesthetic reasons.

One of the most in-demand film-forming systems today is liquid glass; on its basis, the following fire-retardant compositions have been developed:

- Fire retardant varnish [15] is produced on the basis of an aqueous solution of silicates of alkali metals or ammonium, containing methyl or dimethylalkoxy (phenoxy)silanes, alkyl siliconate, as well as polyalcohols, their oxyethyl derivatives, borates, and other auxiliary substances.
- Fire retardant composition for wood coating [16], in addition to liquid glass, contains asbestos, pyrite cinders, dunite, lime, and blast furnace granulated slag.
- Fire protection coating [17] for wood and metals includes liquid glass modified with magnesium hydroxide and urea. As a result of surface treatment with these compositions, the first group of fire protection efficiency is provided.

In contrast to organic binders, silicate compounds do not burn, but they have disadvantages, which include a low tendency to bloating. For this reason, such

a. b.

Fig. 6.2 The mechanism of bloating of fire retardant coatings for wood, I group of fire protection efficiency (weight loss 5–6%): (**a**) fire retardant intumescent paint based on the classical bloating system, polyhydrate compounds, and thermoplastic polymers; (**b**) fire retardant intumescent paint based on the classical bloating system, polyhydrate compounds, and mineral additives

compositions, as a rule, need to be applied in a thick layer on the protected surface. In addition, the use of liquid glass almost completely excludes the use of acid components; increased alkalinity of the environment can in some cases have a negative effect, expressed in an increased tendency of wood to smolder [18].

Using as a binder organic polymeric system can overcome these disadvantages. At the same time, the combustibility of the organic base does not have a negative impact on the fire-retardant properties and acts in a single system with agents of carbonization and gas formation [19].

Compounds that decompose to form inorganic acids usually act as carbonizers, and any low-boiling compound or a substance that decomposes to form low-molecular gaseous products (CO_2, NH_3, etc.) can be a gas-forming agent. For example, coating [20] based on polyvinyl alcohol is known to provide formation of bloated layer with thickness multiple of 30–50 of initial thickness. Based on the study [21], it is stated that as a result of thermolysis of polyvinyl, alcohol formation of graphite-like structures occurs, providing a high degree of thermal insulation protection of the treated material.

Such class of coatings can also include hardenable two-component systems, to which, for example, belongs the composition presented in the patent [22] and including two components. The first component is isocyanate. The second component is a mixture containing from 30 to 60% of tetrafunctional amino alcohol, from 1 to 10% of pigment, from 5 to 20% of ester containing 2 hydroxyl groups, from 20 to 50% of bloating graphite, from 8 to 30% of fire retardant, and from 2 to 5% of thickener. This coating is proposed to be used to protect wood from the effects of fire.

In [23], a study of flame-retardant bloating based on epoxy resin, also containing ammonium polyphosphate, bloating graphite, melamine, zinc borate, bisphenol A, and tetra ethylene tetra amine as a hardener, is presented.

The following three directions in the creation of fire-retardant compositions are more technological and, in most cases, allow to obtain compositions that not only form a physical barrier over the surface of the substrate, but also react with the organic components of wood, providing a certain fire protection effect regardless of the conditions of fire development.

The most studied method of obtaining highly effective flame-retardant compositions for wood of combined action is the targeted synthesis of oligomeric water- or organosoluble compounds from low molecular weight compounds, mainly nitrogen-phosphorus-containing ones. Many works are devoted to this problem, which are discussed in more detail in scientific articles [1, 2].

The following two methods of obtaining flame-retardant compositions are aimed at the use of renewable raw materials or recycling of waste from existing industries. Thus, relatively new and promising ways of obtaining fire retardant compositions for wood are chemical recycling (destruction) of polymeric waste with obtaining solutions of oligomeric compounds, as well as modification of natural polymeric and oligomeric compounds.

As an example of the use of polymer waste for the production of flame-retardant compositions, the work of the research team of the Ural State Forestry Engineering University [24–28] on chemical recycling of carbon-chain polymers containing complex ester groups, such as polycarbonate, polyethylene terephthalate, and polyurethanes, can be presented.

No other foreign and domestic sources containing a study of the flame-retardant effectiveness of products of chemical degradation of polymeric materials in the course of the literature analysis were found; at the same time, there are papers on the use of products of aminolysis of polyesters as inhibitors of salt deposits [29].

An example of the use of natural polymers and oligomers as film-forming bases is the use of polysaccharides to obtain flame-retardant compositions for wood [30–32]. The Institute of Biochemical Physics. N.M. Emanuel Institute of Biochemical Physics of the Russian Academy of Sciences obtained a series of flame retardants for "fire treatment of materials" and fire suppression based on the oxidation products of lignin-containing woodworking waste [33] and a flame-retardant composition for wood based on oxidized starch [4]. Examples of the use of plant raw materials for flame-retardant compositions are found in a number of scientific sources. For example, in [34], it is proposed to use sugars, glycerin, or lactic acid as raw materials for the synthesis of the polymeric base of flame-retardant coatings, and in [35], it is proposed to use polylactic acid and ammonium polyphosphate mixed with lignin or starch as a base for the blowing coating.

In general, we can note the great popularity of flame-retardant systems, which include a complex of different functional components (catalysts of the carbonization process, substances, and compounds – carbon source, gas formers, etc.) that promote coking on the surface of the protected material. When using coke-forming flame-retardant systems, it is potentially possible to achieve a high fire-retardant effect for wooden structures.

The mechanism of coke formation is due to the use of impregnating compositions capable of forming a foamed coke layer up to 15 mm thick in the surface area of wood (Fig. 6.3).

These compositions form a thin functional film on the wood surface and penetrate into the wood structure by 1–3 mm, which determines their coking effect both directly on the surface and interaction with the carbohydrate complex of the wood composite. Compounds working by the mechanism of catalytic dehydration penetrate into the structure of wood to a depth of 3–5, and their action is aimed at participation in the process of carbon formation directly in the wood. In this case, an additional flame-retardant effect is the dilution of the gas phase by inert gases formed as a result of thermal transformations of flame retardants. The system on the basis of P-containing substances, polyhydrate and aromatic compounds, and N-containing gas formers with the consumption of 450 g/m^2 provides timber weight loss no more than 9%, and the composition on the basis of P-containing substances and complex polyhydrate compounds with the consumption of 550 g/m^2 provides timber weight loss no more than 20%.

A priori, similar mechanisms of fire action, aimed at creating a thermal barrier on the protected surface, can be used for wooden structures of long service life.

Complex mechanisms of fire protection action include the use of many known substances and compounds acting as flame retardants, including taking into account the manifestation of the synergistic effect of increasing the flame-retardant effect. For long-term natural aging wood, the issue of fire hazard reduction can be solved by using "classic" flame retardants in combination with reactive compounds, allowing to modify not only the carbohydrate part of wood but also the aromatic complex of wood composite, as well as to change the properties and structure of carbon residue and influence the intensity of charring process.

For the development of such fire protection systems, the most known fire protection mechanisms can be considered, which allow to effectively reduce the indicators of fire danger of wood, such as flammability, flammability, and flame spread on the surface of materials, and also contribute to the reduction of fire danger of wooden structures. These are the main works, which have been performed in the world practice for the last 20 years. The main scientific approaches to reducing the combustibility of wood were defined back in the 1930s–1940s of the last century.

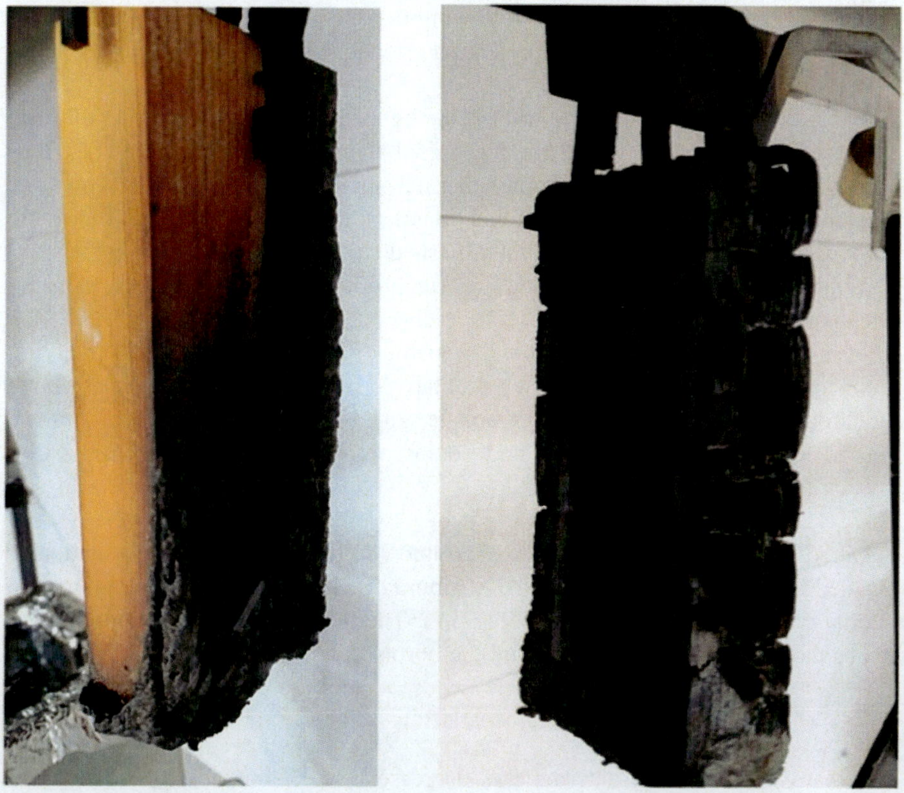

a. b.

Fig. 6.3 Mechanism of coke formation of flame retardants for wood: (**a**) composition based on P-containing substances, polyhydrate and aromatic compounds, and N-containing gas formers; (**b**) composition based on P-containing substances and complex of polyhydrate compounds

So, the work [36] actually predetermined two key mechanisms of reducing the combustibility of wood and wood-based materials associated with the ability of flame retardants to influence the course of gas-phase and solid-phase reactions in the combustion of wood material.

The issue of using complex flame retardants to increase the fire resistance of wooden structures has not actually been considered. Confirmation of the possibility of increasing the fire resistance of wooden structures was actually the only work [37]. The effect of reducing the intensity of charring of wood in this case is crucial in considering the issue of increasing the fire resistance of wooden structures. Experimental confirmation of increasing the fire-resistance limit of wooden structures (beams) by 10% using the surface method of applying flame retardants has been obtained. The use of flame retardants—regulators of the carbon formation process—is a promising direction in the field of fire protection of wood and wood-based materials.

For long-life wooden structures, following the above considerations, the following ways of realization of the fire protection effect can be used:

- By modifying the main components of the lignocarbon complex in order to form, as a result of fire exposure or fire, a charcoal layer with a large number of cross-links, which will achieve a higher thermal stability of the charred layer. For example, in [38] aspects of the application of modifying additives that contribute not only to the formation of spatially meshed structures on the wood surface but also to the foaming of the modification product when exposed to elevated temperatures were considered.
- The use of substances and compounds capable of melting when heated and giving noncombustible residue. Clear representatives of this direction are such flame retardants as potash (potassium carbonate salt), sodium tetraboronoxide (borax), boric acid, and others. When heated, sodium tetraboric acid loses crystallization water to form a foamy residue.

Anhydrous salt melts at temperatures around 740 to 880 °C with the formation of so-called borate glasses. At about 741 °C, borax melts and covers the combustible material with a glassy, non-combustible film. The borax partially decomposes into the metaboron sodium salt $NaBO_2$ and the boron anhydride B_2O_3:

$$Na_2B_4O_7 = 2NaBO_2 + B_2O_3$$

In many respects, this determines a decrease in the flammability of the wood material, as well as possibly reducing the intensity of oxidation of the resulting charcoal and the intensity of coal formation. Thus, in [39] with respect to modern wood, the possibility of modifying samples using borazon compounds (amine borates) was investigated. Weight loss of wood samples with flame retardant at a rate of 150 g/m^2 is no more than 20% under conditions of fire exposure. Thermal analysis revealed a shift of coke oxidation temperature to high temperatures and a twofold reduction in the oxidation rate of the carbon layer compared to the untreated wood sample.

The effect of formation of a glassy thermally stable residue is also known for metaphosphoric acid. In [40], it is noted that metaphosphoric acid is prone to polymerization and formation of a polymer on the surface that protects the wood from ignition and burnout.

The most effective are the fire retardants of complex action in the zone of gas-phase reactions, as well as in the condensed phase. For example, substances from the group of ammonium salts are capable of decomposing into ammonia and free acid when heated. The best-known substances of this group are two-substituted ammonium phosphate (diammonium phosphate) or one-substituted ammonium phosphate (monoammonium phosphate). A feature of ammonium salts is that they lose all ammonia when heated and convert to orthophosphoric acid:

$$(NH_4)_2HPO_4 = 2NH_3 + H_3PO_4$$

Upon subsequent heating, orthophosphoric acid easily melts and protects the surface of the wood from contact with the air and prevents it from igniting.

Upon further heating at about 260 °C, orthophosphoric acid loses water, forming pyrophosphoric acid ($2H_3PO_4) = H_2O + H_4P_2O_7$), which at 300 ° C converts to tetraphosphoric acid ($H_4P_2O_7 = H_2O + 2HPO_3$).

Reducing the flammability of wood impregnated with ammonium salts is mainly due to the fact that when heated, nonflammable fusible phosphoric acids are formed. Ammonium salts of phosphoric acid combine the mechanisms of its action not only in the gas phase but also in the condensed phase. To increase the flame-retardant effect, ammonium salts are used in mixtures with other substances, such as nitrogen-containing compounds.

The effectiveness of these flame retardants is determined by the presence in their composition of phosphorus, which is the most important retardant of flameless (smoldering) combustion of cellulosic materials and its effect in the condensed phase. The influence of phosphorus compounds on the kinetics of gas-phase reactions can be if they, at the temperatures realized on the surfaces of the burning material, are able to pass into gaseous state [41].

Given the importance of ensuring the longevity of wooden structures, it is necessary to solve the complex problem of reducing the fire hazard and increasing the biostability. In this regard, phosphoric acid esters, in particular, dimethylphosphite, were used as an effective biocide and fire retardant. The effectiveness of this flame retardant has been studied in detail for modern wood [42–44]. For example, in [42] it was found that during thermal decomposition of cellulose and wood treated with phosphorus-containing compounds, the yield of volatile products and the thermal decomposition temperature decrease, and the coke yield increases. By means of thermal analysis methods, it was found that as the degree of phosphorylation increases, the flame retardancy of the wood increases. Using a climatic chamber, it was found that the durability of flame-retardant effect of flame retardants is provided by the non-washability of phosphorus-containing compounds. At a consumption rate of 200 g/m^2, the loss of mass is from 21.9% (formulation based on trichloroethyl phosphate) and 15.3% (formulation based on 40% trichloroethyl phosphate and 60% carbon tetrachloride). Based on research results, a class of phosphorus-containing water-soluble firebioprotective compositions with the consumption of their application of 400–500 g/m^2 with delay up to 12 h between layers application was created.

High flame-retardant effect of joint action of dimethylphosphite and organosilicon compounds was shown in [43]. The developed flame-retardant composition at a flow rate of 400 g/m^2 provides a weight loss of the tested sample of 9%. The fireproofing composition reduces the fire hazard indicators of wood and increases the ignition time from 15 to 65–90 s (depending on the type of organosilicon compounds). The methods of thermal analysis established the reduction of weight loss in the temperature range of 300–450 °C in 2.5 times compared to untreated wood and the increase of the coke residue almost in two times.

In [44], it was also shown that compositions based on esters of phosphoric acids can significantly reduce the smoke-forming ability of modern wood samples. It was investigated an increase in fire safety of wood when using the method of surface modification by diethylphosphite. The flame retardant provides an 8.5% weight loss of wood samples, a sixfold decrease in the smoke-forming capacity of wood and a 16-fold decrease in the flame propagation index value. Durability of the developed composition reaches 8 years.

Increase in bioproofness and hydrophobicity of wood when using bioprotective compositions based on dimethylphosphite on small samples of wood with operating life up to 500 years was established in work [45]. In the course of research, modification of surface layers of the samples under study was carried out by dipping them into various modifying solutions for 3 h at solution concentration equal to 10%. During studies by means of electron microscopy, it was found that during a single modification the solution does not penetrate deep into the sample, but accumulates on the surface, in connection with which it is necessary to conduct additional stages of modification. When conducting research on fire resistance, the samples were modified in solutions at a rate of 300 and 500 g/m^2. At the rate of 300 g/m^2, the wood sample becomes hardly flammable. It was found that the continuation of the protective effect depends on the chemical composition of the modifier. When modifying with esters of phosphoric acids and phosphoric acid, the bioprotective effect of the samples increases. The results of thermal analysis established a decrease in the value of mass loss in the temperature range of 150–400 °C. Developed by the results of studies, the biogreening composition based on dimethylphosphite and ethylhydridesiloxane allows to increase the service life of wood in 1.4 times compared with the native wood.

6.2 Study of the Effectiveness of Fire-Retardant Compositions for Wood

The variety of technological methods of impregnation, hardware design of the process, and its driving forces do not allow to form universal classifying attributes, which leads to the existence of many different variants of impregnation methods classification [46–48].

Due to its simplicity and accessibility, the surface impregnation method is the most widespread. Widespread use of this method is provided by the development of theoretical ideas about the mechanism of influence of various fire-retardant compositions on the process of thermal decomposition of wood components, which allows to obtain compositions with high fire protection efficiency even at low consumption (100–200 g/m^2) [3]. The solution is usually applied to the surface of the material or structure by brush, roller, or spraying. Penetration of active substances occurs due to capillary pressure to a depth of up to 2 mm for fresh wood and up to 5 mm for old wood [49, 50].

As part of a comprehensive assessment of the effectiveness of firebioprotective compositions, the base formulation was selected, based on the formulation of a

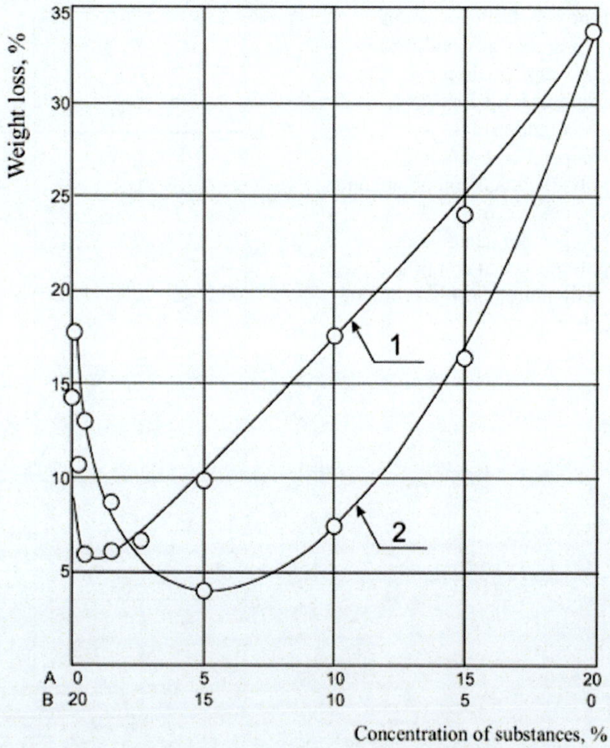

Fig. 6.4 Dependence of weight loss of wood with fire-retardant compositions when tested according to GOST R 53292 (Russia) on the concentration of initial substances: 1: A—polyfunctional compound; B—ammonium phosphate double substituted; 2: A—polyfunctional compound; B—sodium hydroxide

complex composition, which includes both inorganic nitrogen-phosphorus and organic polyfunctional compounds. Studies were conducted in relation to impregnation compositions, which are balanced synergistic mixtures of substances with firebioprotective functions. For the primary assessment of flame-retardant efficiency and optimization of impregnating compositions formulations, the standard methodology according to GOST R 53292 (Russia) was used. It has been established that the greatest flame-retardant effect, i.e., the least weight loss of the sample during the tests, is observed at a certain ratio of the initial components of impregnating compositions. So, for the composition containing polyfunctional compound and ammonium phosphoric acid divalent, the greatest effect was obtained at the content of polyfunctional compound 0.5% in the solution of 20% concentration (Fig. 6.4).

It should be noted that when using the polyfunctional compound alone—the weight loss of wood in a fire test increases, i.e., its thermal destruction increases, but in a mixture with phosphorus-containing retarder of combustion—ammonium phosphate double substituted significantly enhances its fire-retardant effect.

Similar tests of the composition containing a polyfunctional compound and sodium hydroxide (active smoke depressant) showed that the greatest flame-retardant effect is shown when the content of the polyfunctional compound is 5.0% in the solution of 20% concentration. The primary assessment of the fire-protective effect of the impregnating composition by a standard method of GOST R

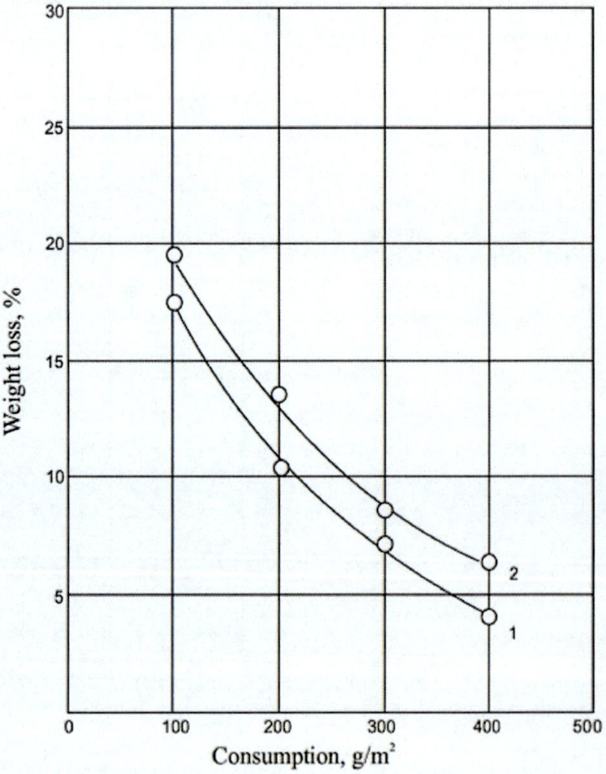

Fig. 6.5 Dependence of weight loss of wood treated with impregnation composition on its consumption: 1—composition containing polyfunctional compound and sodium hydroxide; 2—composition containing polyfunctional compound and ammonium phosphate double substituted

53292 (Russia) showed that already when applying a single layer of the composition on the wood surface at the rate of 100–180 g/m^2, group II fire-protective effectiveness is provided. Increasing the consumption of aqueous composition up to 300–400 g/m^2 allows obtaining group I of fire-protection efficiency (Fig. 6.5). Losses of weight during fire test of wood samples in this case were equal to 6.0–8.0%, respectively.

The test results show that at a flow rate of 300 g/m^2 or more surface treatment with wood compositions achieved group I fire-retardant efficiency (sample loses less than 9.0% weight). Impregnating composition containing sodium hydroxide and polyfunctional compound, with the content of the latter component of 5.0%, has the greatest efficiency—the mass loss is 7.2% at a consumption rate of 300 g/m^2. It is interesting to note that the composition is reactive with the achievement of the maximum synergistic effect of fire protection after 2–3 weeks of application of the flame retardant to the protected wood surface.

To establish the leading mechanism of action of impregnated flame-retardant composition, modern physical and physical-chemical methods of analysis (scanning electron microscopy, dynamic thermogravimetric analysis, infrared spectroscopy) were used. As a result of studies by the method of thermogravimetry, the characteristic temperatures of the maximum decomposition of the initial substances ammonium phosphorus double-substituted and polyfunctional compound were established. Thus, the temperature of maximum decomposition of ammonium

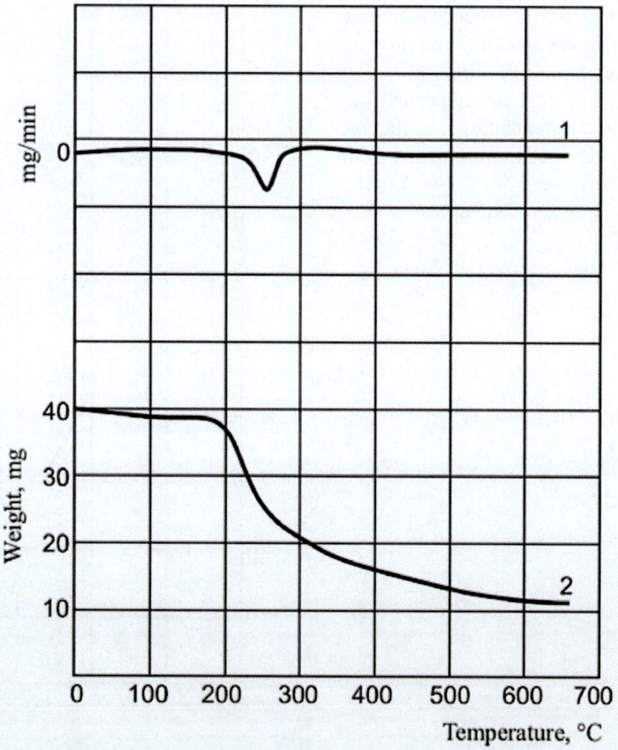

Fig. 6.6 Thermogravimetric decomposition curves of pine with the composition containing ammonium phosphate double-substituted and polyfunctional compound in the air when heated at a speed of 10 deg/min: 1—DTG curve; 2—TG curve

phosphorus divalent is 178 °C; there is a characteristic endothermic effect; at a temperature of 608 °C, this compound has a weight loss of 93%. The polyfunctional compound has a maximum decomposition temperature of 123 °C; there is a characteristic endothermic effect; at 460 °C, this compound has a weight loss of 92.8%. In thermal analysis, the mixtures of these compounds show effects similar to the initial ones, which allow us to conclude that these compounds do not interact with each other.

Figures 6.6 and 6.7 show the curves of thermal analysis of wood with flame retardants.

From the results of thermal analysis of wood with the composition containing ammonium phosphate double-substituted and polyfunctional compound, it is clear that the thermal decomposition has a different character compared to the native wood. Thermal decomposition of wood in the presence of this composition begins in a lower temperature zone compared to the original wood and occurs in several stages:

- Up to 196 °C wood drying and release of physically bound water from the capillaries of wood, the effect of the applied composition is manifested (weight loss—5.8%).
- From 196 to 262 °C occurs release of phosphoric acid. There is an intensive decomposition of wood in the low-temperature area (the temperature of

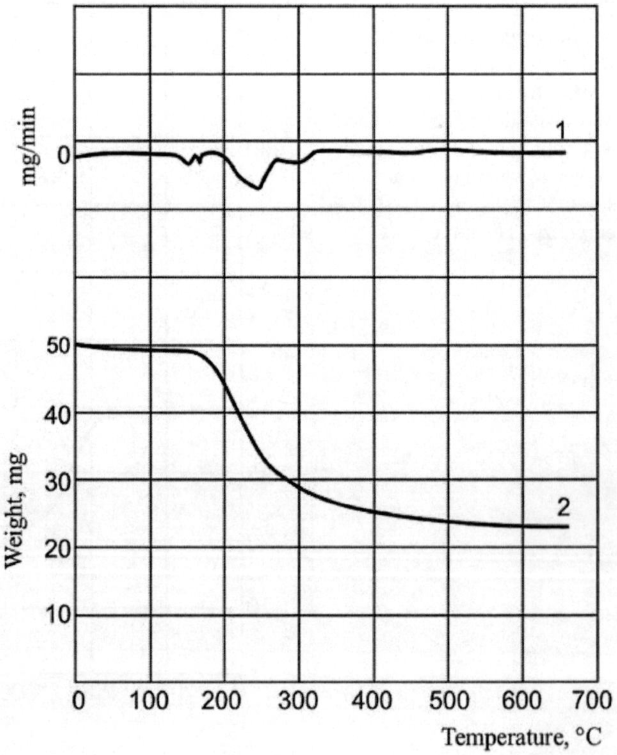

Fig. 6.7 Thermogravimetric decomposition curves of pine with the composition containing sodium hydroxide and polyfunctional compound in the air when heated at a rate of 10 deg/min: 1—DTG curve; 2—TG curve

maximum decomposition is 262 °C), running processes of dehydration, and formation of carbonized layer (weight loss 28.0%).

- From 262 to 620 °C, processes of dehydration, further decomposition of wood complex, and formation of carbonized layer (weight loss 36.2%) occur.

Thermal decomposition of wood in the presence of the composition containing a mixture of ammonium phosphate double-substituted and polyfunctional compound begins in a lower temperature zone compared to the original wood. Temperatures of maximum decomposition, respectively, are 262 and 330 °C; at the same time in the dependence section (*dm*/*dT*) to 200 °C, there is less weight loss compared to the native wood.

At temperatures above 200 °C, weight loss increases and reaches a maximum rate of decomposition at 262 °C. This is explained by the fact that at about 180 ° C, chemically bound orthophosphoric acid is released, and there is an acceleration of dehydration of cellulose.

This process in the presence of orthophosphoric acid reaches the highest rate at a temperature 100 ° C lower than in the decomposition of untreated pulp, while water is released twice as much as in the thermal effect on the original pulp:

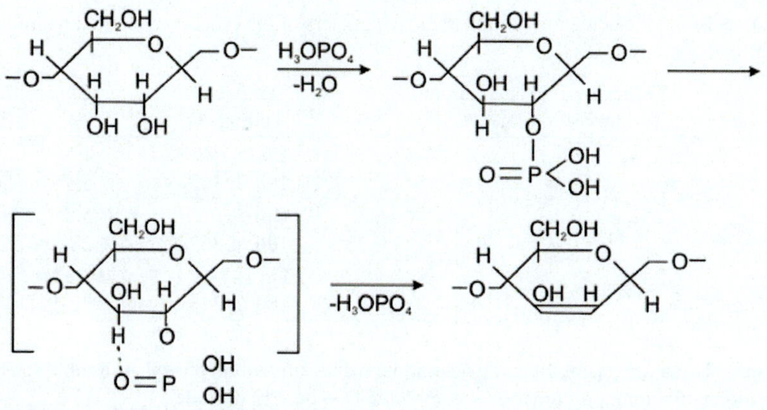

The thermal analysis of wood with the composition containing sodium hydroxide and polyfunctional compound (Fig. 6.7) shows that the thermal decomposition also has a different character compared to untreated wood. We can also distinguish three characteristic areas of thermal decomposition of wood:

- Up to 183 ° C wood drying and release of physically bound water from the capillaries of wood, the effect of the applied composition is manifested (mass loss—3.9%).
- From 183 to 245 °C, intensive decomposition takes place in the low-temperature.

Intensive decomposition occurs in the low-temperature area (maximum decomposition temperature is 245 °C), processes of dehydration, and formation of the carbonized layer (weight loss 22.2%).

- From 245 to 696 °C, dehydration, further decomposition of the wood complex, and building up of the carbonized layer (weight loss 27.9%) occur.

Used modifier-sodium hydroxide promotes the increase of intermicellar spaces in the structure of cellulose. The product of the interaction of alkali with cellulose leads to the formation of a product with a lower degree of crystallinity, and the reaction of thermal decomposition shifts to the low-temperature area, which is accompanied by acceleration of dehydration processes and the yield of carbonaceous residue. In this case, modification of hydroxyl groups of cellulose by reaction of the following kind takes place:

$$[C_6H_7O_2(OH)_3]_n + n\,NaOH \rightarrow [C_6H_7O_2(OH)_2ONa]_n + n\,H_2O$$

At comparison of curves of thermal analysis of wood and wood with proposed compositions, it is seen that the mass loss on the temperature section to 200 °C of wood is 8.5%, and wood with the composition containing ammonium phosphate

Table 6.1 Results of calculation of the kinetic parameters of the process of thermal degradation of wood with flame retardants

Sample name	Temperature of maximum decomposition rate, °C	Temperature section, °C	Δm, %	E_a, kJ/mol
Pine	330	Up to 198 198 to 363	8.5 59.5	31.7 209.0
Pine—composition[a]	262	Up 196 196 to 262	5.8 28.0	51.3 153.9
Pine—composition[b]	245	Up 183 183 to 245	3.9 22.2	63.8 121.3

Note:
[a]Composition containing a polyfunctional compound and ammonium phosphate double substituted;
[b]Composition containing polyfunctional compound—sodium hydroxide

double-substituted and polyfunctional compound and with the composition containing sodium hydroxide-polyfunctional compound is 5.8% and 3.9% respectively, i.e., decreases in 1.5 and 2.2 times.

Thus, the effect of the proposed impregnating compositions affects, first of all, a change in the direction and speed of the main stages of thermal decomposition of wood.

For a more detailed study of the mechanism of flame-retardant action of the compositions, the kinetic parameters of wood thermal degradation with the proposed impregnating compositions at two temperature sections were calculated by the method presented in work [51].

The results of the calculation of the kinetic parameters are presented in Table 6.1.

In the low-temperature region of wood with compositions, the activation energy (E_a) increases, which is indicative of a decrease in the process of wood thermodegradation. In the region of temperatures of maximum decomposition of wood complex components in the presence of compositions, a decrease in activation energy (E_a) is observed. In this area, in the presence of compositions, the area of maximum decomposition shifts to a lower temperature compared to untreated wood. Due to this, there is an acceleration of dehydration processes and yield of the carbonized layer, which explains the high flame-retardant effect of the studied fire-retardant compositions.

The results of thermal analysis and studies of the dynamics of changes in the Fourier-infrared spectra of surface layers of flame-retardant wood samples under heat exposure confirmed the important role of carbon layer formation in reducing the fire danger of wood. When analyzing the Fourier IR spectra in the range of 400–4000 cm^{-1}, attention was paid to the frequency vibration regions of O–H, C–H, and C–O–C bonds in the structural groupings of wood components. Such groupings are the most sensitive to transformations under heat exposure. The increase in the intensity of bands caused by the appearance of $C{=}C_{AROM}$ bonds and the increase in the absorption background testified to the active carbonization of the substrate under the influence of combustion retarder. It was concluded that the

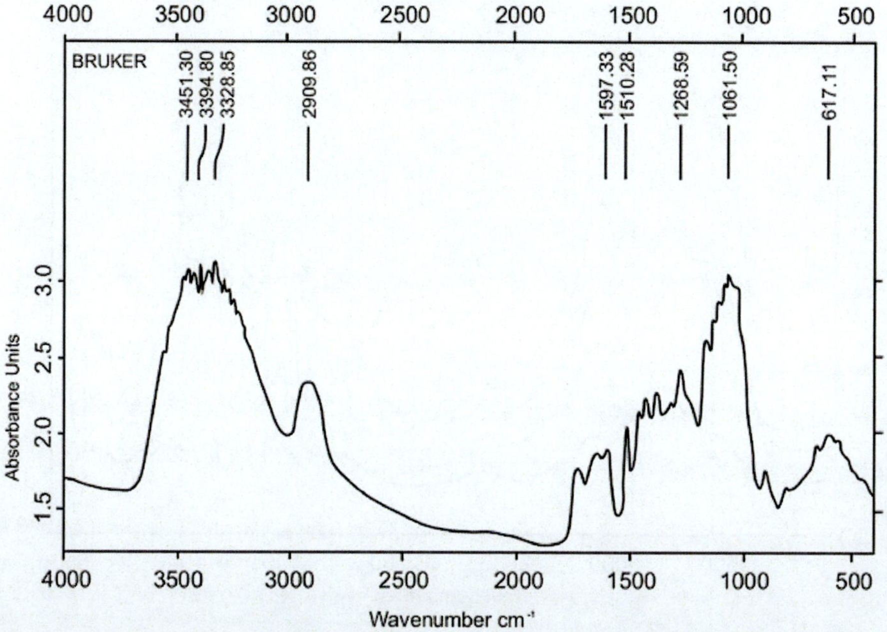

Fig. 6.8 Infrared spectra of pine wood

impregnation composition containing polyfunctional compound and ammonium phosphate double substituted reacts with wood components when heated, accelerates dehydration and cross-linking reactions of macromolecules, and leads to formation of coke layer with improved thermal protection properties (Figs. 6.8, 6.9, and 6.10).

To study the mechanism of flame-retardant effect of the proposed impregnating compositions, replicas of untreated wood surface, as well as wood with the initial substances separately and with the composition containing a mixture of compounds, polyfunctional compound and ammonium phosphate double substituted (magnification of 300 and 800 times) were also performed, Figs. 6.11, 6.12, 6.13, 6.14, and 6.15. The images of wood surface show significant changes in the character of supramolecular formations in the material structure. Untreated wood has pores between individual fibrils or fibers. When treating the wood surface with aqueous solutions of initial components of impregnating compositions, crystalline compounds are observed on the wood surface separately.

As shown by analysis of the replicas obtained from the surface of samples using electron microscopy, after treatment of wood with impregnating composition containing a mixture of compounds (polyfunctional compound and ammonium phosphate double substituted), the structure of the surface layer changes markedly. On the surface of untreated sample, typical elements of coniferous wood structure were found, namely, microfibrils in the form of ribbon-like formations and fibers with fringed pores of round shape. After impregnation of wood with the

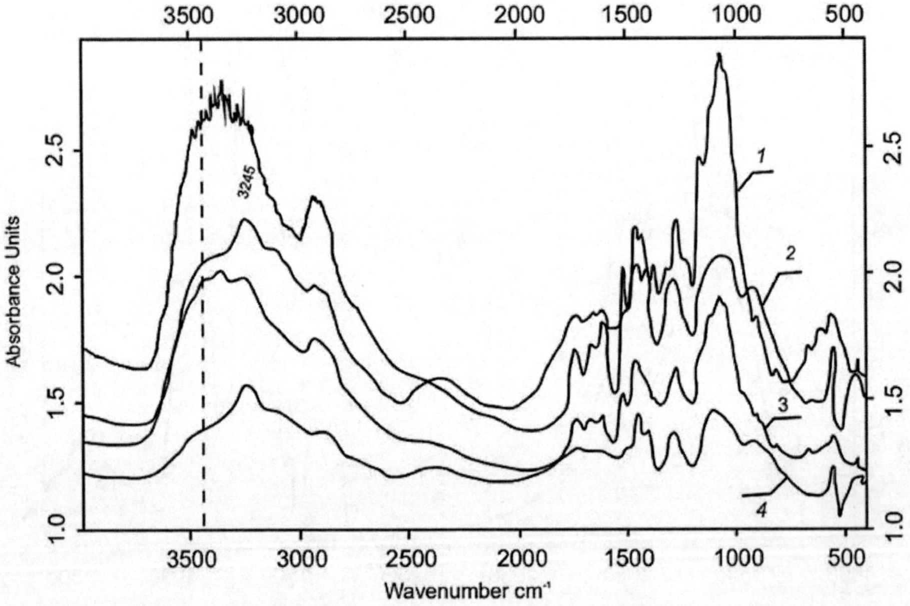

Fig. 6.9 Infrared spectra of wood with polyfunctional compound, ammonium phosphate double substituted, mixture of polyfunctional compound and ammonium phosphate double substituted, mixture of polyfunctional compound, ammonium phosphate double substituted, and boron-containing antiseptic

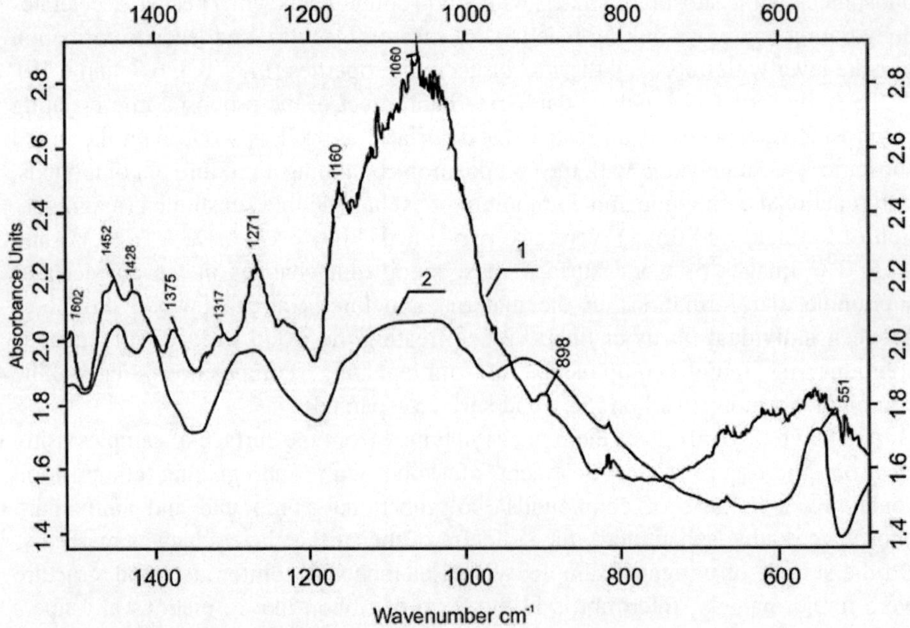

Fig. 6.10 Infrared spectra of wood with ammonium phosphoric acid divalent and ammonium phosphate double-substituted, polyfunctional compound and boron antiseptic

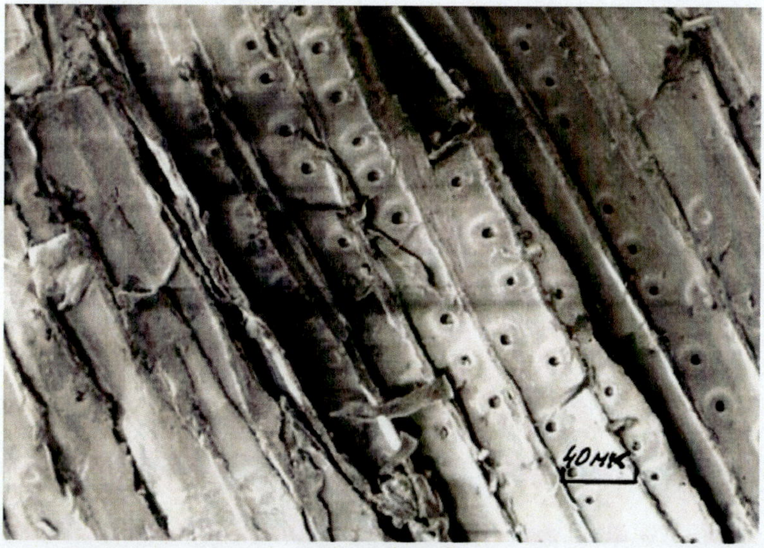

Fig. 6.11 Photo of the original wood surface—magnification of 300 times

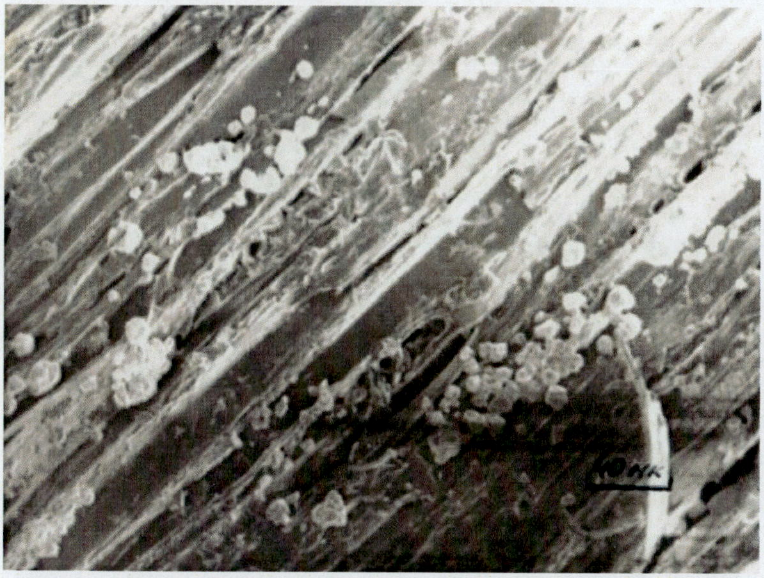

Fig. 6.12 Photo of wood surface with aqueous solution of ammonium phosphate double substituted—magnification of 300 times

composition, pores were no longer visible, and transverse formations—"strands"—were observed between microfibrils. A picture of a kind of spatially cross-linked structure appeared (Figs. 6.13 and 6.14).

From the photos shown, it can be seen that in the case of the studied impregnation composition, containing the original components in the ratio that provides the

Fig. 6.13 Photo of wood surface with aqueous solution of ammonium phosphate double-substituted and polyfunctional compound (19.5/0.5)—magnification of 300 times

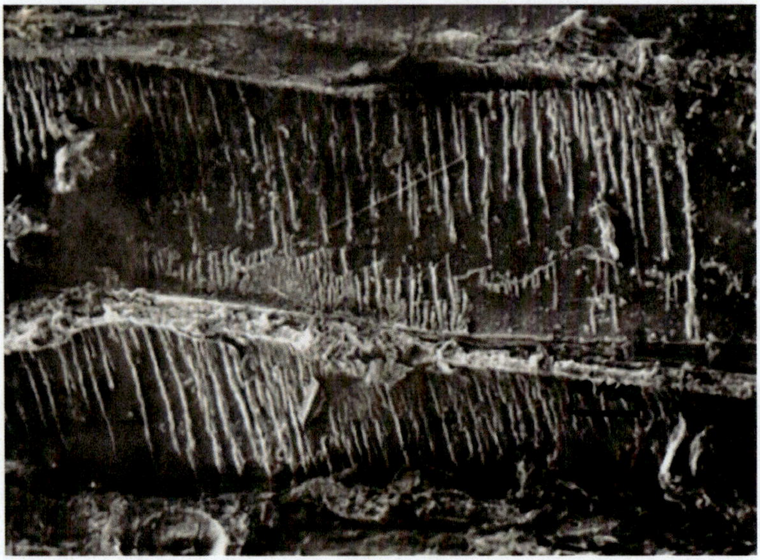

Fig. 6.14 Photo of wood surface with aqueous solution of ammonium phosphate double-substituted and polyfunctional compound (19.5/0.5)—magnification of 800 times

greatest fire-retardant effect, the structure is greatly altered. Between the fibers and fibrils of wood, the appearance of transverse "ties" is found, creating a kind of spatially cross-linked structure. This indicates that even under normal conditions without heating, there is some interaction of reagents with cellulose. It should be

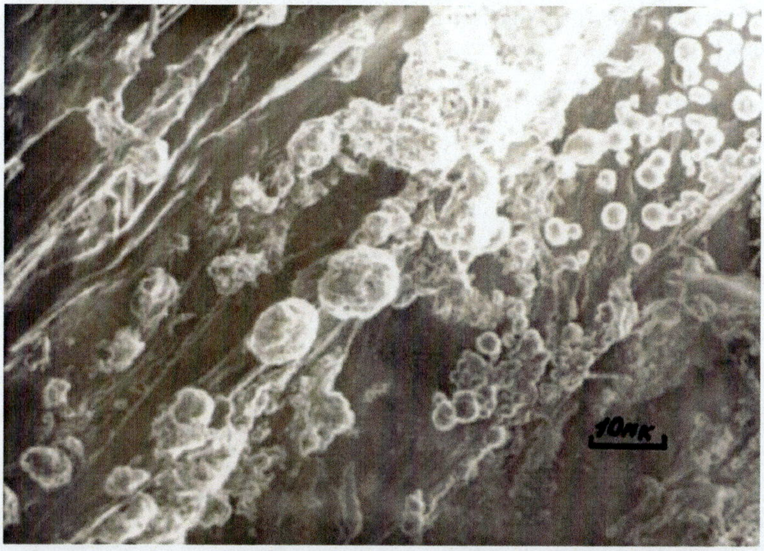

Fig. 6.15 Photo of wood surface with aqueous solution of ammonium phosphate double-substituted and polyfunctional compound (15.0/5.0)—magnification of 800 times

noted that when the content of the polyfunctional compound is increased up to 5% (Photo 6.18), the "mesh" structures on the surface are not observed and the flame-retardant effect decreases. This is apparently explained by the fact that when using an aqueous solution of the polyfunctional compound for wood treatment, there is an increase in degradation of the components of the wood complex. If there is a "deficit" of the polyfunctional compound, the flame-retardant effect of impregnation compositions also decreases, indicating insufficient formation of cross-linked structures on the surface to be protected.

The predominant mechanism of action of the proposed fire-retardant compositions under consideration is associated with the strengthening of the process of charring of the surface layer of wood. Charred dense layer formed on the surface of wood under fire action, in this case, is characterized by greater uniformity and lack of deep cracks compared with the charred surface of untreated pine. This layer is not as porous as in the case of blowing foaming systems. However, it also has the effect of reducing combustibility by reducing the yield of volatile combustible products into the gas phase.

An estimate of the charring rate (β) of wood can be made using the following equation [52]:

$$\beta = q_{\text{net}}/\rho \cdot \Delta H_v, \tag{6.1}$$

where $q_{\text{net}} = q_e - \varepsilon\sigma(T_s^{\ 4} - T_p^{\ 4})$; q_{net}—thermal energy stored by the coke layer, kW/m^2; q_e—density of external radiation heat flux, kW/m^2; T_s—temperature of coke surface, K; T_{ch}—temperature of charring front equal to the temperature of the beginning of intensive decomposition of wood (or temperature of surface of the

Table 6.2 Results of tests of complex flame retardants by the method of determining the flammability of materials

Name composition	Consumption, g/m^2	Weight loss, %	Δt_{max}, °C	Time of reaching Δt_{max}, s	Combustibility group according to GOST 12.1.044 clause 4.3
Composition[a]	300	32.4	69	300	Difficult to ignite
	400	18.5	44	300	Difficult to combust
Composition[b]	300	21.3	55	300	Difficult to combust
	400	15.9	32	300	Difficult to combust

Note:
[a]Composition containing a polyfunctional compound and ammonium phosphate double substituted
[b]Composition containing a polyfunctional compound and sodium hydroxide
Δt_{max}—maximum temperature increase, °C
t_{max}—maximum temperature, °C

sample at ignition), K; ε—radiation coefficient; σ—Stefan-Boltzmann coefficient, $\sigma = 5.67 \cdot 10^{-11}$ kJ/(m^2·s·K^4); ρ—density of surface layer of wood, kg/m^3; and ΔH_v—enthalpy of gasification, kJ/g.

For impregnating compositions containing polyfunctional compound and ammonium phosphate divalent, polyfunctional compound, and sodium hydroxide, the calculated values of gasification heat are 704 and 755 kJ/kg, respectively.

At an external heat flux $q_e = 50$ kW/m^2 when tested according to GOST 30402 (Russia), the charring rate of pine is 1.39 mm/min. For wood with compositions containing polyfunctional compound and ammonium phosphate double-substituted, polyfunctional compound and sodium hydroxide, charring rate is 4.67 and 4.84 mm/min, respectively. The greatest efficiency in reducing the intensity of charring of wood can be found with the use of bloating (intumescent) coatings for fire protection of wood.

In the case of application of bloating coatings based on modified starch with degree of modification $\alpha = 0.5$, wood charring rate is reduced compared with the original wood from 1.39 mm/min to 0.186 mm/min. This may indicate different mechanisms of flame-retardant effect of flame retardants used.

In order to obtain more complete information about the flammability of wood treated with the tested flame retardants, the method of determining the flammability of materials by GOST 12.1.044, paragraph 4.3 (Russia), in accordance with which the duration of fire exposure is increased to 5 min, was applied. The test results are presented in Table 6.2.

Comparison of the results of fire tests by two methods shows that at a rate of 300 g/m^2, the composition containing a polyfunctional compound and ammonium phosphate double substituted, which provides the I group of fire protection efficiency by the standard method (mass loss of the sample less than 9.0%), refers only to a group of materials difficult to flame by the method of determination of the flammability of materials. It seems more reliable to test by the method for determining the flammability of materials, in which, in addition to recording the mass loss, temperature measurements of the flue gases indirectly reflect the potential of heat release during combustion of the sample.

Table 6.3 Effect of flame-retardant impregnation composition containing polyfunctional compound and ammonium phosphate double substituted on flammability and mass burn rate of wood

Sample name	Consumption, g/m^2	q_e, kW/m^2	T_{ign}, s	q_{cr}, kW/m^2	Maximum mass loss rate, [a]g/m^{2}s
Pine timber	–	30	21	12.5	–
	–	40	9		28
	–	50	4		35.7
Pine with impregnation composition containing polyfunctional compound and ammonium phosphate double substituted	300	30	195	22.5	–
		40	39		–
		50	16		21.3
	400	30	600	25.0	–
		40	72		–
		50	29		14.4

Note: [a]$q_e = 50$ kW/m^2

The flammability indicators, as well as the mass burnout rate, were determined under an external heat flux of 18–50 kW/m^2 on the apparatus of GOST 30402 (Russia). Flame spread over the surface of the material, smoke-forming ability, and toxicity of combustion products—according to GOST 12.1.044 (Russia), p. 4.19, 4.18, and 4.20, respectively.

In this work, tests on flammability of wood samples with impregnating compositions were carried out on wood samples 20 mm thick, i.e., thermally thick with the n-value close to the value of 2.

Table 6.3 shows the results of studies of the effect of impregnating fire-retardant composition on flammability and maximum burn rate of pine wood.

Quantitative parameters of flammability, which allow us to compare the effect of fire protection of wood, are the time to ignition of the sample, τ_{ign}, at a given density of external heat flux, q_e; the critical heat flux of ignition, q_{cr}; and the maximum mass loss rate. The surface treatment of wood with a flame-retardant composition increases its resistance to ignition many times over. Not only the period until ignition of samples is prolonged, but also the values of critical heat flux required for their ignition are increased (by 1.5–2 times) (Fig. 6.16).

With increasing consumption of flame-retardant composition, the maximum rate of mass loss during combustion of samples even under fairly harsh conditions of exposure to radiant heat flux ($q_e = 50$ kW/m^2) is reduced by more than two times compared to untreated wood. Thus, flame retardants affect the dynamics of wood pyrolysis, slowing down and reducing the formation and release of volatile combustible products. The quantity and quality of the charcoal layer formed on the surface of the sample is important.

When investigating the process of flame propagation over the surface of flame-retarded wood, the method according to GOST 12.1.044 p. 4.19. This method provides classification of materials by the index of flame propagation, which characterizes the ability of materials to propagate a flame from top to bottom on

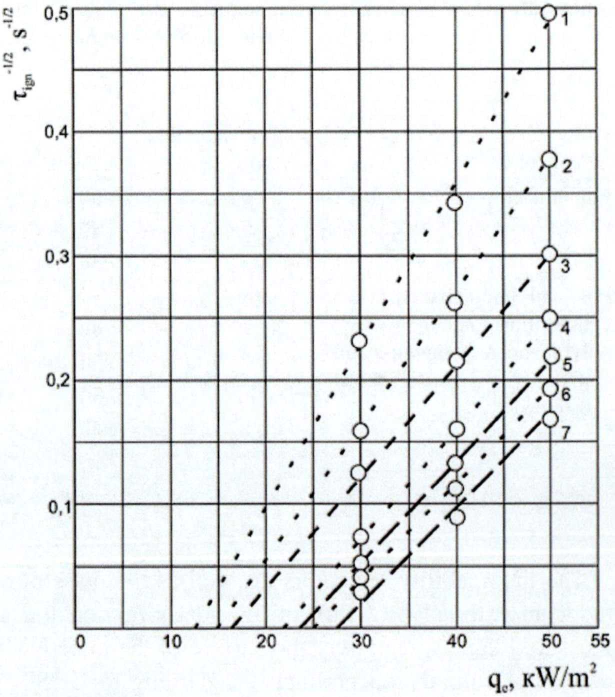

Fig. 6.16 Determination of the critical heat flux density in tests according to GOST 30402 (Russia): 1—pine wood; 2, 4, 6—wood treated with the composition containing polyfunctional compound and ammonium phosphate double substituted, consumption 200, 300, 400 g/m^2; and 3, 5, 7—wood treated with the composition containing polyfunctional compound and sodium hydroxide, consumption 200, 300, and 400 g/m^2

the surface of the sample located at an angle of 30° to the radiation panel under the action of heat fluxes in the range from 12.0 to 32.0 kW/m^2. After processing the results of the experiment, the critical surface heat flux density and flame propagation velocity under the action of different values of heat fluxes were determined. For untreated pine, it is not possible to determine experimentally the critical surface heat flux density for flame propagation, since the flame spreads over the entire length of the sample. Indeed, the minimum sample surface temperature required to maintain the flame spread across the surface of the material ranges from 270 to 370 °C for fiberboard and wood cladding. This temperature can be calculated assuming that the heat loss is mainly due to radiation from the surface.

Figure 6.17 shows the dependences (ln$V - q_e$) for wood and wood impregnated with firebioprotective compositions. We graphically determined that the critical surface heat flux density for pine wood is 5.0 kW/m^2 and, accordingly, the surface temperature is 295 °C, as it follows from the Stefan-Boltzmann equation:

$$T_s = 4\sqrt{q_{cr}}/\varepsilon \cdot \sigma, \mathrm{K} \tag{6.2}$$

where q_{cr}—the intensity of radiant heat flux, kW/m^2; σ—the Stefan-Boltzmann constant, $\sigma = 5.67 \cdot 10^{-11}$, kJ/(m^2·s·K^4); and ε—the relative emissivity or degree of blackness of the body (flame), $\varepsilon = 1.0$.

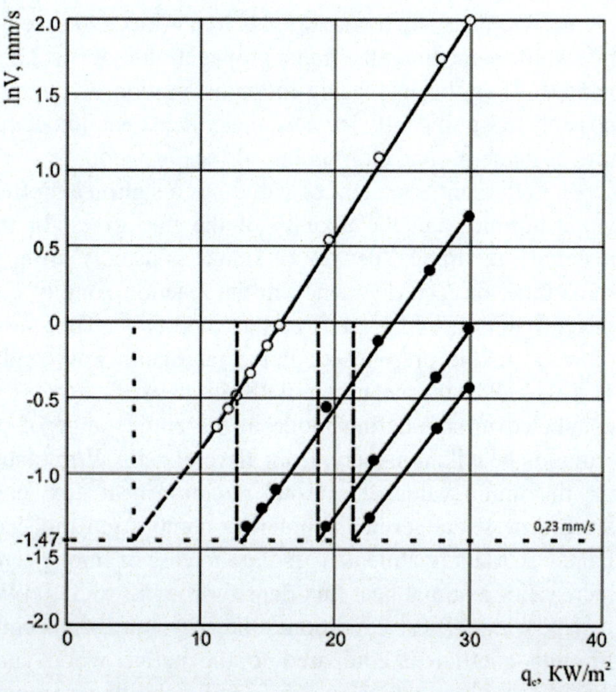

Fig. 6.17 Determination of the critical surface heat flux density for flame propagation according to GOST 12.1.044 (Russia) p. 4.19: 1—pine wood; 2—wood with the composition containing ammonium phosphate double substituted and polyfunctional compound, consumption 200 g/m^2; 3—wood with the composition containing ammonium phosphate double substituted and polyfunctional compound, consumption 300 g/m^2; and 4—wood with the composition containing ammonium phosphate double-substituted and polyfunctional compound, consumption 400 g/m^2

Figure 6.17 shows that the critical heat flux density increases from 13.0 to 22.0 kW/m^2 as the flow rate of the applied composition increases from 200 to 400 g/m^2. It is interesting to note that for the initial wood samples and wood treated with the proposed impregnating compositions, the value of limiting flame spreading speed on the surface remains at the same level: V_{cr} = 0.23 – 0.24 mm/s and is critical (Fig. 6.17).

In this case, the surface temperature at which the process of flame propagation is still possible depends on the effectiveness of the combustion retardant system. Thus, for wood with the composition containing ammonium phosphate double-substituted and polyfunctional compound at the rates of 200, 300, and 400 g/m^2, the experimentally measured temperature is equal to 509, 540, and 576 °C, and the temperature calculated with the value of coefficient of blackness $\varepsilon = 0.75$ is 533, 565, and 602 °C. Thus, the real value of the coefficient of blackness of the surface of pine treated with the composition should correspond to $\varepsilon = 0.95$. If you take for untreated pine also $\varepsilon = 0.95$, the surface temperature of pine at the limit of flame spreading should be 295 °C.

The results of tests to determine the flame propagation index showed that the use of the proposed impregnation compositions significantly reduces the rate of flame propagation on the surface of the material, as well as the values of the flame propagation index. Thus, when applying the composition containing ammonium phosphate double-substituted and polyfunctional compound at a flow rate of 400 g/m^2, the flame propagation rate decreases to a critical value after the passage of the second section, which corresponds to a critical surface heat flux density of 25 kW/m^2. At the same time, the flame propagation index is 2.5, while for wood, this index is 55.0. Thus, by increasing the consumption of impregnating compositions, it is possible to significantly increase the critical heat flux density, as well as to reduce the rate of flame propagation on the surface of wood.

Fire retardant treatment of wood has a significant effect on the smoke density of wood burning and the toxicity of the flue gases. In flame combustion of wood materials, the optical density of smoke is usually lower than in smoldering. At the same time, oxygen deficiency in the reaction zone of coke smoldering leads to an increase in the yield of toxic carbon monoxide. The study of smoke-forming ability of wood in the presence of flame retardants was conducted according to GOST 12.1.044 (Russia) paragraph. 4.18; in the work, studies with flame retardants were conducted in smoldering mode in the entire range of possible values of external radiation heat flux, namely, from 10.0 to 35.0 kW/m^2 and, in flame burning mode, at the maximum value of external radiation heat flux, namely, 35.0 kW/m^2. In the studied most dangerous smoldering combustion mode of wood without and with flame-retardant treatment, a regular increase of smoke generation coefficient with the increase of external heat flux density from 18 to 35.0 kW/m^2 is observed.

Under equal test conditions, the fire-retarded specimens have a lower smoke-forming coefficient compared to the native wood and belong to the group of materials with moderate smoke-forming ability according to the accepted classification. Fireproofing composition containing ammonium phosphate double-substituted and a polyfunctional compound, with application rates of 300 and 400 g/m^2, can reduce the smoke formation coefficient of pine at a heat flow of 35 kW/m^2 by 2 or more times. This effect is achieved due to the fact that the compositions contain in their formulation a mixture of compounds that have synergistic properties in the fire protection of wood. This combination has a stronger effect on reducing smoke formation during wood combustion than the treatment of wood separately with phosphorus-containing flame retardants.

The toxicity of wood smoldering combustion products depends on the external heat flux density. In a wide range of densities of 10–65 kW/m^2 in terms of toxicity of combustion products, native wood belongs to the group of highly hazardous materials (Fig. 6.18, curve 1).

Curve 1 reflects the complex extreme character of the dependence of the toxicity index of wood combustion products on the external heat flux density. In the interval of 10–23 kW/m^2, there is a typical, although weak, dependence of growth of toxicity of combustion products (decrease of toxicity index of combustion products) due to increase of CO concentration. At 23–24 kW/m^2 under test conditions of samples according to GOST 12.1.044, self-ignition of pine timber occurs. As a result, the oxidation of carbon monoxide intensifies, and the toxicity index increases again with increasing intensity of heat flow. Fire-retardant treatment prevents the realization of wood self-ignition and expands the range of smoldering combustion mode. The critical heat flux of self-ignition of wood treated with impregnation composition containing ammonium phosphoric acid divalent and polyfunctional compound at a rate of 300 g/m^2 increases to 44.0 kW/m^2. Surface impregnation of wood with the

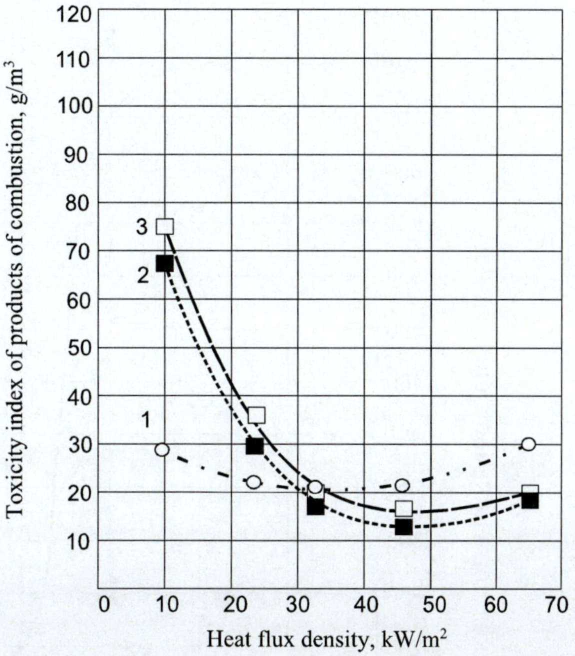

Fig. 6.18 Dependence of the toxicity index of the products of combustion of materials in the smoldering mode on the density of external radiation heat flux: 1—natural pine; 2—wood with the composition containing ammonium phosphate double substituted and polyfunctional compound, consumption of 300 g/m^2; and 3—wood with the composition containing ammonium phosphate double substituted and polyfunctional compound, consumption of 400 g/m^2

fireproof composition containing ammonium phosphate double-substituted and polyfunctional compound provides the transition of wood from the rank of materials highly hazardous in toxicity of combustion products into the group of moderately hazardous at low intensity of heat flux.

When using complex flame retardants containing active smoke depressants, it is possible to significantly reduce the smoke-forming ability of wood (Fig. 6.19).

Flame retardants are water-based compositions containing various alkaline catalysts with component concentration of 20%.

The effect of reducing the smoke-forming ability of pine wood by using flame retardants increases in the following sequence:

sodium carbonate → calcium carbonate → potassium hydroxide → sodium hydroxide.

When using flame retardants based on potassium and sodium hydroxide shifts, the period of flame burning in the region of higher values of heat flux densities with the possible transfer of native wood from the group of materials with high smoke-forming ability to the group of materials with moderate smoke formation. Reducing smoke production in a wide range of heat flux densities when using alkaline catalysts is due to the process of modification of the carbohydrate part of the wood complex, a shift of the thermal decomposition process to the low-temperature region with the formation of carbon residue. The most important aspect in considering the effectiveness of flame retardants with alkaline components is their ability to resist oxidation of the resulting carbon residue. In this connection, it is of interest to study the smoke-

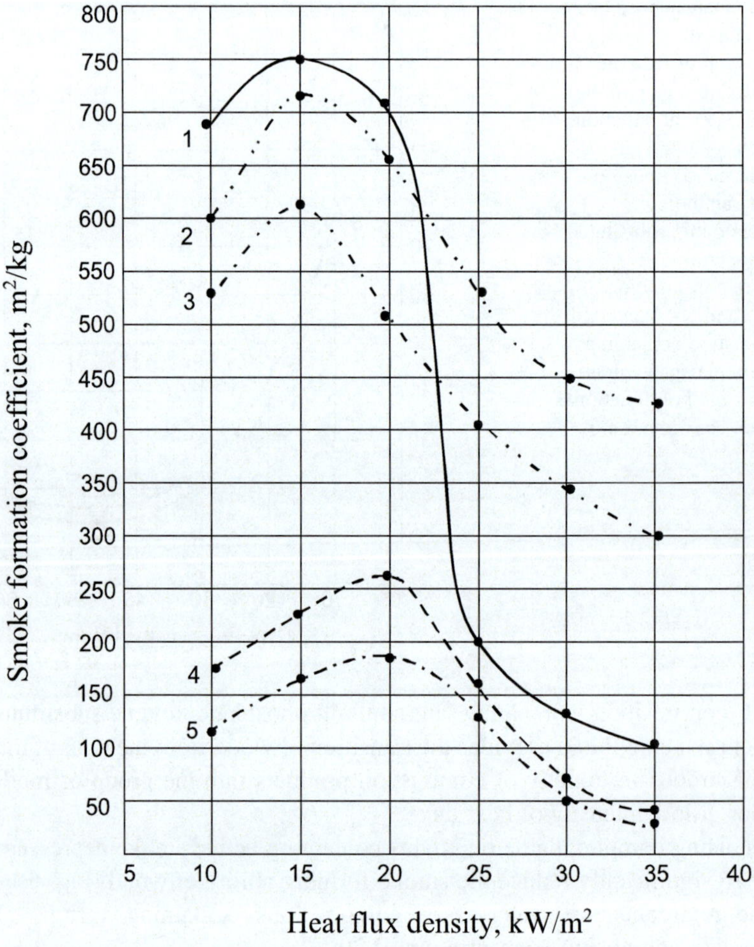

Fig. 6.19 Dependence of smoke formation coefficient when testing pine wood with flame retardants (consumption of 400 g/m^2), containing polyfunctional compound and catalyst of modification process in the ratio that provides the greatest fire protection effect, according to GOST 12.1.044 paragraph. 4.18. on the external radiation heat flux density: 1—pine wood; 2—sodium carbonate; 3—calcium carbonate; 4—potassium hydroxide; and 5—sodium hydroxide

forming capacity of wood with flame retardants depending on the ratio of the polyfunctional compound and various alkaline catalysts of the modification process. Figure 6.19 shows the dependence of smoke formation coefficient of wood with flame retardants on the ratio of the main components of the composition at external radiation heat flux density of 25 kW/m^2.

According to the test results, it was found that the composition containing polyfunctional compound and sodium hydroxide has a greater efficiency in reducing the smoke formation of wood compared to other flame retardants. The highest efficiency of the composition is observed when the ratio of the components is 5: 15, respectively.

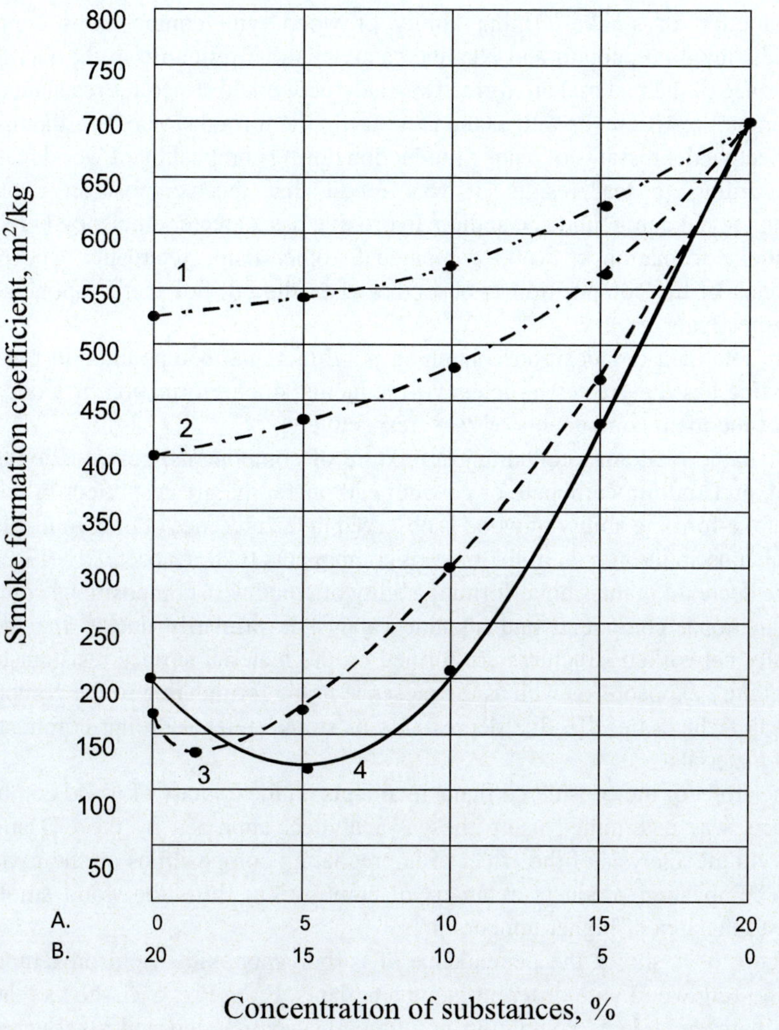

Fig. 6.20 Dependence of smoke formation coefficient of wood with flame retardants on the ratio of the main components of the composition at external radiation heat flux density of 25 kW/m^2: 1—polyfunctional compound—sodium carbonate; 2—polyfunctional compound—calcium carbonate; 3—polyfunctional compound—potassium hydroxide; and 4—polyfunctional compound—sodium hydroxide

Fire retardant composition containing polyfunctional compound and potassium hydroxide has the greatest efficiency in reducing smoke formation of wood in the ratio of the main components of 2:18, respectively. For flame retardants containing a mixture of components, polyfunctional compound, and sodium carbonate or calcium carbonate, the greatest effect in reducing the smoke-forming ability of wood is observed in the presence of only mild alkalis in the aqueous solution, i.e., in the ratio of components 0:20, respectively (Fig. 6.20).

Reduction of smoke-forming ability of wood with compositions containing polyfunctional compound and alkaline catalyst is due primarily to the fact that on the surface of the material are formed spatially networked structures resistant to high temperature exposure, as well as the presence in the formulation of the alkali, which are effective depressants of smoke production during combustion of wood materials.

According to test results, it was found that the composition containing polyfunctional compound and sodium hydroxide has a greater efficiency in reducing the smoke formation of wood compared to other flame retardants. The highest efficiency of the composition is observed when the ratio of the components is 5: 15, respectively.

Fire retardant composition containing polyfunctional compound and potassium hydroxide has the greatest efficiency in reducing smoke formation of wood in the ratio of the main components of 2:18, respectively.

For flame retardants containing a mixture of components, polyfunctional compound, and sodium carbonate or calcium carbonate, the greatest effect in reducing the smoke-forming ability of wood is observed in the presence of only mild alkalis in the aqueous solution, i.e., in the ratio of components 0:20, respectively (Fig. 6.20).

The decrease in the smoke-forming ability of wood with compositions containing polyfunctional compound and alkaline catalyst is primarily due to the fact that spatially networked structures are formed on the material surface resistant to high temperature exposure, as well as the presence in the formulation of the composition of alkali, which are effective depressants of smoke release during combustion of wood materials.

The effect of the considered flame retardants on the toxicity of wood combustion products was determined using the gas-analytical approach by the CO and CO_2 yield. As the analysis of the effect of impregnating compositions on the toxicity of wood combustion products, retarders of combustion, shifts the wood smoldering mode to the area of higher temperatures.

Figure 6.21 shows the dependence of carbon monoxide yield on temperature when testing wood with different flame retardants. From Fig. 6.21 shows that when using flame retardants containing polyfunctional compound and alkaline catalyst with a flow rate of 400 g/m^2 at test temperatures up to 550 °C, the yield of carbon monoxide is less than the yield of the same toxicant when testing the native wood. This is explained by the fact that this composition at temperatures up to 550 °C protects the wood from thermal decomposition and prevents the release of carbon monoxide.

At temperatures above 500 °C, there is spontaneous ignition of untreated wood and a significant decrease in the yield of carbon monoxide, which is due to the active conversion of carbon monoxide to carbon dioxide during the onset of flame wood combustion. Thus, at this temperature site, the carbon monoxide yield for antipyrated wood exceeds the carbon monoxide yield for native wood. The maximum value of carbon monoxide yield for wood with flame retardants is not realized, since heat fluxes greater than 35 kW/m^2 are required for its self-ignition.

When using flame retardants containing polyfunctional compound and alkaline catalyst, with a flow rate of 400 g/m^2 over the entire temperature range, the carbon

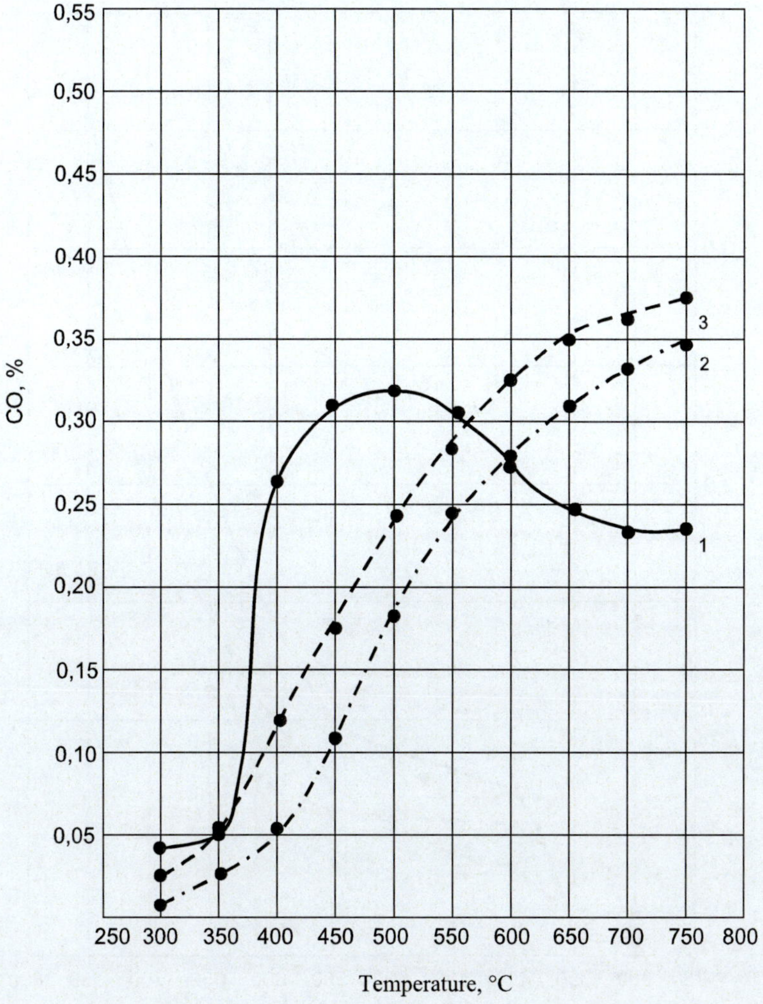

Fig. 6.21 Dependence of carbon monoxide yield on temperature when testing untreated wood and wood with flame retardants: 1—natural wood; 2—wood with flame retardant "polyfunctional compound—sodium hydroxide" with a consumption of 400 g/m^2; and 3—wood with flame retardant "polyfunctional compound—potassium hydroxide" with a consumption of 400 g/m^2

dioxide yield is less than the yield of the same toxicant when testing native wood (Fig. 6.22).

As the results of fire tests show, the values of carbon dioxide yield indicators differ slightly, and the use of these flame retardants reduces the yield of carbon dioxide compared to untreated wood. At the same time, a smaller yield of carbon dioxide provides the composition containing polyfunctional compound and sodium hydroxide in a ratio of 5:15, respectively. For all the samples studied, it was found that the wood self-ignition intensifies the suppression of carbon monoxide and, as a consequence, increases the yield of carbon dioxide.

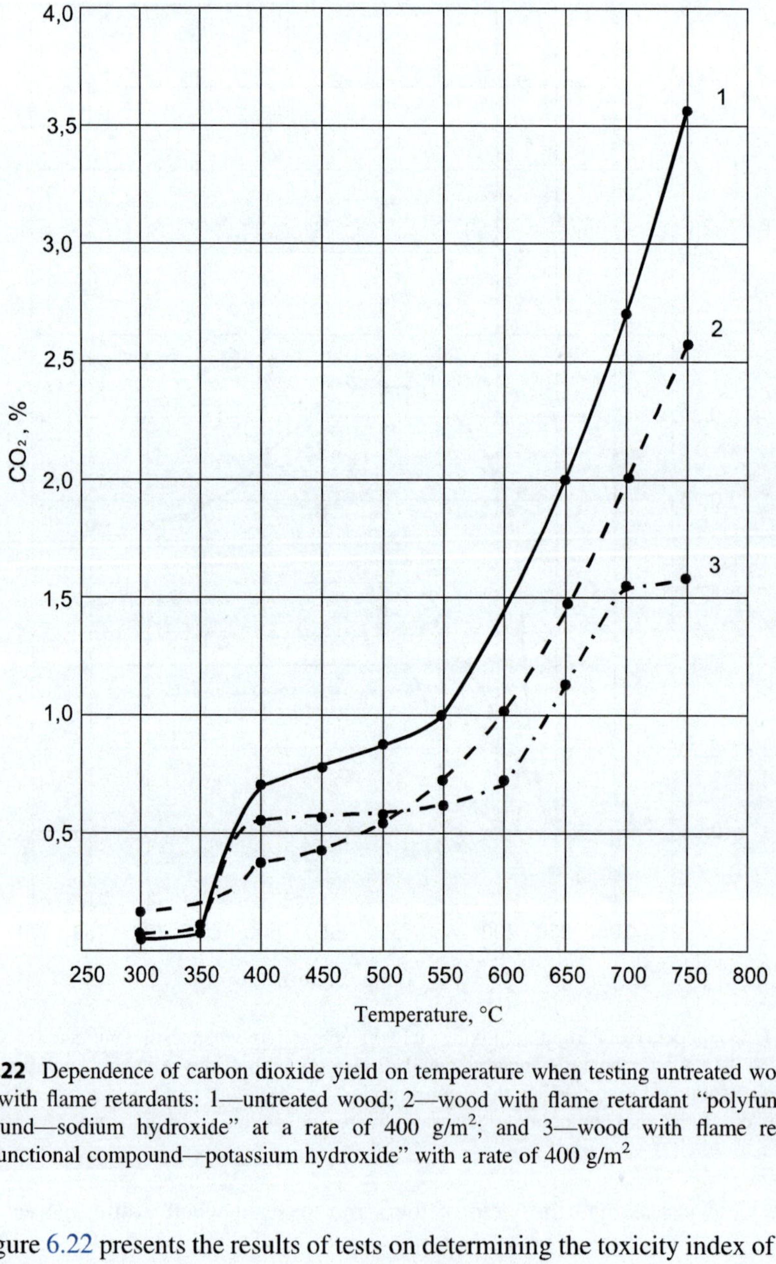

Fig. 6.22 Dependence of carbon dioxide yield on temperature when testing untreated wood and wood with flame retardants: 1—untreated wood; 2—wood with flame retardant "polyfunctional compound—sodium hydroxide" at a rate of 400 g/m^2; and 3—wood with flame retardant "polyfunctional compound—potassium hydroxide" with a rate of 400 g/m^2

Figure 6.22 presents the results of tests on determining the toxicity index of wood combustion products on carbon monoxide yield with flame retardants "polyfunctional compound—sodium hydroxide" and "polyfunctional compound—potassium hydroxide" with a consumption of 400 g/m^2. From Fig. 6.22, it can be seen that for untreated wood the smoldering mode is realized at 500 °C (index of toxicity of combustion products equals 23.8 g/m^3).

The wood treated with flame retardants—polyfunctional compound—sodium hydroxide and polyfunctional compound—potassium hydroxide at the rate of 400 g/m^2, the smoldering regime is realized at temperature of more than 750 °C (index of toxicity of combustion products is 13.5 and 18.5 g/m^3, respectively). Analysis of the effect of flame retardants on the toxicity of combustion products showed that the compositions "sodium hydroxide and polyfunctional compound" and the composition "polyfunctional compound—potassium hydroxide" reduce the toxicity of combustion products, i.e., increase the toxicity of combustion products compared to wood by increasing the yield of carbon monoxide. Despite this, wood and wood with flame retardants are in the same group of toxicity of combustion products and are classified as highly hazardous materials (toxicity index of combustion products ranging from 13 to 40 g/m^3).

The results of fire tests show that for wood treated with flame retardants "polyfunctional compound—sodium carbonate" and "polyfunctional compound—calcium carbonate" at a rate of 400 g/m^2, the smoldering mode is realized at temperatures of 500 and 550 °C, with a toxicity index of 22 and 18 g/m^3, respectively. As can be seen, the toxicity values of the products of combustion of wood and wood with flame retardants differ slightly. Thus, the test results showed that the tested fire-retardant compositions increase the toxicity of combustion products of wood, but do not change its group of toxicity of combustion products—"highly hazardous materials."

In general, the analysis of the effect of impregnating compositions on the toxicity of combustion products of wood showed that these retardants of combustion shift the wood smoldering mode to higher temperatures, increasing the yield of toxic combustion products (Fig. 6.24). The highest yield of carbon monoxide is observed in the area of flameless combustion. So untreated pine has the highest value of CO yield 0.540% at the corresponding mass of the sample at heat flux density of 23.0 kW/m^2. When using the proposed fire-retardant compositions, smoldering mode of pine (the area of maximum yield of carbon monoxide) shifts to the area of more intense heat fluxes (higher temperatures). Compared with untreated pine in the smoldering mode, the yield of carbon monoxide increases. The highest yield of carbon monoxide—0.675% is observed for pine with the composition containing sodium hydroxide and polyfunctional compound in the smoldering mode at a thermal flow of 52.5 kW/m^2 at a flow rate of 400 g/m^2. This, first of all, indicates intensification of the smoldering process and significant expansion of the temperature range of wood smoldering mode towards high q_e. As a result of flame combustion, the intensity of which increases with increasing q_e, the yield of carbon monoxide decreases (Fig. 6.23).

Figure 6.24 shows that impregnation of wood with the proposed compositions at heat flux density up to and including 32.5 kW/m^2 contributes to the reduction of carbon monoxide compared to untreated pine.

Quite often researchers, when studying the effect of combustion retarders on the course of the process of thermal destruction of cellulose materials, operate with such indicator as CO/CO_2 ratio. It is obvious that preference should be given to those

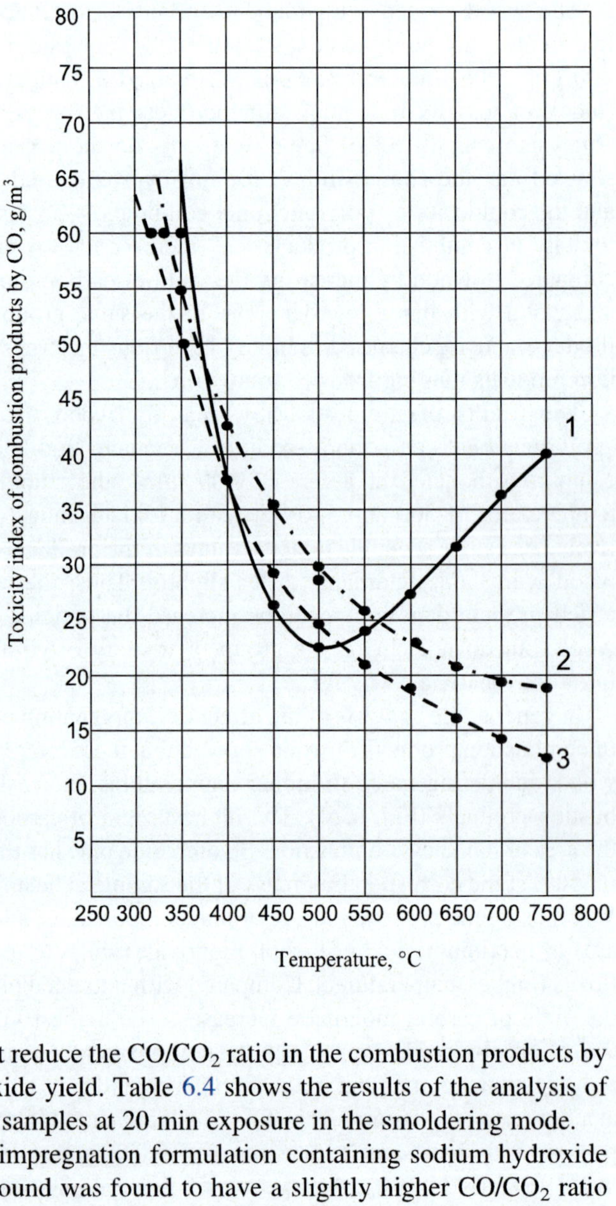

Fig. 6.23 The dependence of the index of toxicity of products of combustion by carbon monoxide on the temperature when testing untreated wood and wood with flame retardants: 1—untreated wood; 2—wood with the flame retardant "polyfunctional compound—potassium hydroxide" with a flow of 400 g/m^2; and 3—wood with the flame-retardant "polyfunctional compound—sodium hydroxide" with a flow of 400 g/m^2

retarders of combustion that reduce the CO/CO_2 ratio in the combustion products by reducing the carbon monoxide yield. Table 6.4 shows the results of the analysis of CO/CO_2 ratio of the tested samples at 20 min exposure in the smoldering mode.

The wood sample with impregnation formulation containing sodium hydroxide and a polyfunctional compound was found to have a slightly higher CO/CO_2 ratio than the wood with impregnation formulation containing ammonium phosphate and a polyfunctional compound. There is a general tendency for the relative value of CO/CO_2 to increase with increasing heat flux density applied to the surface of the sample. This is due, first of all, to intensification of the process of carbonization of wood.

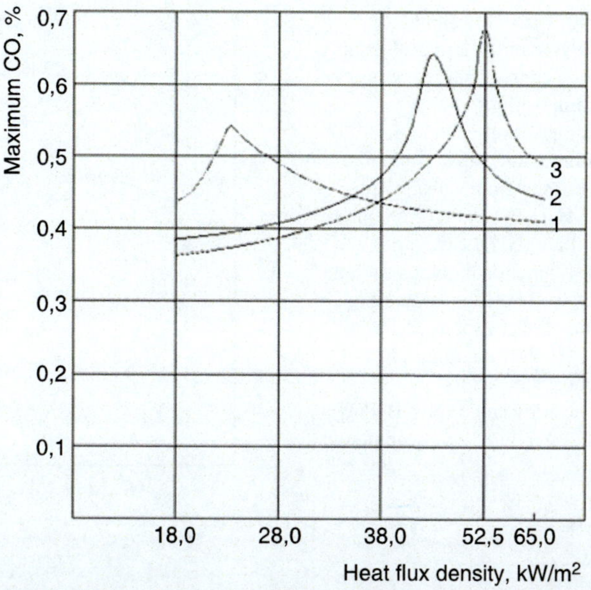

Fig. 6.24 The dependence of the maximum yield of carbon monoxide for the exposure time of 20 min on the heat flux density in the tests: 1—native pine; 2—pine with the composition containing ammonium phosphate divalent and polyfunctional compound (consumption 400 g/m^2); and 3—pine with the composition containing sodium hydroxide—polyfunctional compound (consumption 400 g/m^2)

Table 6.4 Analysis of CO/CO_2 index of untreated wood and wood with impregnation compositions in the mode of flameless combustion

Sample name	q_e, kW/m^2	CO	CO_2	CO/ CO_2
Pine	23.0	0.54	0.75	0.72
Pine with a formulation containing polyfunctional compound and ammonium phosphoric acid divalent	44.0	0.63	0.85	0.74
Pine with composition containing polyfunctional compound sodium hydroxide	52.5	0.68	0.89	0.77

6.3 Study of the Effectiveness of Fire-Retardant Compositions for Long-Term Natural Aging Wood

To establish the effectiveness of fire-retardant impregnating compositions for wood of long natural aging, we conducted studies of fire-retardant compositions containing a complex of substances and compounds, including the traditional phosphorus-nitrogen-containing flame retardants: $NH_4H_2PO_4$, $(NH_4)_2SO_4$, $C_2H_7PO_3$, and polyfunctional compounds.

For the primary assessment of the effectiveness and mechanism of flame-retardant effect, a standard method for assessing the flame-retardant effectiveness of compositions and substances for wood according to GOST 53292 (Russia) was used.

According to the results of fire tests, it was found that the proposed flame-retardant impregnation compositions with a surface application rate of not less than 400 g/m^2 provide Group I fire protection efficiency (weight loss less than 9.0%) (Fig. 6.25).

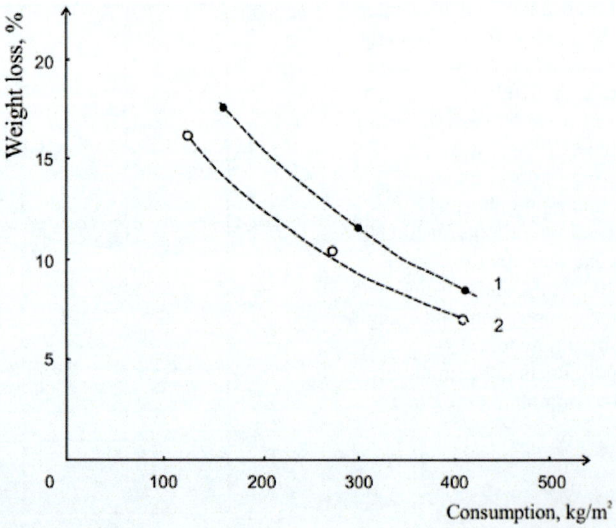

Fig. 6.25 Results of assessment of fire protection efficiency of impregnating compositions: 1—composition containing ammonium salts (ammonium phosphate one-substituted ($NH_4H_2PO_4$) and (ammonium sulfate $(NH_4)_2SO_4$) and polyfunctional compound and 2—composition containing dimethylphosphite ($C_2H_7PO_3$), ammonium phosphate acid one-substituted ($NH_4H_2PO_4$), and polyfunctional compound

It should be noted that in the process of fire test visually on the surface of the fencing, wooden structures with fire protection observed characteristic effects accompanying the process of their charring. These effects largely characterize the mechanism of flame-retardant action of the developed flame retardants. In the case of using a composition containing ammonium salts (ammonium phosphate one-substituted ($NH_4H_2PO_4$) and (ammonium sulfate $(NH_4)_2SO_4$) and a polyfunctional compound, characteristic foam bubbles, which retained their integrity until the completion of the fire test, could be distinguished on the surface of the carbon layer during the experiment (Fig. 6.26).

For the composition containing dimethylphosphite ($C_2H_7PO_3$), ammonium phosphate one-substituted ($NH_4H_2PO_4$) and polyfunctional compound during the test, it was possible to visually fix the formation of a protective glassy layer on the surface of the sample, which also locally preserved to the end of the experiment (Fig. 6.27).

Thermal analysis methods make it possible to efficiently and accurately evaluate the behavior of various materials under high-temperature heating, to identify the individual stages of thermal degradation, and to identify their combustion mechanisms, using microsamples (grains).

It was important to find out the possibility of assessing the effectiveness of various flame retardants for long-term natural aging wood using thermal analysis methods (thermogravimetry (TG), differential thermogravimetry (DTG), differential scanning calorimetry (DSC), as well as establishing key parameters of the thermodestruction process of the studied samples, allowing a predictive assessment of the fire hazard of wooden structures with fire protection.

Samples for the study were microcuts, which were selected by cutting with a special cutter a thin surface layer of wood treated with fire-retardant impregnation compositions at a rate of not less than 400 g/m^2. The research was carried out on a Du Pont 9900 thermoanalyzer using a TGA-951 thermobalance. Conditions for thermoanalytical experiments: heating rate 20 °C/min; atmosphere—nitrogen up to 500 °C then air; gas flow rate—50 ml/min; and weight of the sample 4–7 mg.

Fig. 6.26 General view of a wood sample with the composition containing ammonium salts (ammonium phosphate one-substituted ($NH_4H_2PO_4$) and (ammonium sulfate $(NH_4)_2SO_4$) and polyfunctional compound after fire tests according to GOST 53292 (Russia)

Conditions of thermal analysis of the objects under study were chosen so as to reveal changes in the course of thermal analysis dependences as much as possible and to determine by characteristic points the corresponding regularities of the main processes of thermal destruction in different environments.

For the analysis of DSC curves of the studied samples, a high-precision thermal analyzer of Q600 type was used. The experimental conditions were as follows: heating rate 5 °C/min; atmosphere—nitrogen up to 500 °C; and then air.

Thermal analysis of the samples under study in a nitrogen atmosphere was carried out with a subsequent change of the medium to air at 500 °C at a rate of 20 °C/min. The obtained TG and DTG curves are shown in Fig. 6.28.

Thermogravimetric analysis under the conducted conditions showed that the samples of materials have three pronounced intervals of degradation:

in the first interval (30–150 °C), the moisture yield occurs in the range of 3.5–5.79% by mass;

the second interval (150–450 °C) actively destroys the actual wood of pine, with a loss of mass of about 60–70%, while treatment with the composition based on the system $NH_4H_2PO_4$–$(NH_4)_2SO_4$ and composition based on dimethylphosphite ($C_2H_7PO_3$) shifted the maximum decomposition reaction to lower temperatures, including the amplitude (from 24.9%/min to 23.6 and 5.7%, respectively).

Fig. 6.27 General view of a wood sample with the composition containing dimethylphosphite ($C_2H_7PO_3$), single-substituted ammonium phosphate ($NH_4H_2PO_4$), and polyfunctional compound

At the third stage, after changing the nitrogen atmosphere to air, coke oxidation occurs; the rate of coke oxidation when using flame-retardant compositions decreases from 28.6 to 20.6 and 4.72%/min for compositions based on the system $NH_4H_2PO_4$–$(NH_4)_2SO_4$ and the composition based on dimethylphosphite ($C_2H_7PO_3$), respectively.

It attracts attention that the flame-retardant composition on the basis of dimethylphosphite ($C_2H_7PO_3$) actively modifying the components of wood composite significantly changed the basic characteristics of the thermal destruction process (Table 6.5).

The stage in the temperature range 150–450 °C underwent the greatest changes. A significant shift of the DTG maximum up to 283 °C is observed, as well as a noticeable decrease in sample mass loss from 70.05 to 38.5%.

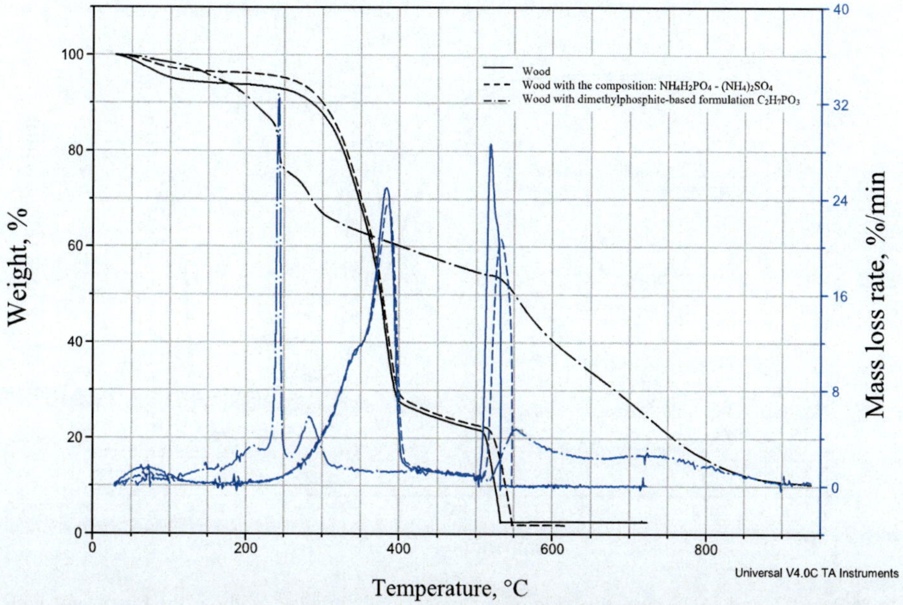

Fig. 6.28 TGA and DTG curves of modern pine wood samples without treatment and with treatment with the composition based on the system: $NH_4H_2PO_4$–$(NH_4)_2SO_4$ and the composition based on dimethylphosphite ($C_2H_7PO_3$) (20 °C/min, atmosphere—nitrogen to 500 °C then air)

Table 6.5 Results of thermal analysis of modern wood with flame retardants

Characteristics	Specimen name: Pine native	Pine with composition based on system: $NH_4H_2PO_4$–$(NH_4)_2SO_4$	Pine with composition based on dimethylphosphite ($C_2H_7PO_3$)
In the range of 30–150 °C			
Weight loss, %	5.79	3.5	4.18
In the range of 150–450 °C			
Weight loss, %	70.05	71.6	38.5
Maximum DTG T, °C/A, % $\cdot min^{-1}$	383/24.9	385/23.6	283/5.7
In the range of 450–600 °C			
Speed of coal oxidation, % $\cdot min^{-1}$	28.6	20.6	4.72

If we analyze the TG and DTG curves, we can testify that the high weight loss of the wood sample with the composition based on dimethylphosphite ($C_2H_7PO_3$) is observed at a temperature of about 215 °C, due to the active course of dehydration and carbonization of the wood material. It can be concluded that this composition is characterized by the classical mechanism of "catalytic dehydration." Of great interest is the effect of fire-retardant impregnating compositions on the peculiarities of thermal decomposition of long-term natural aging wood.

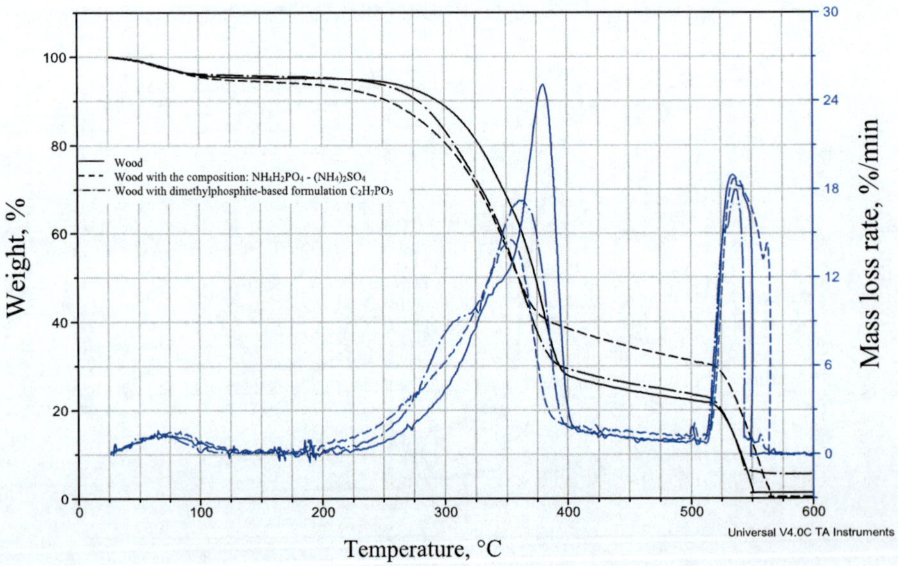

Fig. 6.29 TG and DTG curves of long-term pine wood samples (without treatment and with treatment with the composition based on the system: $NH_4H_2PO_4$–$(NH_4)_2SO_4$ and with the composition based on dimethylphosphite ($C_2H_7PO_3$) (20 °C/min, atmosphere—nitrogen up to 500 °C then air))

Figure 6.29 shows TG and DTG curves of pine wood samples of long-term service life (81 years of operation) without treatment and with treatment with composition based on the system $NH_4H_2PO_4$–$(NH_4)_2SO_4$ and composition based on dimethylphosphite ($C_2H_7PO_3$).

Analysis of TG and DTG curves of long-term natural aging wood shows the presence of three pronounced degradation intervals, as in the case of modern wood samples. Temperature stages have similar intervals. Visually by the change of TG and DTG curves, we can conclude that the composition based on the system $NH_4H_2PO_4$–$(NH_4)_2SO_4$, given the classical ideas about the effectiveness and mechanism of fire-retardant action of flame retardants, is more effective compared to the composition based on dimethylphosphite ($C_2H_7PO_3$). Earlier start of the carbonization process results in the temperature range of 150–450 °C to the lowest mass loss (60%). The main thermal characteristics are presented in Table 6.6.

Taking into account the peculiarities of behavior of long-term wooden structures which are characterized by a high degree of thermal damage and significant heat release during combustion, it is important to reduce not only the mass loss by shifting the process of coal formation to the low-temperature area but also the properties and structure of the formed carbon, its oxidation, and calorific capacity. In this regard, the use of a composition based on dimethylphosphite ($C_2H_7PO_3$) for long-term natural aging wood is more preferable in terms of its directional ability to reduce the oxidation rate of carbon residue (reduction of oxidation rate from 18.97 to 17.97%/min).

Table 6.6 Results of thermal analysis of long-lived wood (81 years) with flame retardants

Characteristics	Specimen name Pine (81 year of use)	Pine (81 year of use) with composition on the basis of system: $NH_4H_2PO_4$–$(NH_4)_2SO_4$	Pine (81 year of use) with composition based on dimethylphosphite ($C_2H_7PO_3$)
In the range of 30–150 °C			
Weight loss, %	4.78	5.6	4.28
In the range of 150–450 °C			
Weight loss, %	70.85	60.0	69.7
Maximum DTG T, °C/A, %·min^{-1}	380/25.1	353/14.5	362/17.2
In the range of 450–600 °C			
Speed Of coal oxidation, %·min^{-1}	18.97	18.74	17.97

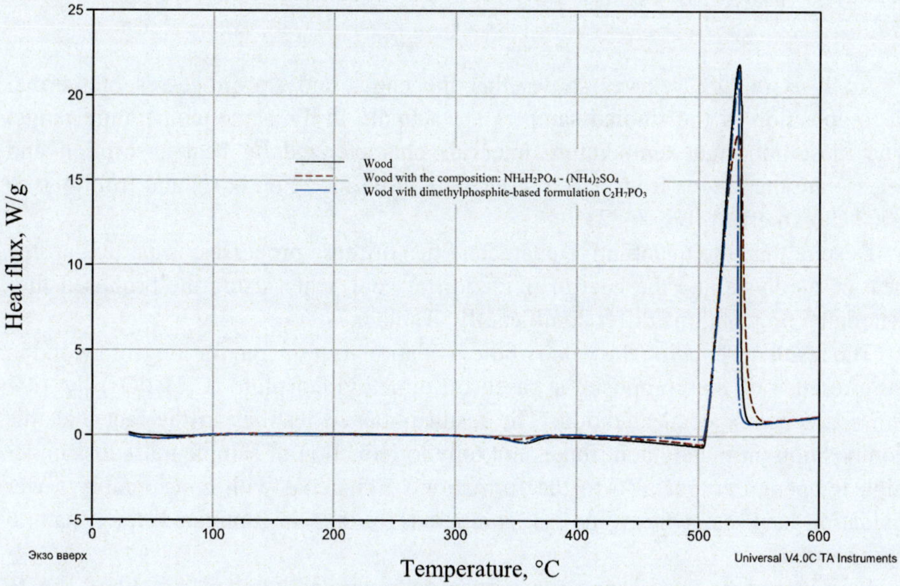

Fig. 6.30 DSC curves of long-life pine wood samples (without treatment and with treatment with the composition based on the system: $NH_4H_2PO_4$–$(NH_4)_2SO_4$ and with the composition based on dimethylphosphite ($C_2H_7PO_3$) (5 °C/min, atmosphere—nitrogen up to 500 °C then air))

In order to establish the possibility of reducing the thermal oxidation of the formed coal in the presence of flame, retardants were taken DSC curves using thermal analyzer "Q600." DSC curves of the studied samples are shown in Fig. 6.30.

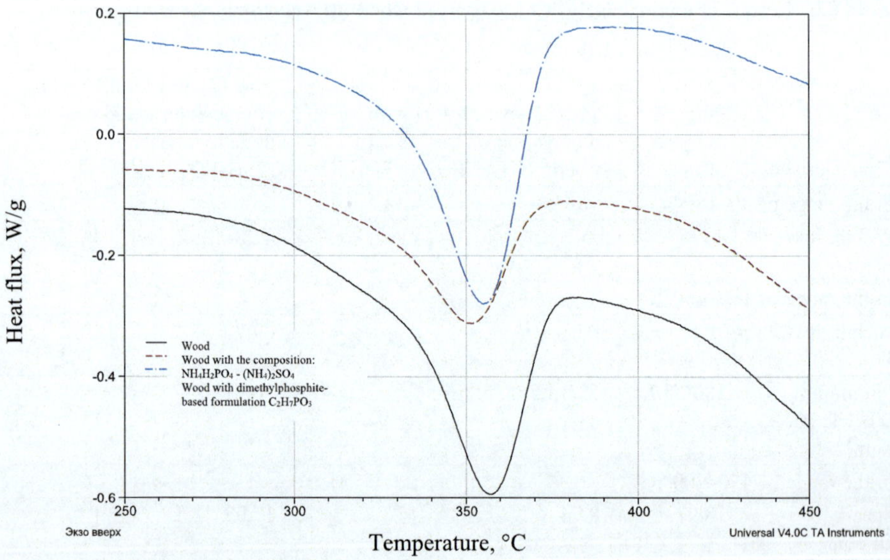

Fig. 6.31 DSC curves of long-life pine wood samples (without treatment and with treatment with the composition based on the system: $NH_4H_2PO_4$–$(NH_4)_2SO_4$ and with the composition based on dimethylphosphite ($C_2H_7PO_3$) in the interval 250–450 °C (atmosphere—nitrogen))

Analysis of DSC curves shows that the endo- and exo-processes of thermal decomposition of the studied samples are actually in the same temperature range. The most important temperature intervals characterized by heat absorption and release are the ranges from 250 to 400 °C (endo effect, Fig. 6.31) and from 480 to 600 °C (exo effect, Fig. 6.32).

Despite this, the quantitative characteristics of these processes, in particular, the heat of pyrolysis and the heat of oxidation of coal, when using the proposed fire-retardant compositions differ significantly (Table 6.7).

The result of reducing the value of the oxidation heat of coal for long-life wood in the presence of the composition based on dimethylphosphite ($C_2H_7PO_3$) by 1.27 times deserves special attention. The result obtained testifies to the fact that this composition ultimately contributes not only to reduction of sample mass loss under high temperatures but also to the formation of charcoal with considerably lower oxidation and calorific value as compared to that of untreated long natural aging wood.

It is important that during the longer operation of wooden structures (up to 200 years of operation), this composition retains its effectiveness in reducing the oxidation rate of the formed charcoal residue (Fig. 6.33).

Thermogravimetric analysis of 200 years old pine wood has shown that the composition based on the system $NH_4H_2PO_4$–$(NH_4)_2SO_4$ and the composition based on dimethylphosphite ($C_2H_7PO_3$) actively modify the wood components.

The compositions contribute to acceleration of carbonization processes of wood material, with maximum DTG shifts to lower temperatures from 353 °C for long

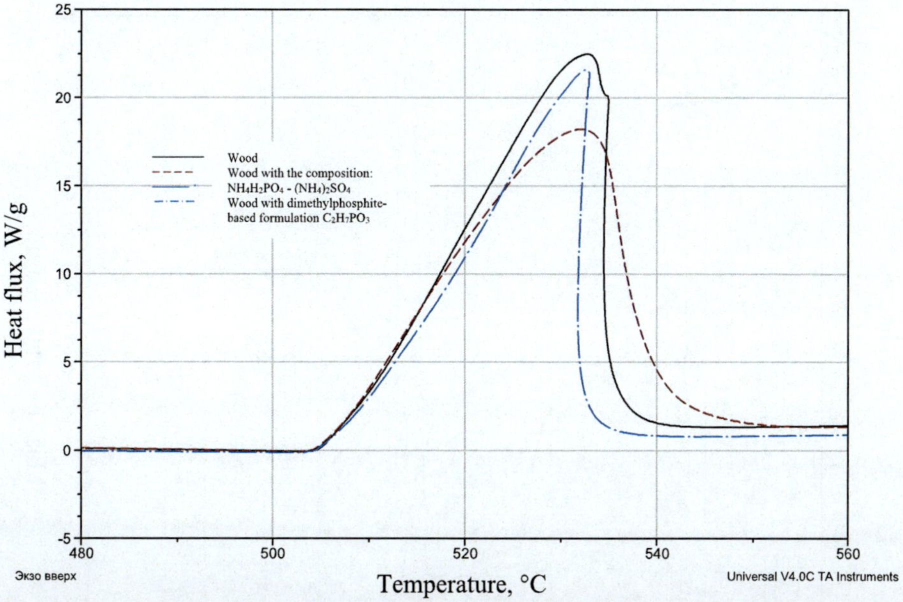

Fig. 6.32 DSC curves of long-life pine wood samples (without treatment and with treatment with the composition based on the system: $NH_4H_2PO_4$–$(NH_4)_2SO_4$ and with the composition based on dimethylphosphite ($C_2H_7PO_3$) in the interval 480–560 °C (atmosphere—air))

Table 6.7 Results of differential-scanning calorimetry of long-term wood samples with flame retardants

Characteristics DSC	Sample name		
	Pine (81 year of use)	Pine (81 year of use) with composition on the basis of system: $NH_4H_2PO_4$–$(NH_4)_2SO_4$	Pine (81 year of use) with composition based on dimethylphosphite ($C_2H_7PO_3$)
Heat of pyrolysis, J/g (nitrogen)	−164.6	−102.5	−170.3
Heat of coal oxidation, J/g (air)	+4315	+4442	+3374

natural aging wood without fire protection to 309 and 278 °C for the same wood with fire protection composition based on the system $NH_4H_2PO_4$–$(NH_4)_2SO_4$ and composition based on dimethylphosphite ($C_2H_7PO_3$), respectively (Table 6.8).

At the same time, the appearance of additional DTG peaks in the interval 150–450 °C and a decrease in the rate of degradation are observed. In addition, the temperature range of oxidation expanded towards higher temperatures. Modification of the wood structure with the used compositions led to a sharp decrease in the coal oxidation rate from 27.5 to 19.2 (composition based on the system $NH_4H_2PO_4$–$(NH_4)_2SO_4$ and 4.7%/min (composition based on dimethylphosphite ($C_2H_7PO_3$)).

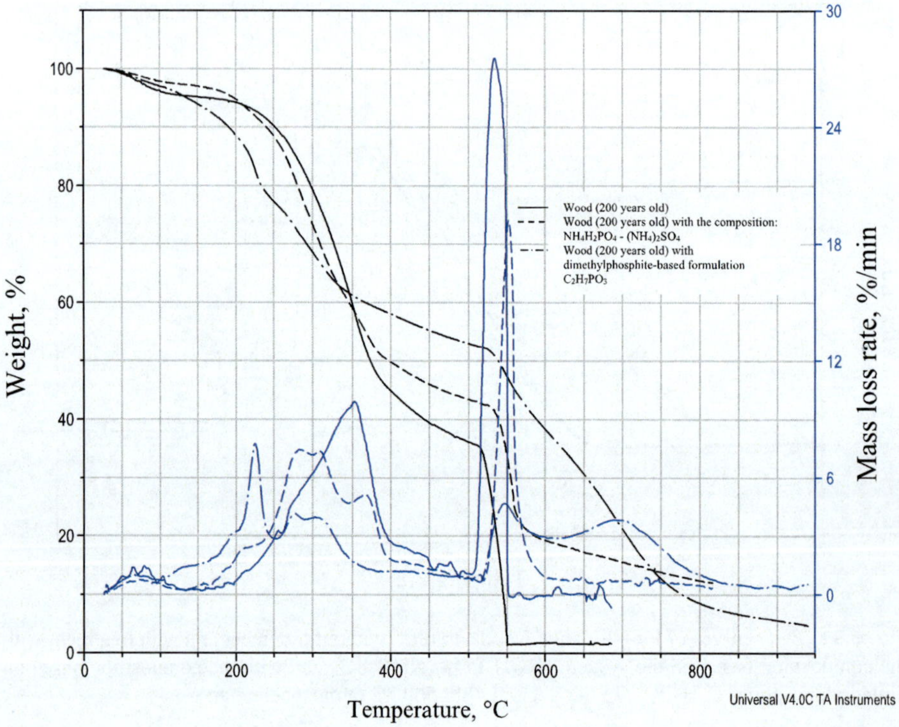

Fig. 6.33 TG and DTG curves of wood samples with a lifetime of 200 years (without treatment and with treatment with the composition based on the system: $NH_4H_2PO_4$–$(NH_4)_2SO_4$ and with the composition based on dimethylphosphite ($C_2H_7PO_3$) (20 °C/min, atmosphere—nitrogen up to 500 °C then air))

Changing conditions and duration of operation of wooden structures, as well as the use of various compositions and coatings for fire protection of wood, affects the properties of the resulting charcoal residue. The results of studies of the properties and structure of coal layers formed during thermal decomposition and combustion of wood largely explain the differences in the values of charring parameters and thermophysical characteristics of the samples under study. To study the charcoal structures of wood samples by vapor sorption methods, we took samples of modern and artificially aged wood, as well as charcoal samples from the surface layer of wood with different flame-retardant compositions (composition based on potassium oxide hydrate (KOH) and a polyfunctional compound in the ratio of components 18: 2 in aqueous solution, composition based on the system $NH_4H_2PO_4$–$(NH_4)_2SO_4$, composition based on dimethylphosphite ($C_2H_7PO_3$), and BB-11 (boric acid—10%, sodium tetraborate—10%, water—the rest).

For research by method of vapor sorption, the following conditions were observed. The adsorbent was a fine carbon powder (0.3–0.5 mm) from the surface carbon layer of the studied samples. Benzene was chosen as the adsorbate. The choice of this substance is explained by the fact that it has a high surface tension and can wet the surface of wood capillaries. Obtained charcoal adsorption curves for natural wood and wood treated with flame retardants (at $t = 20$ °C) are shown in Fig. 6.34.

Table 6.8 Results of thermal analysis of long-term wood (200 years) with flame retardants

Characteristics	Sample name		
	Pine (200 years in service)	Pine (200 years of use) with composition based on system $NH_4H_2PO_4$–$(NH_4)_2SO_4$	Pine (200 years of use) with composition based on dimethylphosphite ($C_2H_7PO_3$)
In the range of 30–150 °C			
Weight loss, %	4.84	3.0	6.0
In the range of 150–450 °C			
Weight loss, %	55.73	51.0	38.7
Maximum DTG T, °C/A, %·min^{-1}	353/9.9	309/7.3	278/4.1
In the range of 450–600 °C			
Speed of coal oxidation, % ·min^{-1}	27.5	19.2	4.7

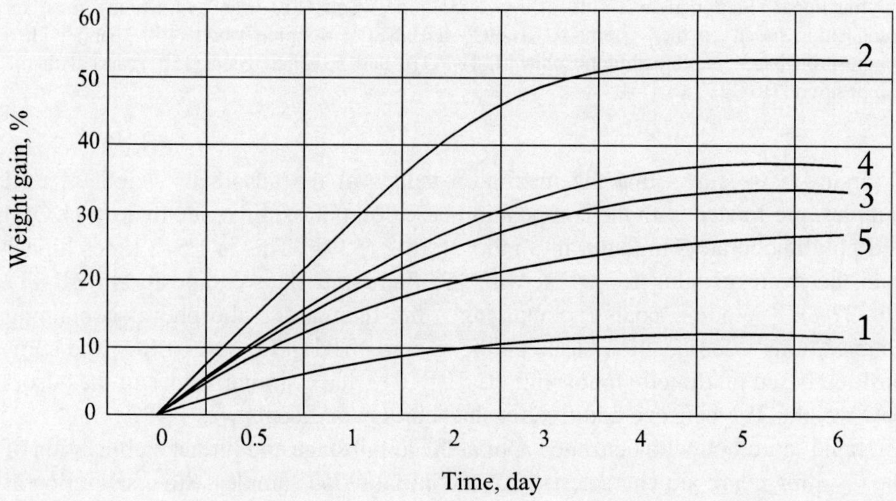

Fig. 6.34 Kinetic curves of benzene vapor sorption at temperature $t = 20$ °C on the surface of pine timber after testing in the presence of flame-retardant systems: 1—pine wood; 2—pine wood with a composition based on potassium oxide hydrate (KOH) and a polyfunctional compound in a ratio of 18:2; 3—pine wood with a fire retardant composition based on the system: $NH_4H_2PO_4$–$(NH_4)_2SO_4$; 4—pine wood with a composition based on dimethylphosphite ($C_2H_7PO_3$); and 5—pine wood with the composition "BB-11"

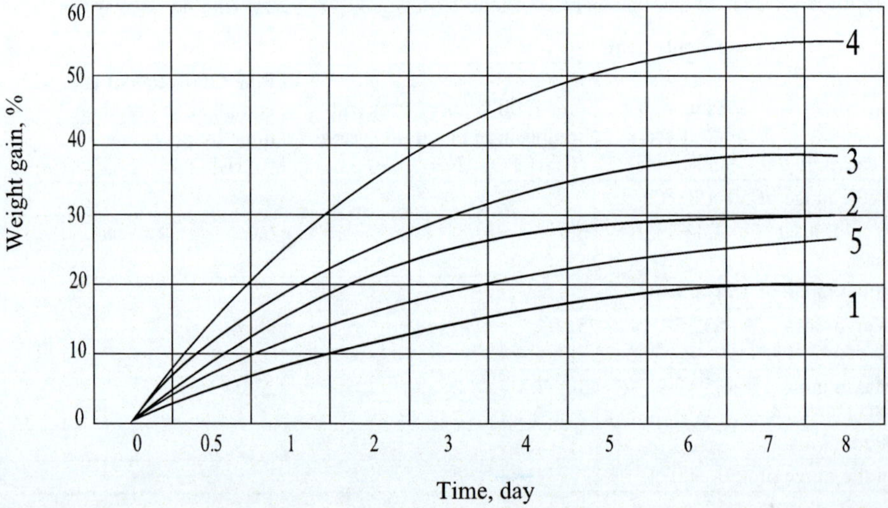

Fig. 6.35 Kinetic curves of benzene vapor sorption at temperature $t = 20$ °C on the surface of pine wood (150 years) after testing in the presence of flame retardants: 1—pine wood (150 years); 2—pine wood (150 years) with the composition based on potassium oxide hydrate (KOH) and polyfunctional compound in a ratio of 18: 2; 3—pine wood (150 years) with a fire retardant composition based on the system: $NH_4H_2PO_4$–$(NH_4)_2SO_4$; 4—pine wood (150 years) with a composition based on dimethylphosphite ($C_2H_7PO_3$); and 5—pine wood (150 years) with the composition "BB-11"

Figure 6.34 shows that the maximum values of the adsorbate weight of coal samples, pre-treated with the composition based on potassium oxide hydrate (KOH) and polyfunctional compound in a ratio of 18:2 (53%). This is 1.5–2 times higher than the average value for wood with the flame-retardant composition "BB-11" (26–37%). Wood coals containing fire-retardant phosphorus-containing compositions—composition based on the system $NH_4H_2PO_4$–$(NH_4)_2SO_4$ and composition based on dimethylphosphite ($C_2H_7PO_3$)—have similar values of the adsorbate weight. The benzene additive for the source wood coals was 12.8%.

Rapid saturation with benzene vapor at the initial stage and further stabilization of the resulting curve are characteristic of the initial wood samples. Slow saturation at the initial stage (1 h), further intensification of the process, and uniform saturation are characteristic of modified charcoal samples. Complete saturation of investigated samples by benzene vapors was observed in 6 days.

Figure 6.35 shows kinetic curves of benzene vapor sorption on the surface of coals of artificially aged pine wood (150 years old) with the selected flame retardants (at temperature $t = 20$ °C).

The test results show that the benzene vapor absorption period for artificially aged wood is longer in time compared to modern wood. This may indicate a more developed structure of the carbon layer for long-lived wood.

The results of research by the method of vapor sorption, as well as the calculated characteristics of the sorption parameters of charcoal, confirm the abovementioned assumption about the possible decrease in the efficiency of applied fire-retardant compositions and coatings for wood materials and products with a long service life. According to the results of the study, the characteristics of the porous structure of coals were calculated: the total pore volume ($V_{\sum}$), the average pore radius (d_{av}), and the specific surface of coals (S_{sp}).

The total pore volume was determined by the formula

$$V_{\sum} = \frac{\Delta g}{g\rho_s}, (\mathrm{sm/g}), \tag{6.3}$$

where Δg—the adsorbent gain due to absorption of benzene, g; g—the adsorbent weight, g; and ρ_s—the density of benzene at 20 °C (0.876 g/sm^3).

Specific surface area was estimated by the formula

$$S_{sp} = a_m N_A \omega_m 10^{-20}, \mathrm{m}^2/\mathrm{g} \tag{6.4}$$

where a_m—the weight gain, mol/g; N_A—the Avogadro number (6.02214179 (30) × 1023 mol^{-1}); and ω_m—the area of the area occupied by one molecule (49 A^2).

The average pore radius was estimated by the formula

$$d_{av} = 2 \times 10^4 \frac{V_{\sum}}{\rho}. \tag{6.5}$$

Table 6.9 shows the results of the study of the porous structure of coke formed during the fire combustion of pine wood with different flame-retardant treatment with the following characteristics: pore diameter (d_{av}), total pore volume ($V_{\sum}$), specific surface (S_{sp}), as well as the absorption capacity of carbon residues (a_m).

Fireproofing of modern and artificially aged wood significantly changes the porous structure of the carbon layer formed during combustion on the surface of the material. First of all, this manifests itself in an increase in their absorption capacity, a decrease in the average diameter of the pores, and, accordingly, an increase in the specific surface. The degree of change in the indicated parameters of the porous structure depends on the type of fire-proofing treatment of wood, as well as on its artificial aging.

Especially revealing is the comparison of data on two fire-retardant systems providing Group I efficiency according to GOST R 53292 (Russia standard): the composition based on dimethylphosphite ($C_2H_7PO_3$) and the composition containing potassium oxide hydrate and a polyfunctional compound. Artificially aged wood with fire-retardant composition on the basis of dimethylphosphite ($C_2H_7PO_3$) forms during burning a coke, which in comparison with the coke of treated modern natural wood has increased its absorptivity from 18.48% to 26.76%, decreased by 21% in pore diameter, and increased the specific surface by 58.5%.

Fire-retardant impregnation of modern wood with a composition containing potassium oxide hydrate and a polyfunctional compound led to the formation of coke with the smallest pore diameter (12.80 nm) and a developed specific surface.

Table 6.9 Effect of flame-retardant treatment of wood on the characteristics of charcoal layers

Sample name	Adsorbate—benzene			
	a_m, %	V_Σ, sm^3/g	d_{av}, nm	$S_{уд}$, m^2/g
Modern pine	12.8	0.4726	25.03348	377.5744
[a]Pine, 150 years old	16.2	0.5332	21.56388	814.7575
Pine modern with the fireproofing composition on the basis of the system: $NH_4H_2PO_4$–$(NH_4)_2SO_4$	17.95	0.5123	19.76785	935.3324
Pine modern with composition with dimethylphosphite ($C_2H_7PO_3$)	18.48	0.6812	18.51378	1038.637
Pine modern with composition based on potassium oxide hydrate (KOH) and polyfunctional compound	21.81	0.6327	12.80586	1554.857
Pine modern with composition BB-11	16.96	0.7088	20.17486	882.7961
Pine (150 years) with fire retardant composition based on the system: $NH_4H_2PO_4$–$(NH_4)_2SO_4$	20.65	0.5781	16.11560	1434.879
Pine (150 years) with dimethylphosphite-based composition ($C_2H_7PO_3$)	26.76	0.5234	14.65459	1645.746
Pine (150 years) with a composition based on potassium oxide hydrate (KOH) and a polyfunctional compound	17.67	0.4992	19.89676	904.5326
Pine (150 years) with the composition BB-11	16.53	0.3968	20.56678	823.7534

[a]Note: Artificial aging pine with a service life of 150 years was obtained under special conditions according to the methodology [38] and reproducible cyclic environmental conditions

However, when treating artificially aged wood, its effect on the characteristics of the porous structure of the formed coke turned out to be less significant.

The average pore diameter of the coke became equal to 19.9 nm, and the specific surface increased only 1.1 times compared to the sample without fire-retardant impregnation. Minimal differences were found in the parameters of the porous structure of coxes when using the fire-retardant treatment with the composition BB-11 for modern and artificially aged wood. The formation of coxes with fine porous structure leads to reduction of their thermal conductivity and improvement of thermal protection properties. The presence of combustion retardant elements—phosphorus in the coke structure—contributes to the reduction of oxidation ability and slows down the burnout of the carbon layer on the wood surface. The results obtained once again emphasize that the effectiveness of flame-retardant effect of combustion retardants used in various impregnating compositions for wood depends on the direction of chemical reactions and the contribution of physical processes in the combustion of materials.

6.4 Fire Hazards of Long-Lived Wooden Structures with Fire Protection

6.4.1 Effectiveness of Flame Retardants in Reducing the Fire Hazard of Long-Term Naturally Aged Wood

This section presents the results of fire tests to evaluate the effect of fire-retardant compositions on the fire hazard of wood of different life cycles. The estimation of fire hazard was carried out in relation to samples of modern and 81-year-old pine wood in the presence of fire-retardant compositions based on the system $NH_4H_2PO_4$–$(NH_4)_2SO_4$ and the composition based on dimethylphosphite ($C_2H_7PO_3$). Fire retardants with a rate of surface application of 200–400 g/m^2 were applied to the surface of wood samples.

The study of the effectiveness of flame-retardant compositions in reducing the flammability of long-term natural aging wood under the influence of external radiation heat flux was carried out on the installation according to GOST 30402 (Russia).

Figure 6.36 shows the results of studies of the effect of the external radiation heat flux density on the ignition time of wood with a lifetime (81 years) with flame-retardant composition based on the system $NH_4H_2PO_4$–$(NH_4)_2SO_4$, depending on the consumption of its surface application. As it is shown by the results of fire tests, this flame-retardant composition effectively increases the values of ignition time of

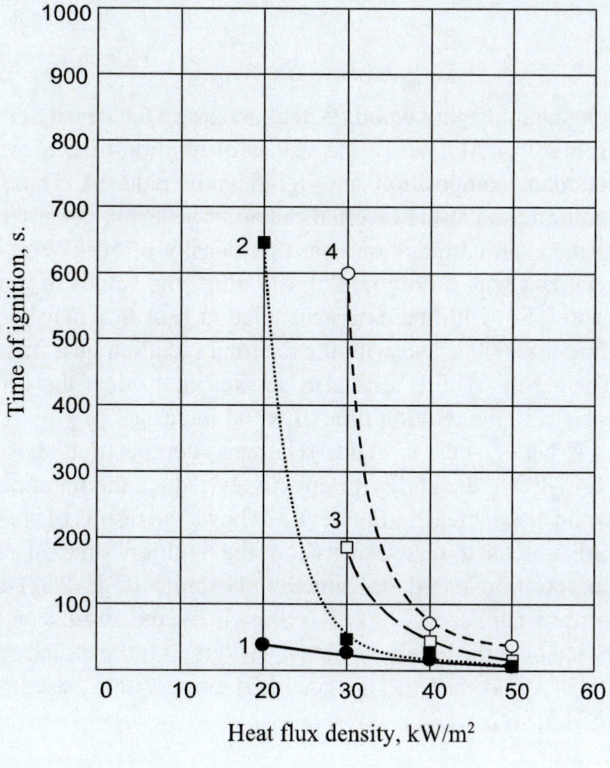

Fig. 6.36 Influence of the external radiation heat flux density on the time of ignition of wood (81 year of operation) with fire retardant composition on the basis of the system $NH_4H_2PO_4$–$(NH_4)_2SO_4$ depending on the consumption of its application: 1—native pine with the density of 558 kg/m^3, humidity 12.2%; 2—pine with composition, consumption 200 g/m^2; 3—pine with composition, consumption 300 g/m^2; and 4—pine with composition, consumption 400 g/m^2

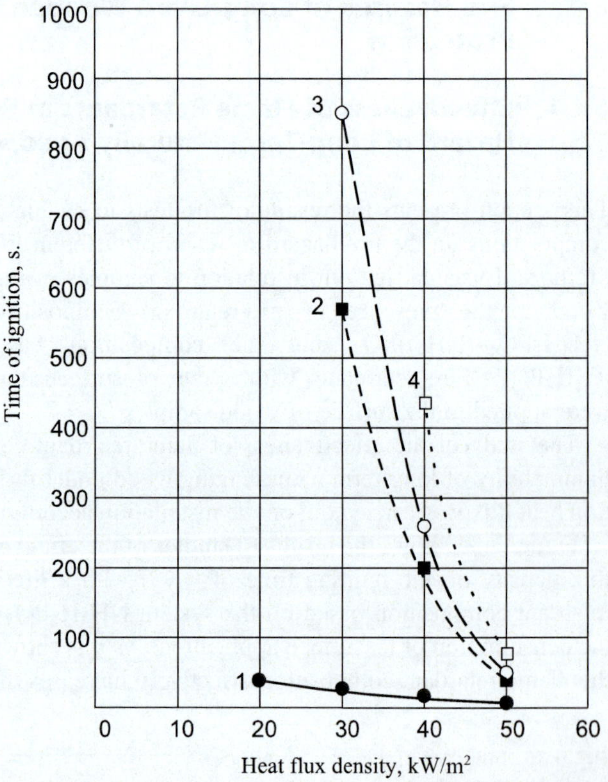

Fig. 6.37 Effect of the external radiation heat flux density on the ignition time of wood (81 year of operation) with fire retardant composition based on dimethylphosphite ($C_2H_7PO_3$) with different rates of its application: 1—native pine with a density of 558 kg/m^3, humidity 12.2%; 2—pine with a composition of 200 g/m^2; 3—pine with a composition, consumption of 300 g/m^2; and 4—pine with a composition, consumption of 400 g/m^2

long natural aging wood. When increasing the density of external radiation heat flux from 20 to 50 kW/m^2, the values of ignition time of wood and wood with flame-retardant composition are significantly reduced. Thus, for the native wood, the ignition time at the external radiation heat flux density of 30 kW/m^2 was 31 s and at the external radiation heat flux density of 50 kW/m^2–11 s. When using a flame retardant with a flow rate of 400 g/m^2, the values of ignition time increase significantly. So, with the external radiation heat flux density of 30 kW/m^2, the ignition time was 600 s, and with the external radiation heat flux density of 50 kW/m^2 29 s. The results of fire tests also showed that when the flow rate of the composition increases, the ignition time of wood increases.

When using a flame-retardant composition based on dimethylphosphite ($C_2H_7PO_3$), the ability to effectively reduce the flammability of long natural aging wood also increases. Figure 6.37 shows the results of studies of the effect of external radiation heat flux density on the ignition time of wood with flame-retardant composition based on dimethylphosphite ($C_2H_7PO_3$) on the consumption of its surface application. As it is shown by the results of fire tests, the composition based on dimethylphosphite ($C_2H_7PO_3$) is more effective in reducing the flammability of wood than the fire-retardant composition based on the system $NH_4H_2PO_4$–$(NH_4)_2SO_4$.

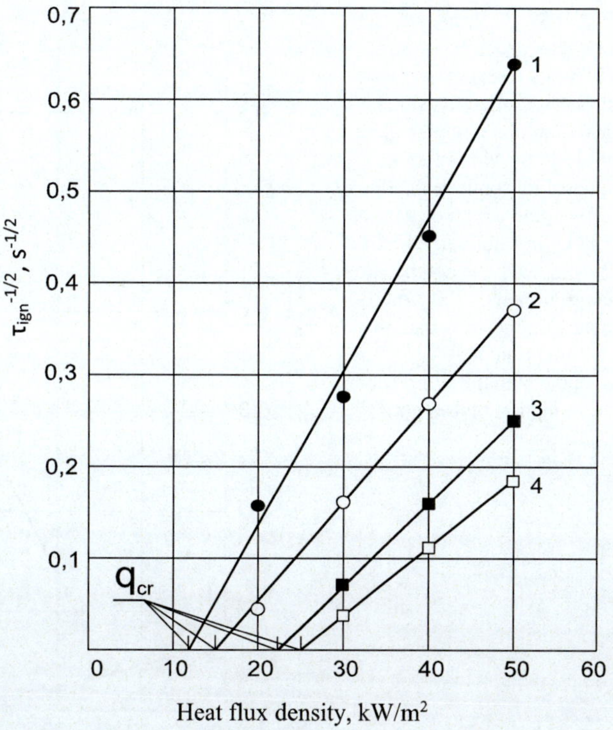

Fig. 6.38 Critical surface heat flux density of wood (81 year of operation) with fire-retardant composition based on the system $NH_4H_2PO_4$–$(NH_4)_2SO_4$ depending on the consumption of its application: 1—native pine, with a density of 558 kg/m^3, humidity 12.2%; 2—pine with the composition, consumption of 200 g/m^2; and 3—pine with the composition, consumption of 300 g/m^2; 4—pine with the composition, consumption of 400 g/m^2

The use of flame retardants not only increases the period until the ignition of wood samples but also increases the values of the critical surface heat flux density required for their ignition. The critical heat flux density of ignition of wood samples was determined by linear extrapolation of experimental values of the ignition time as a function of the external radiation heat flux density. This dependence was plotted in the coordinates $\tau^{-1/2} - q_e$ used to analyze the flammability of long-term natural aging wood samples (81 years of operation).

Figures 6.38 and 6.39 show the dependences of the critical surface heat flux density of wood with flame-retardant compositions on the consumption of surface application.

For natural wood, the value of the critical surface heat flux density of ignition is 14.1 kW/m^2, and for wood with fire-retardant compositions based on the system $NH_4H_2PO_4$–$(NH_4)_2SO_4$ and on the basis of dimethylphosphite ($C_2H_7PO_3$) at the rate of surface application of 400 g/m^2, the value of the critical surface heat flux density of ignition is 25.0 and 32.5, respectively.

The most important indicators of fire hazard of materials are those characterizing the loss of mass during fire exposure under given conditions of fire tests. The most important characteristic, reflecting the rate of mass loss of a material during fire exposure, is the mass rate of burnout. Using an additional device for measuring the mass of the sample during fire tests on an installation according to GOST 30402 (Russia), the effect of a variety of fire-retardant composition and the external

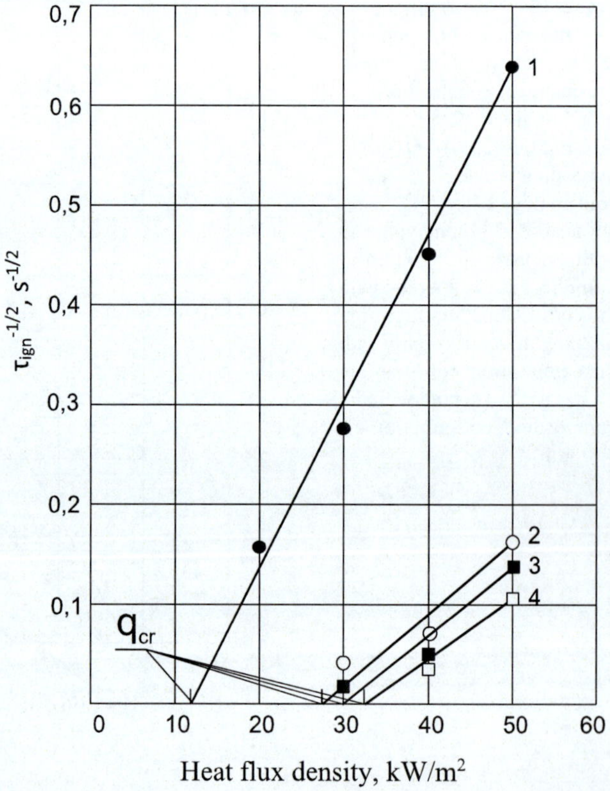

Fig. 6.39 Critical surface heat flux density of wood (81 years of operation) with fire-retardant composition based on dimethylphosphite ($C_2H_7PO_3$) with different rates of its application: 1—native pine with a density of 558 kg/m^3, humidity 12.2%; 2—pine with the composition, consumption of 200 g/m^2; 3—pine with the composition, consumption of 300 g/m^2; and 4—pine with the composition, consumption of 400 g/m^2

radiation heat flux density on the mass rate of wood was studied. As shown by the results of tests on samples, under conditions with an additional source of ignition, there are certain regularities in the change of mass burnout rate for wood with flame-retardant materials under the action of external radiation heat flux.

In Fig. 6.40, the results of researches of mass burnout rate at burning of samples of natural wood and wood with flame-retardant compositions based on the system $NH_4H_2PO_4$–$(NH_4)_2SO_4$ and based on dimethylphosphite ($C_2H_7PO_3$) with the surface application rate of 400 g/m^2 under the influence of the external heat flux density of 35 kW/m^2 arc presented.

Efficiency of flame-retardant compositions in reducing the mass burnout rate of long-term natural aging wood samples was conducted by the maximum value of mass burnout rate and time to reach this maximum value. As the results of fire tests are showed, the composition based on the system $NH_4H_2PO_4$–$(NH_4)_2SO_4$ is less effective in reducing the mass burnout rate of wood (81 year of operation) at the initial stage of fire tests compared to the composition based on dimethylphosphite ($C_2H_7PO_3$), which is due to the characteristic mechanism of fire protection action of these compositions. Thus, for pine wood of long natural aging without flame retardant, the maximum mass burnout rate is 21.1 g/m^2·s, and the time to reach this maximum value was 144 s. In contrast, flame retardants significantly reduce these fire hazard characteristics of wood.

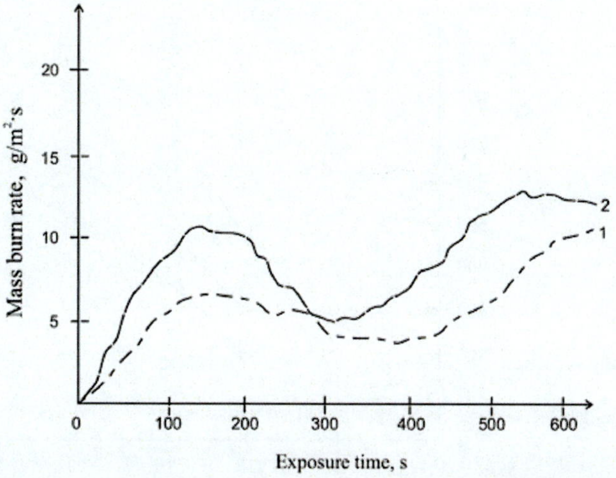

Fig. 6.40 Mass burn-up rate during combustion of samples of native wood with fire-retardant compositions at the heat flux density of 35 kW/m^2: 1—pine (81 year of operation) with fire retardant composition based on the system: $NH_4H_2PO_4$–$(NH_4)_2SO_4$ (consumption 400 g/m^2); 2—pine (81 year of operation) with fire retardant composition based on dimethylphosphite ($C_2H_7PO_3$) (consumption 400 g/m^2)

Table 6.10 Results of the assessment of the main indicators of fire danger of wood with fire protection

Type of wood	Q_{low}, kJ/g	D_{max}, m^2/кg	[a]Ψ, g/m^2·s	q_{cr}, kW/m^2
Long-life pine timber (81 years) ($\rho = 554$ kg/m^3) with the following system: $NH_4H_2PO_4$–$(NH_4)_2SO_4$ (consumption 400 g/m^2)	14.7	230	10.9	25.1
Long-life pine timber (81 years) ($\rho = 554$ kg/m^3) with formulation on the basis of dimethylphosphite ($C_2H_7PO_3$) (consumption 400 g/m^2)	13.1	140	6.7	27.1

[a]Values of mass burn rate were determined at a heat flux density of 35 kW/m^2 at the main stage of the combustion process of a wood sample

For wood with a flame-retardant system based on the system $NH_4H_2PO_4$–$(NH_4)_2SO_4$ with a surface application rate of 400 g/m^2, the maximum mass burnout rate at the first stage of burning of samples is 10.9 g/m^2·s. For wood with flame-retardant composition on the basis of dimethylphosphite ($C_2H_7PO_3$) with a similar flow rate, the maximum mass burnout rate is 6.7 g/m^2·s.

In Table 6.10, the characteristics of flammability and mass burnout rate of wood depending on the type and consumption of flame-retardant composition and the density of external radiation heat flux on the installation according to GOST 30402 (Russia) are summarized.

An important revealed effect is the ability of fire-retardant compositions to reduce the smoke-forming ability of wood regardless of the period of operation by shifting the mode of maximum smoke formation towards higher heat fluxes up to 25 kW/m^2 (Figs. 6.41 and 6.42).

The obtained dependences have a pronounced extreme character, due to the transition of pre-ignition combustion mode to flame combustion mode. Application

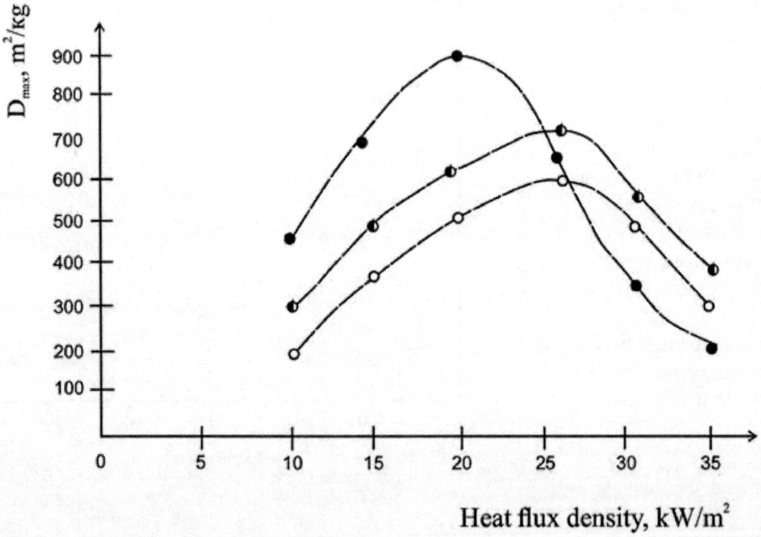

Fig. 6.41 Variation of wood smoke formation coefficient with the value of heat flux density: 1—modern wood native; 2—modern wood with the composition based on dimethylphosphite ($C_2H_7PO_3$); and 3—modern wood with the composition based on the system: $NH_4H_2PO_4$–$(NH_4)_2SO_4$

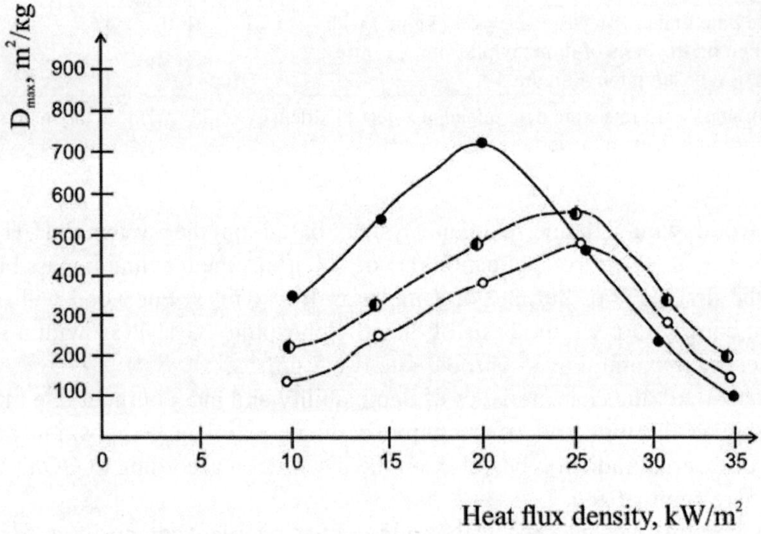

Fig. 6.42 Variation of wood smoke formation coefficient with the value of heat flux density: 1—natural wood (81 year of operation); 2—wood (81 year of operation) with the composition based on dimethylphosphite ($C_2H_7PO_3$); and 3—wood (81 year of operation) with the composition based on the system: $NH_4H_2PO_4$–$(NH_4)_2SO_4$

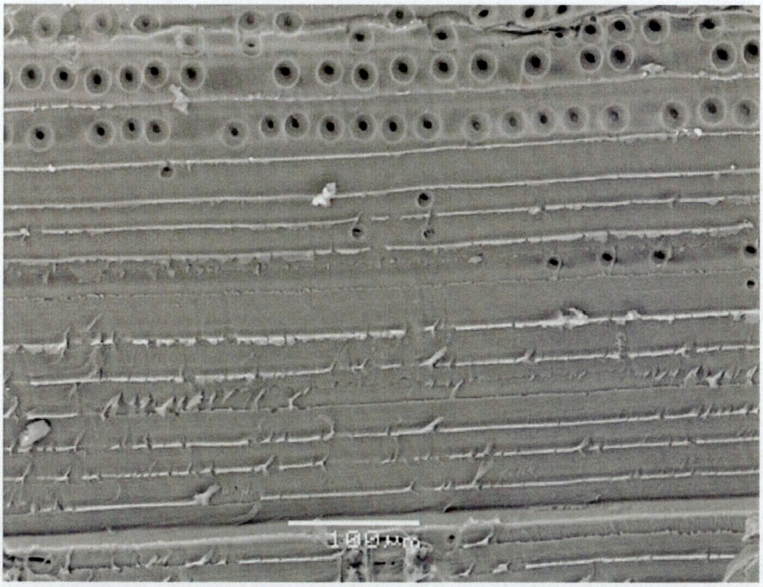

Fig. 6.43 Surface replicas of a modern wood sample with a composition based on the system: $NH_4H_2PO_4$–$(NH_4)_2SO_4$

for fire protection of wood composition based on dimethylphosphite ($C_2H_7PO_3$) allows to reduce significantly the value of wood smoke formation coefficient of long-term natural aging (81 years of operation) and to transfer the material from the group with high smoke-forming ability to the group of materials with moderate smoke-forming ability according to GOST 12.1.044 (Russia).

As the analysis of replicas obtained from the surface of the examined samples by means of electron microscopy showed that after treatment with phosphorus-nitrogen-containing impregnating composition based on the system $NH_4H_2PO_4$–$(NH_4)_2SO_4$, the structure of the surface layer of wood modern visibly changes (Fig. 6.43).

On the surface of a cut of an untreated wood sample, microfibrils in the form of ribbon-like formations and fibers, between which there are fringed pores of round shape, were found. After impregnation of wood with compositions with the optimal ratio of components, pores are no longer visible, and cross-linking between microfibrils is observed. A picture of a kind of spatially cross-linked structure appears.

For modern wood samples with the composition based on dimethylphosphite ($C_2H_7PO_3$), a characteristic denser microfibrillar structure with the presence of a thin salt layer of composition on the modified surface of the sample is also visible (Fig. 6.44).

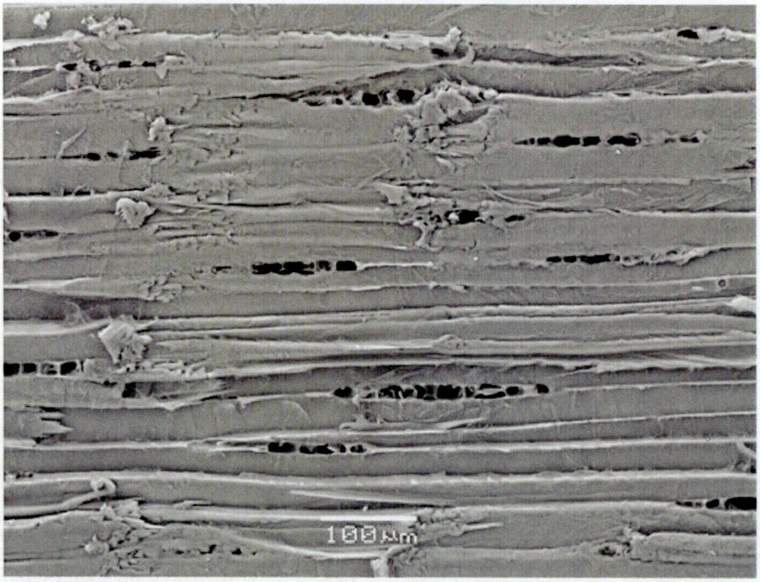

Fig. 6.44 Surface replicas of a modern wood sample with a composition based on dimethylphosphite ($C_2H_7PO_3$)

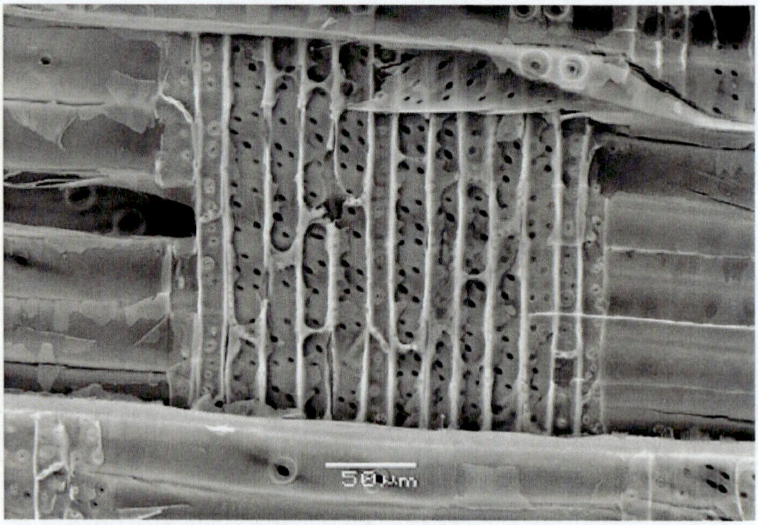

Fig. 6.45 Surface replicas of a wood sample (81 years of operation) with a composition based on the system: $NH_4H_2PO_4$–$(NH_4)_2SO_4$

Similar effects of wood surface modification with the use of fire-retardant impregnating compositions were found for samples of long natural aging (81 years of operation) (Figs. 6.45 and 6.46).

Fire-retardant impregnating compositions fully preserve the primary texture of wood. During the experimental evaluation of the thermal conductivity coefficient of

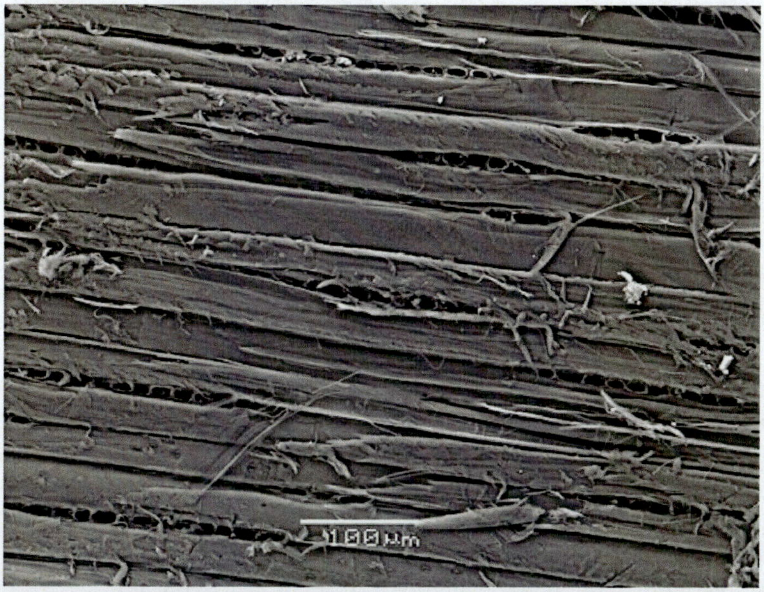

Fig. 6.46 Surface replicas of a wood sample (81 years of operation) with a composition based on dimethylphosphite ($C_2H_7PO_3$)

Table 6.11 Values of thermal conductivity coefficient for wood with fire-retardant compositions (across fibers) of different service life

Sample name	Thermal conductivity coefficient, W/m·K
Modern pine wood	0.096
Long-life pine timber 81 years	0.107
Modern pine with a $NH_4H_2PO_4$–$(NH_4)_2SO_4$ system	0.094
Modern pine with dimethylphosphite ($C_2H_7PO_3$)	0.095
Long-life pine timber with a system: $NH_4H_2PO_4$–$(NH_4)_2SO_4$	0.094
Long-life pine with dimethylphosphite ($C_2H_7PO_3$) formulation for 81 years	0.093

wood with fire-retardant compositions, there was found a slight decrease in the values of this indicator when using fire-retardant impregnating compositions based on the system $NH_4H_2PO_4$–$(NH_4)_2SO_4$ and based on dimethylphosphite ($C_2H_7PO_3$) (Table 6.11).

When using a composition based on dimethylphosphite ($C_2H_7PO_3$) for fire protection of long-term wood, the thermal conductivity coefficient decreases by 1.15 times. Apparently, flame retardants are able to modify the components of the wood complex to form a surface "monolithic" heat-resistant layer, which has lower thermal conductivity values than the native wood. Modification of the surface layer, in all likelihood, will provide a high effect not only in reducing the fire risk indicators of wooden structures of long-term operation but also in increasing their bioproofness. The most effective to protect wood from biodegradation, wood-destroying fungi of Coniophora puteana species, bluetongue fungi, mold, etc. is a fire-retardant composition based on dimethylphosphite ($C_2H_7PO_3$), containing phosphorous acid esters as the active working substance.

6.4.2 Fire Hazard of Long-Life Wooden Structures with Fire Protection

It was important to find out the possibility of reducing the fire hazard of wooden structures of different life cycles by using effective fire-retardant impregnation compositions (flame retardants). It was necessary to determine the degree of the developed flame retardants' influence on the peculiarities of wood structures' behavior during a long-life operation under the conditions of fire tests equivalent to the developed fire.

Indicative is the ability of flame retardants to restrain the process of charring of wood of long-term aging under conditions of fire or heat exposure.

Table 6.12 shows the comparative results of effectiveness of fire-retardant compositions on the intensity of charring process of wood of long-term artificial aging under conditions of standard fire temperature regime.

The results presented in Table 6.12 show that the use of flame retardants for wood, in turn, has a significant effect on the charring parameters of the material. While flame retardants increase the ignition delay time of the material at standard fire temperatures, slow the rate of charring, and reduce the thickness of the charred layer, wood aging reduces the effectiveness of flame retardants.

Table 6.12 Effect of artificial aging of pine wood on the effectiveness of fire retardants in reducing the charring process of wood material under conditions of standard fire temperature regime

Wood sample	δ_c, mm	V, mm/min	ρ_c, кg/m^3	τ_{ign}, min: s	T_s, °C
Modern pine	20	0.95	268	3: 44	732
Pine, 150 years old	26	1.5	221	5:00	758
Pine modern with composition based on system: $NH_4H_2PO_4$–$(NH_4)_2SO_4$	8	0.41	217	6:15	605
Pine, 150 years old with composition based on system: $NH_4H_2PO_4$–$(NH_4)_2SO_4$	15	0.77	220	6:52	670
Modern pine with composition based on modified polysaccharides with oxidation degree $\alpha = 0.5$[a]	7	0.47	242	7:14	–
Pine, 150 years old with composition on the basis of modified polysaccharides with oxidation degree $\alpha = 0.5$[a]	12	0.64	225	7:36	–

Note: [a]Compositions based on modified polysaccharides have an intumescent mechanism of flame-retardant action

It is interesting to note that treating the surface of modern wood with a flame retardant based on the system $NH_4H_2PO_4$–$(NH_4)_2SO_4$ leads to a significant reduction in the surface temperature of the char layer compared to that without treatment. This means that the flame retardant should slow down the rate of burnout of the coal layer under fire conditions. However, with the same treatment of artificially aged pine, the surface temperature of the charcoal layer is higher than that of the fire-retarded modern wood specimen.

In general, compared with the composition based on the system $NH_4H_2PO_4$–$(NH_4)_2SO_4$, foaming coating based on starch oxide with $\alpha = 0.5$ reveals a higher efficiency. A layer of foamed coke with high thermal insulating properties presents a barrier to heat transfer to the wood surface and reduces the likelihood of pyrolysis and charring.

To establish the effectiveness of the developed flame-retardant compositions in reducing the fire hazard of enclosing wooden structures, fire tests according to GOST 30403 (Russia) were carried out.

Fire retardant compositions were applied to the surface of wooden structures in three layers (consumption not less than 400 g/m^2) with a break between applications of 1 day. After application of the third layer, the treated wood was assembled into a structure of 2400x1300 mm and condensed for 1 month before testing.

To conduct fire tests to assess the effectiveness of the investigated fire-retardant compositions on the fire hazard of wooden structures, pine wood structures of different life span were prepared, including those with fire-retardant impregnation compositions (Fig. 6.47).

Fig. 6.47 Fragments of a long-life (81-year operation) timber sheathing wall with a fire-retardant composition based on the system $NH_4H_2PO_4$–$(NH_4)_2SO_4$ (**a**) and a fire-retardant composition based on dimethylphosphite ($C_2H_7PO_3$) (**b**)

According to the existing methodology, the test furnace design is divided into a fire chamber and a heat chamber, each of which has two control zones for the dynamics of temperature changes, including measurements of average temperatures in each of the chambers and in the sectional volume of the structure (20, 40, 60, 80, and 100 mm from the surface exposed to fire exposure) in the geometric center of the structure in the fire chamber and in the control zone in the heat chamber. The duration of the fire exposure for all types of the enclosing wooden structures was 45 min.

When analyzing the intensity of temperature build-up (Figs. 6.48 and 6.49) during fire tests for both modern and long-life timber structures with fire-retardant impregnation compositions, it can be noted that characteristic differences are observed only during the initial stage of tests.

The results of fire tests are primarily affected by the small amount of flame retardant in the surface layer of the structure. The main scientific concept associated with the use of various fire-retardant compositions for wood of long-term natural aging is related to the possibility of influence of flame retardants on the features of formation of carbon residue, its structure and properties during the first minutes of fire exposure in order to not only reduce the fire risk properties of wooden structures at the initial stage but to preserve thermal stability and structural integrity in a prolonged fire.

The proposed scientific concept is reflected in the results of conducted fire tests. Nevertheless, despite the insignificant amount of fire retardant in the surface layer of the structures, the effect of fire-retardant compositions in the reduction of temperature compared to the structures made of modern wood is manifested throughout the fire tests. Regardless of the period of operation throughout the fire test, the compositions are able to reduce the temperature in the fire chamber of the installation within 20–80 °C, i.e., in general, one can testify to a significant reduction in the overall heat effect.

The effectiveness of fire-retardant compositions is also evident in the assessment of ignition time of modern fencing wooden structure under conditions of standard fire temperature regime. It was visually found that the ignition of a sample of modern wooden structure occurs 2.5 min after the start of the standard temperature exposure. The fire-proof treatment of wooden structures with compositions based on the system $NH_4H_2PO_4$–$(NH_4)_2SO_4$ and on the basis of dimethylphosphite ($C_2H_7PO_3$) enables to increase the resistance of structures to ignition by 1.4 and 1.8 times, respectively. Visual observations of the condition of the surface of the tested structure show that carbonization and subsequent charring of the surface layer of the wooden structure with fire protection on average by 1–2 min ahead of similar processes for untreated wood, which agrees with the "classical" theory of reduction of combustibility of wood.

For long-term structures, the fire protection efficiency of the composition based on the system $NH_4H_2PO_4$–$(NH_4)_2SO_4$ decreases. At some time intervals of testing, the temperature value is comparable to the thermal effect for a sample of long-term wooden structure without flame-retardant treatment.

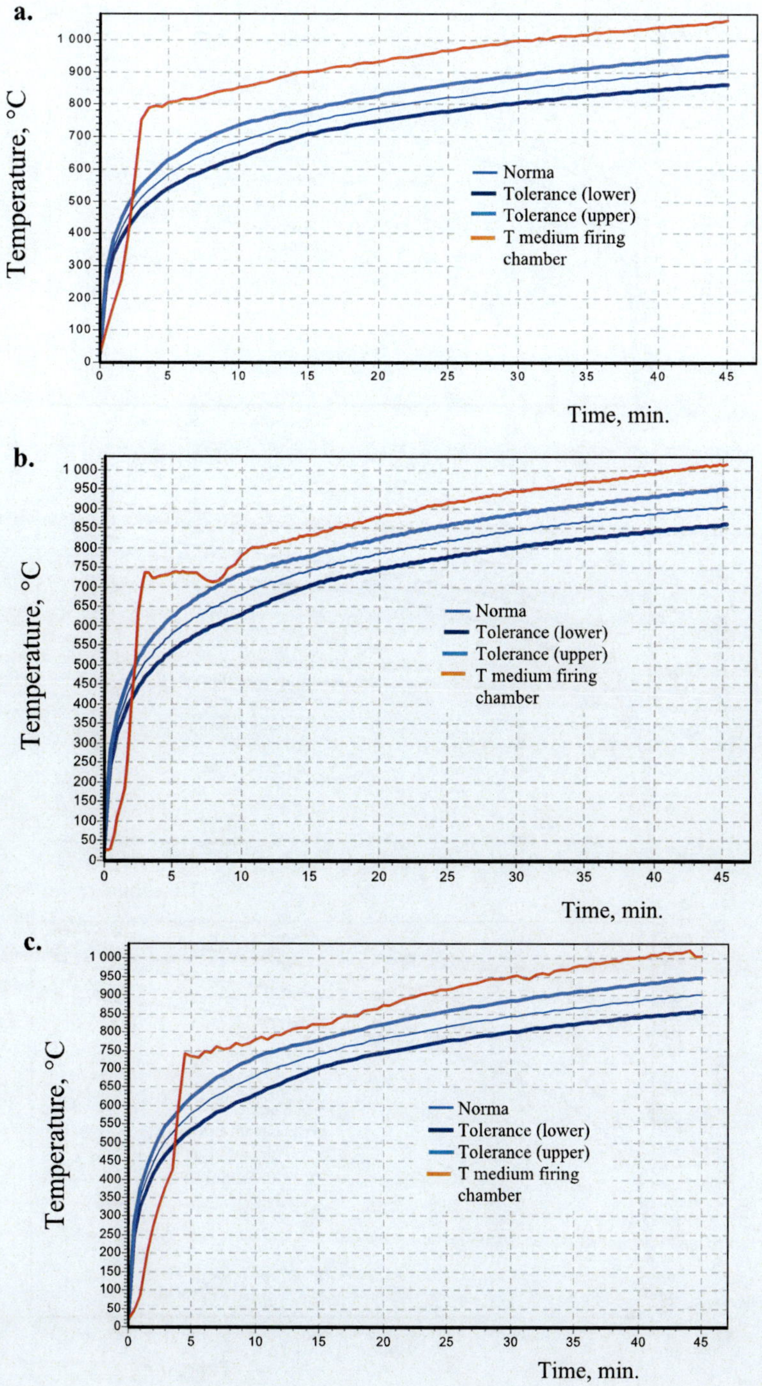

Fig. 6.48 Dynamics of temperature rise in the fire chamber during tests: (**a**) wooden structure without fire protection; (**b**) wooden structure with fire protection composition based on the system $NH_4H_2PO_4$–$(NH_4)_2SO$ (consumption 400 g/m^2); (**c**) wooden structure with fire protection composition based on dimethylphosphite ($C_2H_7PO_3$) (consumption 400 g/m^2)

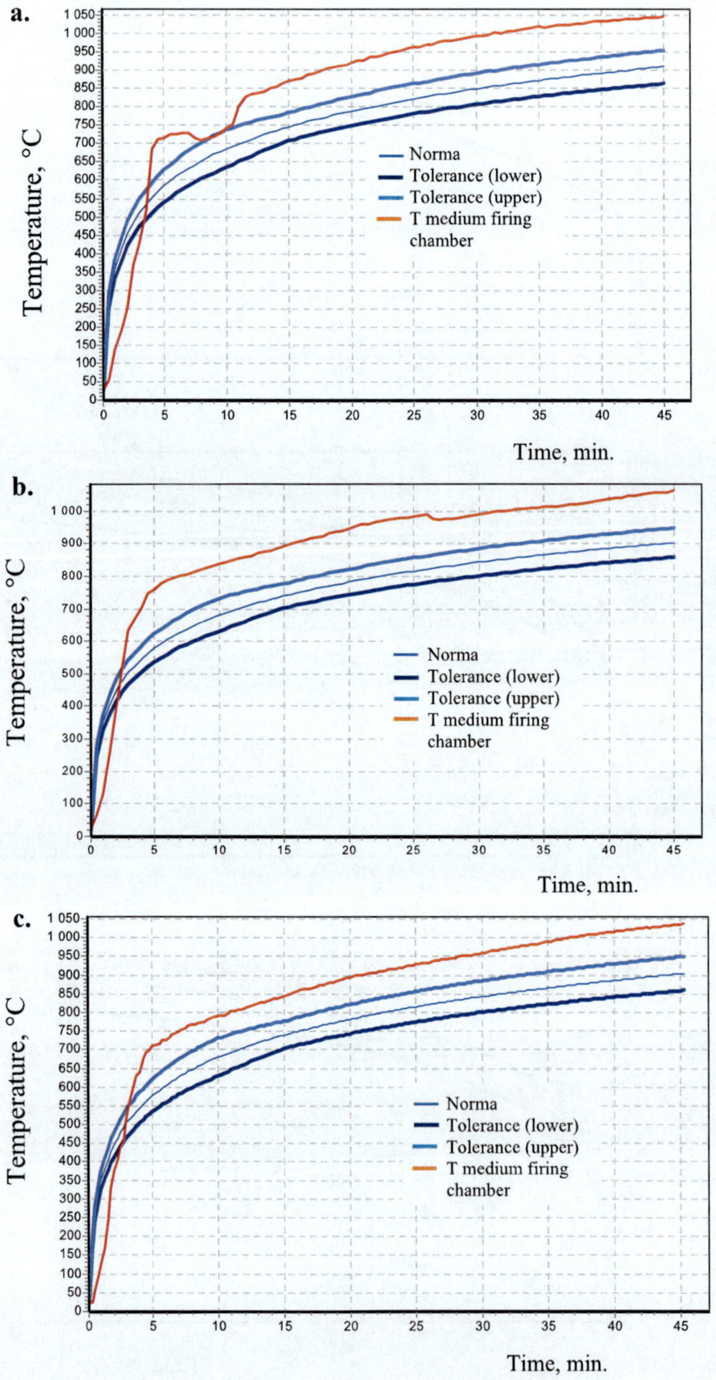

Fig. 6.49 Dynamics of temperature rise in the fire chamber during tests of pine structures with a long service life (81 years of operation): (**a**) wooden structure without fire protection; (**b**) wooden structure with a fire-retardant composition based on the system $NH_4H_2PO_4$–$(NH_4)_2SO$ (consumption 400 g/m^2); (**c**) wooden structure with a fire-retardant composition based on dimethylphosphite ($C_2H_7PO_3$) (consumption 400 g/m^2)

In comparison, the effectiveness of the fire-retardant composition based on dimethylphosphite ($C_2H_7PO_3$) in reducing the temperature values for the wooden structure with a service life of 81 years is maintained. Throughout the fire tests, the temperature value in the fire chamber is on average 50 °C lower compared to the test results of the native wood structure.

Apparently, it is the ability of the flame-retardant composition based on the system $NH_4H_2PO_4$–$(NH_4)_2SO_4$ to reduce the oxidative and calorific value of the carbon layer, which actually forms in the first minutes of the fire test.

As in the case of tests of a modern wood envelope with fire-retardant compositions, an earlier onset of characteristic darkening and carbonization of the surface layer of the sample was visually detected for structures with a life span of 81 years. This was also confirmed by the results obtained using thermal analysis techniques.

A similar effect of flame retardants in reducing the temperature increase during fire tests was established in the heat chamber of the installation for both the sample of modern wooden structure and the long-life structure with flame retardants.

Figure 6.50a–c shows curves of temperature changes during fire tests for a wooden structure with a service life of 81 years. Fire protection effect of compositions, as well as in case of direct fire exposure, is visually evident at the initial stage of heat exposure. Five minutes after the start of the test, the temperature in the heat chamber when testing an untreated wooden structure reached 500 °C (Fig. 6.50a), while the one treated with compositions based on the system $NH_4H_2PO_4$–$(NH_4)_2SO_4$ reached 460 °C (Fig. 6.50b) and on dimethylphosphite ($C_2H_7PO_3$) reached 450 °C (Fig. 6.50c).

It is important in the development of fire-retardant compositions is not only their ability to reduce the fire hazard of wooden structures but also to increase the fire resistance of structures. Of particular interest is the assessment of the effect of the developed fire-retardant compositions on the intensity of charring process—the main process determining the fire resistance of wooden structures. Average values of charring rates of wooden structures of different service life in fire and heat chambers of the facility are presented in Table 6.13.

As it can be seen from the presented results, the fire-retardant composition is the most effective in reducing the rate of charring of wooden structures of different service life. Thus, values of charring rates for samples of wooden structures with the composition based on dimethylphosphite ($C_2H_7PO_3$) in the fire chamber of the installation are lower in 1.26–1.29 times compared with the structures without fire protection.

Fireproofing composition based on the system $NH_4H_2PO_4$–$(NH_4)_2SO_4$ is less effective in reducing the intensity of charring of wooden structures. So, a decrease in the rate of charring of structure samples in the presence of composition on the basis of system $NH_4H_2PO_4$–$(NH_4)_2SO_4$ is 1.07 times lower compared with the structure without fire protection and 1.1 times lower compared with the wooden structures of long-term operation.

The analysis of the presented data shows that the rate of charring of wooden structures in the presence of fire-retardant compositions is not a constant value and

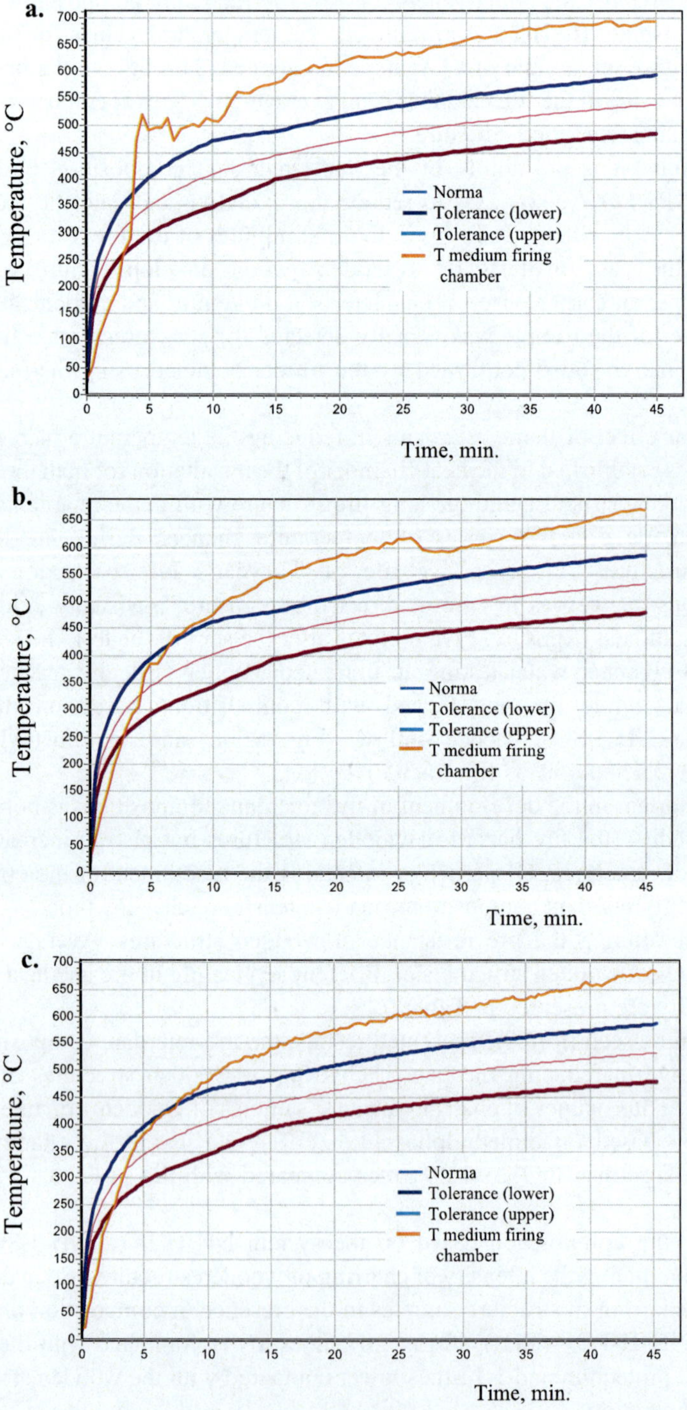

Fig. 6.50 Dynamics of temperature rise in the heat chamber during tests of pine structures with (**a**) untreated wooden structure; (**b**) wooden structure with fire-retardant composition based on the system $NH_4H_2PO_4$–$(NH_4)_2SO$ (consumption 400 g/m^2); (**c**) wooden structure with fire retardant composition based on dimethylphosphite ($C_2H_7PO_3$) (consumption 400 g/m^2)

Table 6.13 Values of charring rates of wooden structures of different service life

Location for measuring the charring depth[a]	Modern fencing wooden structure (pine wood)			Fencing wooden structure of long service life (81 years of operation)		
	–	Composition with composition based on the system: $NH_4H_2PO_4$–$(NH_4)_2SO$	Composition with composition based on dimethyl phosphite ($C_2H_7PO_3$)	–	Composition with composition based on the system: $NH_4H_2PO_4$–$(NH_4)_2SO$	Composition with composition based on dimethyl phosphite ($C_2H_7PO_3$)
Heat chamber	0.61	0.52	0.41	0.75	0.69	0.61
Fire chamber	0.72	0.67	0.57	1.06	0.95	0.82

Note: [a]The average values of charring rate in the chambers of the unit based on the results of at least 30 measurements of the charring depth in each chamber are presented

depends on the type of flame retardant and the mechanism of its fire-retardant action. This must be taken into account when calculating the fire-resistance limits of wooden structures of different service life and the choice of fire-retardant compositions.

Visual analysis of the degree of thermal damage to long-life wooden structures indicates that after prolonged fire tests (45 min), the fire protection effect of compositions is maintained (Fig. 6.51). In the first case (Fig. 6.51a), foamy formations are observed on the surface of the carbon layer, and in the second case (Fig. 6.51b), the presence of a glass-like layer can be identified actually on the entire surface.

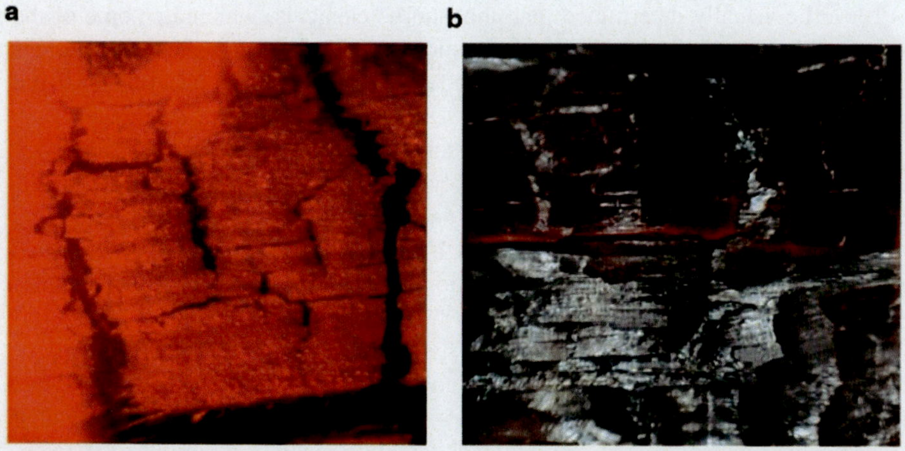

Fig. 6.51 Structure of the carbon layer of a long-lasting wooden structure with fire-retardant impregnation compositions based on the system $NH_4H_2PO_4$–$(NH_4)_2SO_4$ (**a**) and based on dimethylphosphite ($C_2H_7PO_3$) (**b**)

In order to qualitatively establish the degree of thermal damage of the examined samples of wooden structures, as well as the integrity of the carbon layer, the method of electron microscopy was used in the work. First of all, it was necessary to understand how the structure of the carbon layer changes for wooden structures with the increase of their service life.

Figure 6.52 shows replicas of the carbon residue surface for samples of modern wooden structure (a) and wooden structure with a service life of 81 years (b).

The presented replicas of the surfaces of coals indicate that the charcoal layer of long-lived wood is characterized by a more developed and open inner surface compared to the sample of modern wood. This largely explains the increased oxidative capacity of charcoal for wood structures of long life.

The results obtained by electron microscopy allowed us to establish characteristic changes in the structure of the carbon residue of the samples under study in the presence of flame-retardant compositions.

As an example, Fig. 6.53a, b shows replicas of the surface of the carbon layer of the samples of fireproof wooden structures of long-lasting operation after fire tests with flame retardants.

The charred dense layer formed on the wood surface during the test is characterized by greater homogeneity and lack of a pronounced "honeycomb-like" open structure compared to the charred surface of untreated wood. The composition based on the system $NH_4H_2PO_4$–$(NH_4)_2SO_4$ promotes the formation of a spatially cross-linked structure of the charcoal layer, and the composition based on dimethylphosphite ($C_2H_7PO_3$) is characterized by the formation of a glassy heat-resistant layer on the charcoal surface.

According to the results of the study by energy dispersive X-ray spectroscopy, it was found that after a long fire exposure (45 min) in the formed coal layer, the presence of phosphorus elements was detected (Fig. 6.54).

It can be visually determined that phosphorus elements, which are some of the most well-known inhibitors of the combustion and smoldering process of cellulose and wood materials, are evenly distributed in the structure of coal (points of red color).

The combination of the above mechanisms with the long-term thermal stability of phosphorus elements allows to considerably reduce its free specific surface of the coal bed structure, to reduce its oxidative and calorific capacity. The presence of a significant amount of phosphorus elements in the coal bed structure is also confirmed in the analysis of energy spectra of radiation of the investigated coal bed samples (Fig. 6.55).

Figure 6.55 shows that in addition to the traditional elements characteristic of charcoal, such as carbon (83.07–88.97%) and oxygen (8.03–11.56%), an increased content of the elements of phosphorus (0.45–1.67%) was found. The natural content of elements of phosphorus in wood pine charcoal is usually not more than 0.016%. In addition, traces of the elements K, Ca, and Na were identified in insignificant amounts.

The detected ability of flame retardants to change the structure of the charcoal layer makes it more resistant to burnout and oxidation, which has a favorable effect a

b

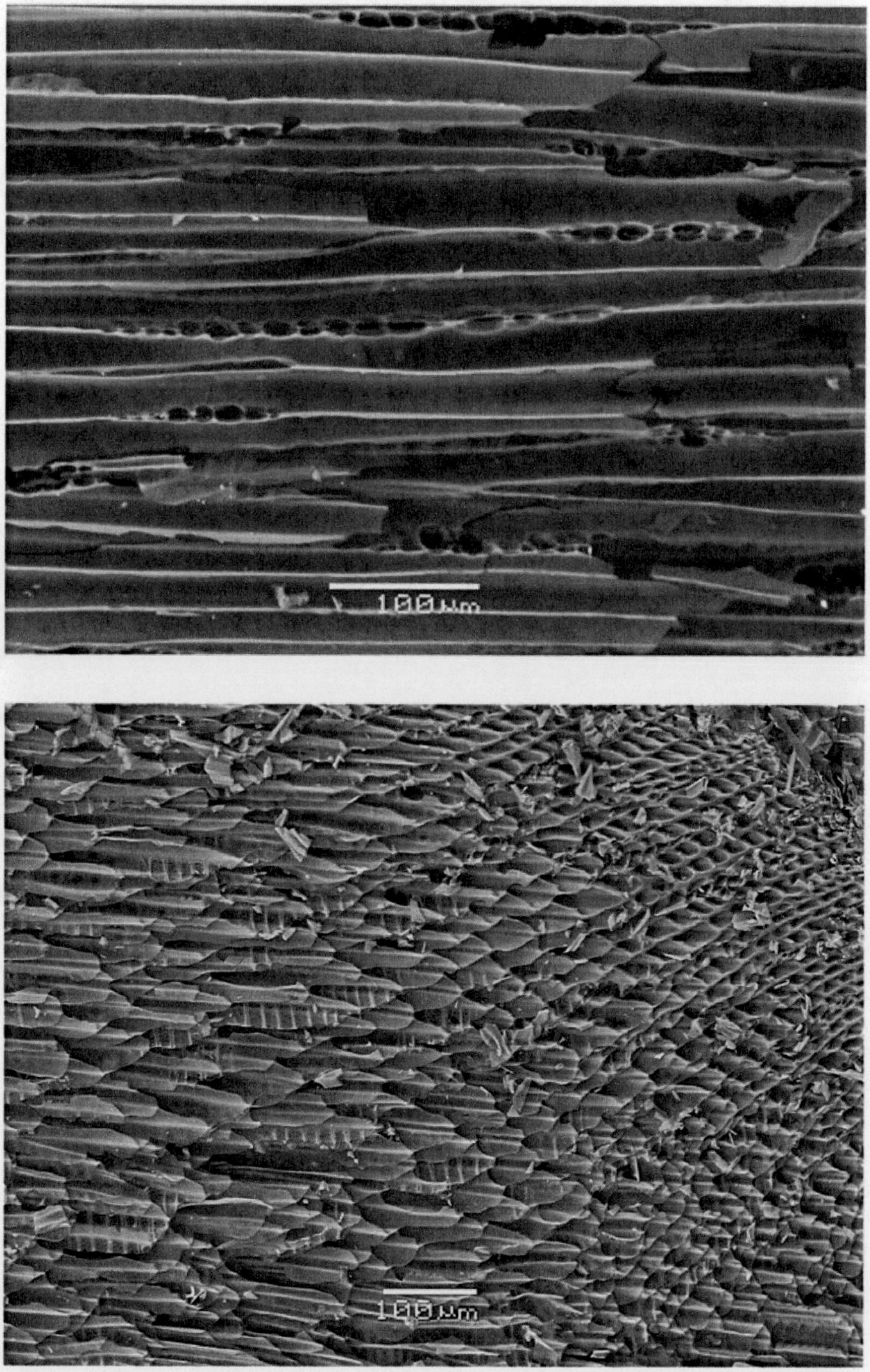

Fig. 6.52 Carbon layer structure of a modern timber envelope (**a**) and a timber structure with a lifetime of 81 years (**b**) at the time of turning off the gas burner (magnification ×500 times)

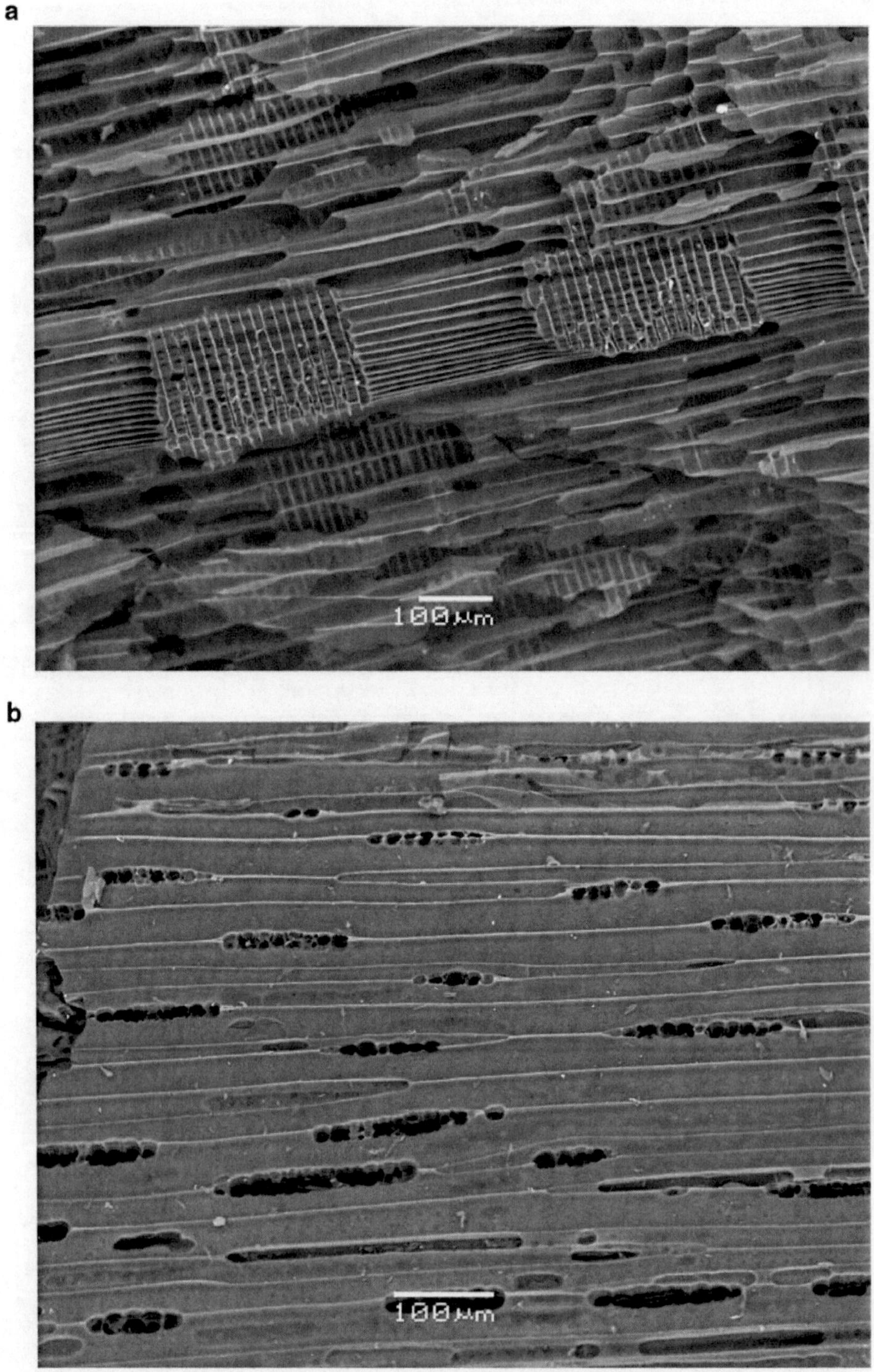

Fig. 6.53 The structure of the carbon layer of samples of long-term wooden building envelopes after fire tests with a fire-retardant composition based on the system $NH_4H_2PO_4$–$(NH_4)_2SO_4$ (**a**) and with a fire-retardant composition based on dimethylphosphite ($C_2H_7PO_3$) (**b**) (×500 times magnification)

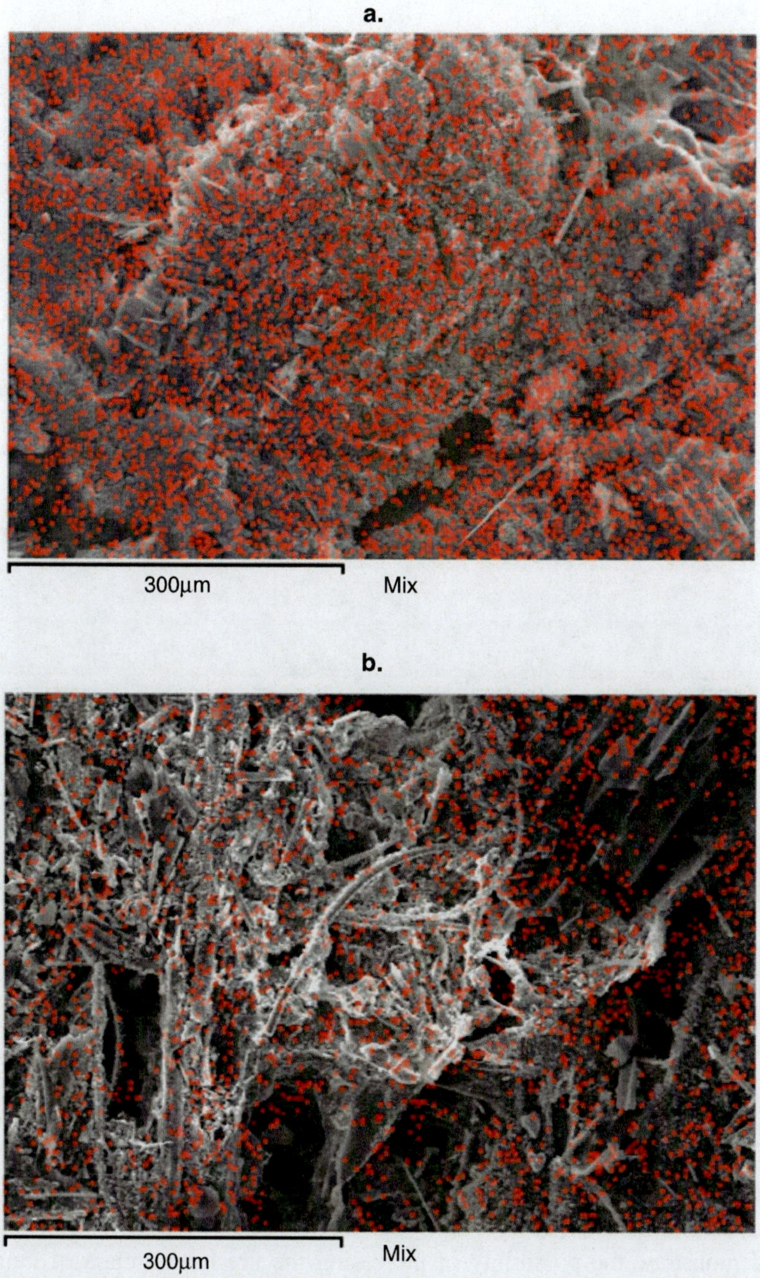

Fig. 6.54 Results of photoregistration of phosphorus elements in the structure of the carbon layer of the samples of long-term wooden building envelopes after fire tests with the fire-retardant composition based on the system $NH_4H_2PO_4$–$(NH_4)_2SO_4$ (**a**) and with the fire-retardant composition based on dimethylphosphite ($C_2H_7PO_3$) (**b**) (magnification ×200 times)

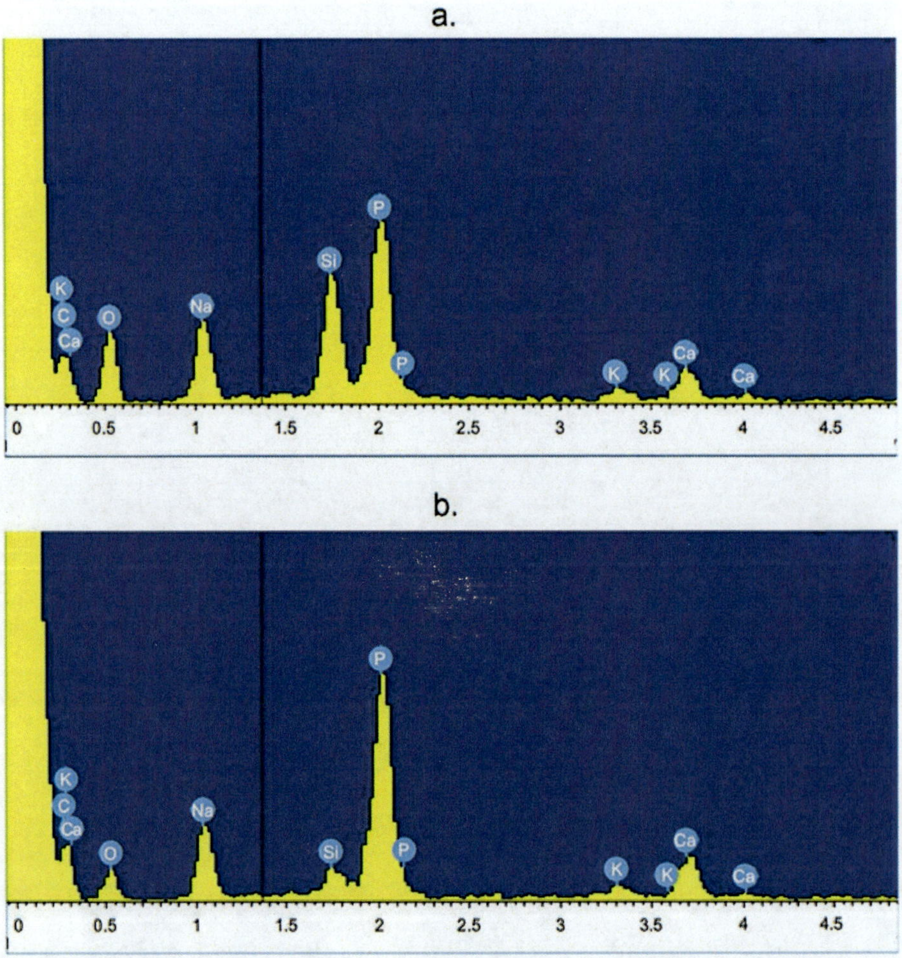

Fig. 6.55 Energy spectra of the carbon layer of samples of long-term wooden building envelopes after fire tests with fire-retardant impregnation compositions based on the system $NH_4H_2PO_4$–$(NH_4)_2SO_4$ (**a**) and based on dimethylphosphite ($C_2H_7PO_3$) (**b**)

on reducing the thermal effect during fire tests. This circumstance is decisive for fire protection of wooden structures of long-term operation, fire hazard and fire resistance of which is largely characterized by properties and structure of carbon layer, its high oxidation, and calorific capacity.

To demonstrate the possibility of increasing the fire hazard class of long-term wooden structures by means of flame-retardant impregnating compositions, fire tests have been conducted according to GOST 30403 (Russia) for 15 min. At the same time, the consumption of compositions was increased to 650–720 g/m^2.

Fire tests were conducted to determine the fire hazard class of wooden structures with fire protection, as well as to evaluate the value of temperature increase in fire and heat chambers, and the degree of thermal damage and charring intensity (depth and charring rate) of the tested samples under the conditions of accepted fire tests.

Temperature changes during fire tests are shown in Fig. 6.56a, b.

The results of estimation of temperature increase along the thickness of the wooden structure sample are shown in Figs. 6.57 and 6.58.

During the tests, the temperature values in the fire chamber did not exceed the allowable values defined by GOST 30247.0 and in the heat chamber GOST 30403.

The following changes were visually recorded during the tests:

5, 10, 15 min—when bringing the burning torch to the end parts of the structure, no flashes and no burning of emitted products from the thermal effect on the sample were observed.

The main results of the tests are as follows:

1. The presence of thermal effect from combustion of gases emitted from the sample was not recorded.
2. The presence of flame combustion of gases, emitted at thermal decomposition of materials of the sample, lasting more than 5 s, was not fixed.
3. The presence of burning melt, flowing from the ends of the sample within the control zone, during the tests was not fixed.
4. Damage to the specimen to a depth of more than 2 mm in the control zone has not been recorded.

After removal of the structure from the test bench, the following was observed:

- In the fire chamber, charring over the entire area of the fire chamber
- In the heat chamber, charring over most of the zone of the heat chamber to a depth of more than 2 mm and the size of damage to the structure of 750 mm

According to the results of fire tests, the values of depth and rate of charring of the samples under study, enclosing wooden structures of long-term service life with fire protection, were determined. Experimental data are presented in Table 6.14.

According to the results shown in Table 6.10, there is an overall trend towards a 1.1–1.2 times reduction in the charring rate of timber structures when using dimethylphosphite ($C_2H_7PO_3$)-based flame retardant. The sample of a long-life wooden structure (81 years of operation) with fire retardant composition based on dimethylphosphite ($C_2H_7PO_3$) at the rate of its surface application of 710 g/m^2 according to GOST 30403 belongs to fire hazard class K2(15).

The proposed concept of application of flame-retardant compositions shows itself consistent according to the results of the submitted studies.

Using methods of thermal analysis, it was shown that in the development and application of flame retardants for long-life wood, the determining characteristics are the rate and the heat of oxidation of carbon residue. The developed compositions allow to effectively reduce these characteristics, especially the flame-retardant composition on the basis of dimethylphosphite and modifying additives. So, the oxidation heat of carbon residue for wood (81 years of operation) with this flame retardant

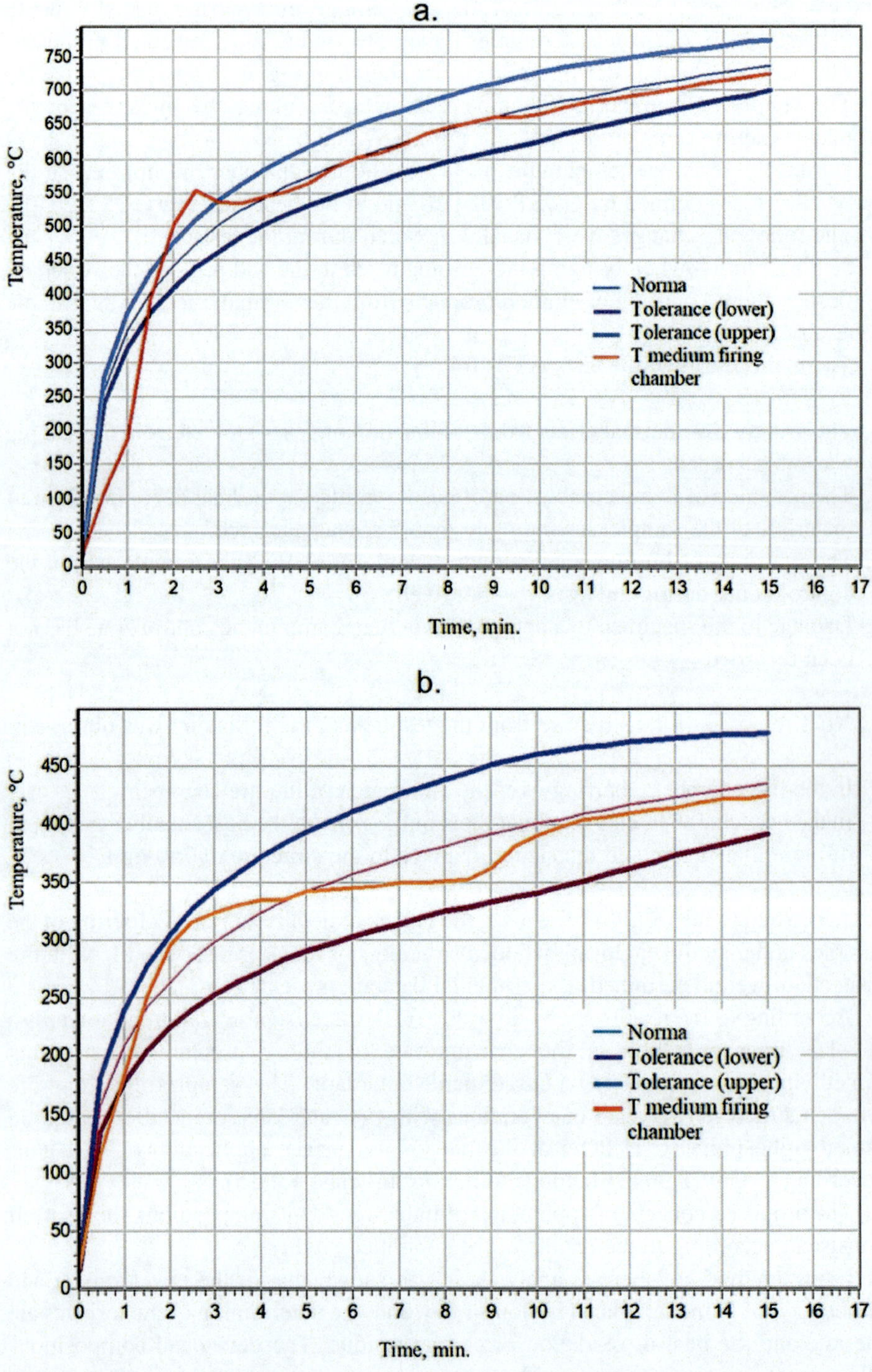

Fig. 6.56 Dynamics of temperature increase in the fire (**a**) and thermal (**b**) chambers during tests of wooden structures with a lifetime of 81 years with fire-retardant composition based on dimethylphosphite ($C_2H_7PO_3$) (surface application rate of 710 g/m^2)

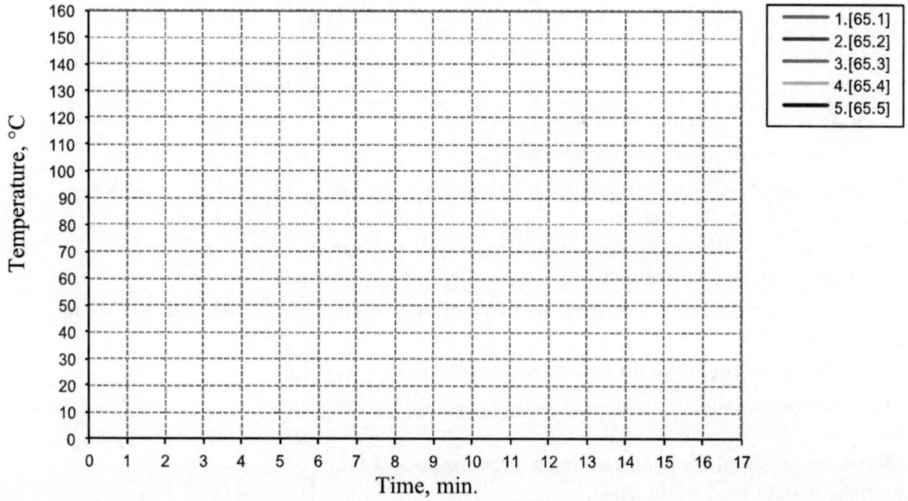

Fig. 6.57 Temperature change in the geometric center of the heated surface of the sample design

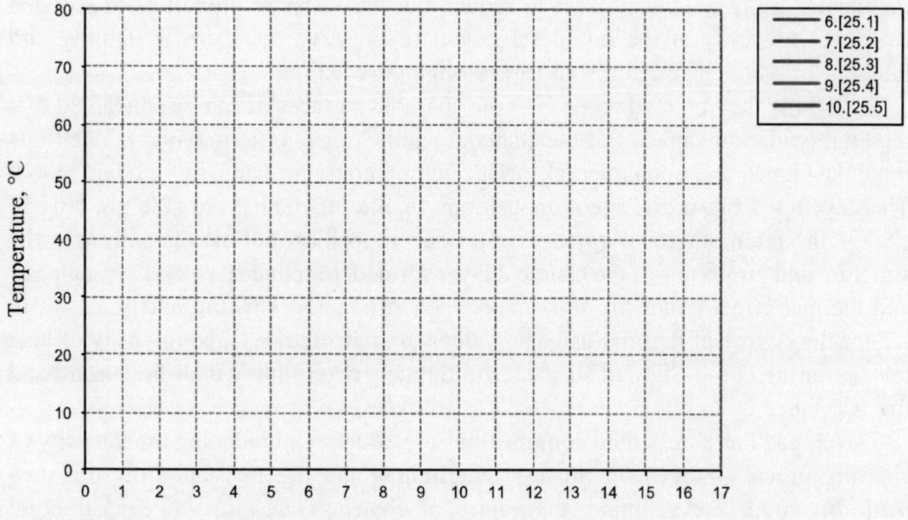

Fig. 6.58 Temperature change in the control zone for the sample design

is reduced by 1.27 times compared with that of modern wood. It is shown that with increasing the service life of wooden structures up to 200 years, the effectiveness of this flame retardant in reducing the specified characteristics of the thermal analysis remains.

It has been shown under conditions of a prolonged fire action in accordance with GOST 30403 (Russia) that fire-retarding compositions are able to influence the

Table 6.14 Charring parameters of samples of wooden structures of long-term operation

Measurement place	Charring depth, mm	Charring rate, mm/min
Long-life wooden construction (81 years of use)		
Bottom beam (0.10 m from the bottom level of the unit)	19	0.95
Thermocouple installation location in the fire chamber (0.75 m from the bottom level of the unit)	15	0.88
Thermocouple installation location in the heat chamber (1.8 m from the bottom level of the unit)	8	0.4
Long-life wooden structure (81 years of operation) with fire retardant composition based on dimethylphosphite ($C_2H_7PO_3$)		
Bottom beam (0.10 m from the bottom level of the unit)	16	0.87
Thermocouple installation location in the fire chamber (0.75 m from the bottom level of the unit)	14	0.81
Thermocouple installation location in the heat chamber (1.8 m from the bottom level of the unit)	7	0.37

peculiarities of coal residue formation, its structure, and properties during the first minutes of a fire action in order to reduce the fire-resistant properties of wooden structures not only at the initial stage, but to preserve the thermal stability and integrity of wooden structures under prolonged fire action.

Analyzing the obtained replicas of the charcoal surfaces, it can be concluded that the high oxidation capacity of the charcoal residue for long-term wood is due to its more developed and open internal surface compared to the sample of modern wood. The developed fire-protective compositions (flame retardants) are able not only to change the main stages of wood thermal decomposition but also to influence the structure and properties of the charcoal layer formed, to reduce its oxidative capacity and thermal effect under fire tests of wooden structures. This ultimately allows to reduce the degree of thermal damage, values of charring rates, and intensity of heat release during combustion of structures, ultimately determining their fire hazard and fire resistance.

Developed flame-retardant compositions are effective in reducing the intensity of charring process—the main process determining the fire resistance of structures made of wood. For example, the values of charring rates for wooden structure samples with the composition based on dimethylphosphite ($C_2H_7PO_3$) in the fire chamber of the installation are lower in 1.26–1.29 times as compared to the structures without fire protection. The ability of the proposed flame-retardant compositions to reduce the rate of charring is due to the mechanism of their flame-retardant action. It was established that the developed fire-protective compositions are able to influence the peculiarities of coal residue formation, its structure, and properties and to reduce the heat effect (by 20–80 °C) throughout the experiment and the intensity of charring of wooden structures (by 1.07–1.29 times) in the conditions of fire tests according to GOST 30403 (Russia), despite the small amount of fire retardant in the surface layer of the structure.

The effectiveness of fire-retardant compositions is also manifested in the reduction of the ignition time of wooden structures under the standard fire temperature conditions. Fireproofing treatment of wooden structures with compositions makes it possible to increase resistance of structures to flammability by 1.4–1.8 times. Mechanism of fire protection action of compositions is manifested not only in the change of basic stages of wood thermal decomposition, but also in their ability to influence the structure and properties of the charcoal layer formed, including its oxidative capacity and the value of heat effect in conditions of fire tests of wooden structures. This is extremely important because these characteristics are closely related to the degree of thermal damage, the rate of charring, and the intensity of heat release, ultimately determining their fire danger and fire resistance.

The developed fire-protective compositions contain not only "classical mechanisms" of fire protection, which are effective for modern wood, but they are also able to influence the character of the course and intensity of the oxidation process of the carbon layer of wooden structures. The compositions' ability to block the formation of an open "honeycomb-like" structure with the formation of a spatially cross-linked sub-foam structure in one case and a glass-like layer in the other case allows not only to reduce the intensity of charring but also to effectively reduce the rate and heat of oxidation of the carbon layer. Efficiency of these mechanisms is confirmed by results of large-scale tests according to GOST 30403 (Russia). Reducing the values of charring rate of wooden structures allows providing the required indicators on fire danger and fire resistance of wooden structures.

References

1. Balakin, V. M., & Polishchuk, E. Y. (2008). Nitrogen-phosphorus-containing flame retardants for wood and wood composite materials (Literary review). *Fire and Explosion Safety, 17*(2), 43–51.
2. Balakin, V. M., Polishchuk, E. Y., & Rukavishnikov, A. V. (2010). Fire protective compositions and coatings based on aminoaldehyde oligomers (Literary review). *Fire and Explosion Safety, 19*(4), 22–27.
3. Afanasyev, S. V., & Balakin, V. M. (2012). *Theory and practice of fire protection of wood and wood products: Monograph Samara* (p. 138).
4. Aseeva, R. M., Vakhrushev, L. P., Lomakin, S. M., Nesterov, I. M., Sakharov, P. A., Serkov, B. V., & Sivenkov, A. B. (2015). Thermo-oxidative decomposition of a new fire retardant for wood – "Frackfire". *Technologies of Technological Safety, 1*(59), 29–34.
5. Koteneva, I. V., Sidorov, V. I., & Kotlyarova, I. A. (2012). *Patent of the Russian Federation № 2011120392/13, 23.05.2011. Firebioprotective composition "Xylostat +"*. Russian patent № 2465129. Bulletin number 30.
6. Saleh, A. I. (2012). *Fireproofing composition for the treatment of wood.* Patent of the Russian Federation № 2011101296/13, 13.01.2011. Russian patent № 2469843. Bulletin No. 20.
7. Krivtsov, V., Maksimenko, N. A., Maksimenko, S. A., & Melnikov, N. O. (2013). *Method of obtaining firebioprotective composition. Patent of the Russian Federation № 2011136737/13, 06.09.2011*. Russian patent number 2486051. Bulletin number 35.
8. Lukyanenko, N. A., Lukyanenko, K. N., & Verenkova, E. M. 2015. *Composition for firebioprotective impregnation of wood. Russian patent number 2538256*. Patent of the Russian Federation № 2012129756/05, 16.07.2012. Bulletin No. 3.

9. Farooq, H. (2008). *Fire retardant composition.* USA Patent No. 20080258121, 23.04.2007. United States Patent No. 11/738,662.
10. Tianhong, G. (2015). *Novel wood fire retardant.* Patent CN № 20131591595, 22.11.2013. Patent China № 104647510(A).
11. Tianhong, G. (2015). *Inorganic fire retardant for wood fire retardation treatment, and preparation method thereof.* Patent CN № 20131592211, 22.11.2013. Patent China № 104647529 (A).
12. Qingdao Tongchuang energy saving environmental Protection Engineering Co Ltd. (2015). *Low-cost and high-efficiency and composite wood fire retardant and preparation method thereof.* Patent CN № 20131602350, 26.11.2013. Patent China № CN104647544(A).
13. Qingdao jiayiyang industry and trade Co Ltd. (2015). *Phosphorus-nitrogen-containing wood fire retardant and preparation method thereof.* Patent CN № 20131614910, 28.11.2013. Patent China № 104669382(A).
14. Shifang midee wood industry Co Ltd. (2015). *Environment-friendly and fireproof wood fire-retardant liquid.* Patent CN № 20131720580, 24.12.2013. Patent China № 104723427(A).
15. Levichev, A. N., Pavlyukovich, N. G., Kazievm M. M., & Valetsky, P. M. (2006). *Fire-retardant composition for wood treatment.* Patent of the Russian Federation No. 2005116003/04, 26.05.2005. Russian patent No. 2299229. Bulletin No.14.
16. Patent of Russian Federation № 2007102966/03, 20.11.2006. (2008). *A composition for wood covering.* Patent of Russia № 2329985. Bulletin No. 21.
17. Vershinin, S. N., Vershinina, T. S., & Vershinin K.S. (2010). *Fireproof composition for wood and metal.* Patent of Russian Federation № 2008150674/03, 22.12.2008. Russian patent № 2381197. Bulletin No. 4.
18. Tripolitsyn A.A. (2015). *Development of flame retardant composition to create a polymer coating of wood materials.* Dissertation for the degree of Candidate of Technical Sciences: 05.17.06. Tripolitsyn Andrey Aleksandrovich, Saint-Petersburg, p. 134.
19. Serkov, B. B., Aseeva, R. M., & Sivenkov, A. B. (2014). Physico-chemical basis of combustion and fire hazard of wood (part 2). *Technologies of Technosphere Safety, 1*(41).
20. Shatalin, S. S., & Mnatsakanov, S. S. (2013). *Fire protection coatings based on polyvinyl alcohol.* Materials of the All-Russian Scientific-Practical Conference "Past - Present - Future of St. Petersburg State University of Cinema and Television". Saint-Petersburg, pp. 122–125.
21. Shatalin, S. S. (2015). *Features of thermolysis of polyvinyl alcohol in flame retardant compositions.* Dissertation for the degree of Candidate of Technical Sciences: 05.17.06. Shatalin Sergey Sergeevich. Saint-Petersburg, p. 84.
22. Ulcar, J., & Csaba, T. (2015). *Fire resistant coating.* Patent USA № 9187674, 17.11. United States of America Patent No. 13/083734.
23. Faiz, A., Sami, U., Wan, F. M., Hammad, A., & Zaiful, Z. (2013). Non-toxic mineral based intumescent fire retardant coating for structural applications. *International Journal of the institute of Materials Malaysia, 1*, 137–156.
24. Balakin, V. M., & Krasilnikova, M. A. (2015). Nitrogen-phosphorus-containing fire retardant compositions based on the products of degradation of polyethylene terephthalate by aliphatic di- and polyamines. In *Proceedings of the V International Conference-School on Chemistry and Physical Chemistry of Oligomers "Oligomers-2015"*, Volgograd, pp. 162–165.
25. Balakin, V. M., Galliamov, A. A., & Postnikov, S. V. (2015). Synthesis and properties of nitrogen-phosphorus-containing flame retardant compositions based on the products of aminolysis of polyurethanes. In *Collection of abstracts of the V International Conference-School on Chemistry and Physical Chemistry of Oligomers "Oligomers-2015"*, Volgograd, pp. 135–138.
26. Islentyev, S. V., & Balakin, V. M. (2015). Phosphorus-containing flame retardant composition for wood materials based on nitrogen-containing products of polycarbonate degradation by aliphatic amines. In *Collection of abstracts of the V International Conference-School on Chemistry and Physical Chemistry of Oligomers "Oligomers-2015"*, Volgograd, pp. 113–118.

27. Starodubtsev, A. V., & Balakin, V. M. (2015). Fireproof compositions based on the products of degradation of complex aliphatic polyethers monoethanolamines. In *Collection of abstracts of the V International Conference-School on Chemistry and Physical Chemistry of Oligomers "Oligomers-2015"*, Volgograd, p. 252.
28. Balakin, V. M., Balakin, S. A. V., Kichanov, V. E., & Krasilnikova, M. A. (2013). Structure and properties of products of degradation of polyethylene terephthalate with diethanolamine and triethanolamine. *Journal of Plastic Mass, 6*, 3–5.
29. Katsadze, V. A., & Vinogradov, D. A. (2007). Volumetric impregnation of wood by centrifugal method. *Forest Journal, 3*, 82–85.
30. Grigoriev, G. V. (2013). *Perfection of the centrifugal method of impregnation of deciduous wood*. Dissertation for the degree of Candidate of Technical Sciences: 05.21.01. Grigoriev Gleb Vladimirovich, p. 147.
31. Saenko, A. I. (1997). *Increase of fire protection of wood by deep impregnation with antipyrenes based on electrohydraulic effect*. PhD thesis: 05.26.03. Saenko Alexander Ivanovich, p. 142.
32. Kunitskaya, O. A., & Burmistrova, S. S. (2015). Experimental studies of impregnation of wood by means of water hammer. *Systems, 4*, 103–108.
33. Varfolomeev, S. D., Lomakin, S. M., Gorshenev, V. N., Sakharov, P. A., Sakharov, A. M., & Demin, V. L. (2011). *Antipyrene, method of its production, method of fireproofing materials and method of extinguishing the seat of fire*. Patent of the Russian Federation № 2009125336/05, 02.07.2009. Russian patent number 2425069, Bulletin No 1.
34. Daniliuc, A., Deppe, B., Deppe, O., Friebel, S., Kruse, D., & Philipp, C. (2012). New trends in wood coatings and fire retardants. *European Coating Journal, 07*, 20–25.
35. Reti, C., Casetta, M., Duquesne, S., Delobel, R., Soulestin, J., & Bourdigot, S. (2009). Intumiscent biobased-polylactid films to flame retard nonwovens. *Journal of Engineered Fibers and Fabrics, 4*(2), 33–39.
36. Tidemann, B. G., & Sciborsky, D. B. (1940). *Chemistry of combustion* (p. 215). Publishers of People's Commissariat of the Russian Federation.
37. Artsybasheva, O. V., Garbuz, M. V., Serkov, B. B., & Sivenkov, A. B. (2013). On improving the fire resistance of wooden structures using flame retardants. *Technologies of Technological Safety, 1*, 46.
38. Sivenkov, A. B. (2015). *Influence of physical and chemical characteristics of wood on its fire hazard and fire protection efficiency*. Dissertation for the degree of Doctor of Technical Sciences: 02.00.06. Sivenkov Andrey Borisovich [N.N. Semenov Institute of Chemical Physics RAS], p. 289.
39. Kotlyarova, I. A. (2012). *Modification of cellulose and wood by functional borazot compounds*. Dissertation for the degree of Candidate of Technical Sciences: 05.17.06. Kotlyarova Irina Aleksandrovna. State University, p. 146.
40. Aseeva, P. M., & Zaikov, G. E. (1981). *Reducing the flammability of polymeric materials* (p. 64).
41. Aseeva, P. M., & Zaikov, G. E. (1981). *The burning of polymer materials* (p. 280). Nauka.
42. Pokrovskaya, E. N. (1993). *Phosphorylation and silylation of wood*. Dissertation for the degree of Doctor of Technical Sciences: 02.00.08, 05.21.03. Pokrovskaya Elena Nikolaevna. Moscow Engineering and Construction Institute. V., Moscow Engineering and Construction Institute n.a. V.V. Kuibyshev, p. 51.
43. Kobelev, A. A. (2012). *Development of complex fire and moisture protective composition based on compounds that provide surface modification of wood*. Dissertation for the degree of Candidate of Technical Sciences: 05.26.03, Academy of State Fire Service of Russia, p. 162.
44. Portnov, F. A. (2016). *Development of smoke suppressors, providing reduction of fire hazardous properties of wood*. Dissertation for the degree of Candidate of Technical Sciences: 05.26.03. Portnov Fyodor Aleksandrovich, Moscow State Construction University, p. 157.
45. Koteneva, I. V. (2004). *Increase of biostability and hydrophobicity of wood by surface modification with phosphorus- and organosilicon compounds*. Dissertation for the degree of

Candidate of Technical Sciences: 05.23.05. Koteneva Irina Vasilyevna, Moscow State University of Civil Engineering, p. 133.
46. Lagendorf, G., & Eichler, X. (1982). *Dehumidification of wood: Translation from German* (p. 144). Forest Industry.
47. Kalninsh, A. J. (1971). *Preservation and protection of timber: A handbook* (p. 315). Forest Industry.
48. Kunitskaya, O. A. (2015). *Substantiation of directions of diversification of processing low-grade wood at complex timber enterprises with the use of innovative technologies: Monograph*, p. 205.
49. Kulakov, V. S., Krasheninnikova, N. N., Sivenkov, A. B., Serkov, B. B., & Demidov, I. A. (2012). Reducing the fire danger of wooden building structures by deep impregnation of wood by firebioprotective composition KSD-A (grade 1). *Fire and Explosion Safety, 21*, 35–42.
50. Nigmatullina, D. M., Polishchuk, E. Y., Sivenkov, A. B., & Stenina, E. I. (2017). Study of the effectiveness of deep impregnation of wood with different fire protection systems. In *Proceedings of the 5th International Scientific and Practical Conference "Roitman Readings"*, pp. 60–63.
51. Papkov, V. S., & Slonimsky, G. L. (1966). Microthermogravimetric analysis of thermodegradation of polymers. *High-Molecular Compounds, VIII*(1), 80–87.
52. Mikkola, E. (1991). Charring of wood based materials. In *Proceeding of 3 International symposium on fire safety science, Edinburgh*, pp. 547–556.

Conclusion

Over the centuries of human history, a vast amount of experience in the creation and application of natural, artificial, and synthetic building materials has been accumulated. Significant development of industrial fields has been observed in recent centuries, which is associated with the development of new industrially important chemical technologies. This was facilitated by epochal discoveries in science and technology all over the world. For example, progressive results have been obtained for the construction field, which is experiencing the need to use building structures made of various materials, including the use of wood-based materials.

Wood as a building material has a timeless importance in the practice of reliable and durable construction. To date, there are a number of unique examples of longevity of objects involving wooden structures. In this regard, it should be noted that there is a significant stock of buildings and structures in the world with a service life of several decades to several hundred years. These buildings have a significant time period of operation, exceeding, in many cases, the normative lifetime of their service. The development of this area in recent decades has led to the emergence of unique buildings and structures of medium- and high-storey as well as high-rise facilities using wooden structures. At the same time, in the conditions of designing, construction, and operation of objects with the use of such structures, the most important priority task is to ensure fire safety.

In the present book, an emphasis is laid on the study of peculiarities of fire behavior of wooden structures during their long-term operation. Extensive information is presented about the fire hazard of wood of long natural aging and wooden constructions of long-term exploitation as well as about their fire protection.

The authors of the monograph, to a certain extent, summarize the results of research with regard to long natural aging wood. The main concept of the presented scientific materials is that as a result of long-term natural aging in wood, irreversible complex physical and chemical processes occur, leading to changes in the macro- and microstructure, chemical and elemental composition, and operational properties of wood, which affects the fire danger and fire resistance of wooden structures, as well as the choice of type of fire protection. In fact, for the first time in such a statement, the issues of influence of external environmental factors or conditions of

S. Andrey Borisovich et al., *Fire Hazard and Fire Resistance of Wooden Structures*,
https://doi.org/10.1007/978-3-031-24074-4

natural aging on the characteristics of fire danger and fire resistance of wooden structures are considered. The specifics of wood reaction to heat and fire are considered from the standpoint of changes in its physicochemical, thermochemical, and thermophysical properties in conditions of different time periods of operation of wooden structures. So, for example, the analysis of the influence of the content of the main components in the chemical composition of wood on the value of the lower heat of total combustion of wood, which is a thermodynamic characteristic of a particular substance, clearly confirms the validity of the adopted concept. Thus, it is revealed the ability to increase the fire hazard of wooden structures during their long-term operation, which is noticeably reflected in the increase in heat release during the burning of wood of long-term natural aging, as well as the degree of its thermal damage (the intensity of charring).

For the first time, a new type of threat or danger to humans in conditions of fire occurrence in buildings and constructions, having wooden structures of long-term operation, was established, namely the potential reduction of the level of fire safety of a building or structure, caused by changes in physical, chemical, thermal, thermodynamic, fire hazardous properties of wood, staying in conditions of long-term exposure to the environment.

The authors of the book sought to systematize and summarize the results of the basic experimental and theoretical works available in the scientific press, which allow to give a general idea of the change in the structure, composition, and properties of wood, as well as the relationship of these changes with patterns of its combustion process in different conditions of fire exposure (fire). In general, according to the state of works in the field of wood fire hazard research and modeling of fires in buildings and constructions with wood as a building construction material, it can be testified that the operating life is actually not taken into account when assessing the fire hazard of wooden structures and the features of fire scenarios development. For this reason, the assessment of fire hazard characteristics and fire resistance of wooden structures of different service life and modeling of fire hazards is carried out to a greater extent using the indicators of modern wood. The obtained body of experimental data presented in the monograph can be a certain basis for the study of the influence of long-term processes, occurring in wood as a result of its long natural aging, on the features of the behavior of wooden structures in fire conditions and the efficiency of various ways and kinds of their fire protection. In this regard, undoubtedly, the studies of the degree of change in some important indicators of fire danger of wood in the course of its natural and artificial aging are key.

A significant contribution to the science of the development of ideas about the behavior of long-term natural aging wood in fire conditions is associated with the establishment of the increase in the intensity of its charring, as well as abnormal heat release during its combustion, which is actually decisive in predicting fire scenarios and calculation assessment of the fire-resistance limits of wooden structures. For different conditions of fire exposure, a significant number of wood samples of different species and species, as well as from different climatic regions, it was found that with the increase of service life of wooden structures, there is a potential increase of charring rate by 1.2 and more times, depending on the service life of wooden structures and conditions of fire exposure (fire). This is caused, among other things, by the increasing contribution of aromatic components of lignin in intensification of charring process, as well as by increasing values of wood combustion heat. Data on the charring rate of wooden structures depending on their service life were obtained under the influence of different fire temperature conditions, including standard fire temperature conditions. This fact determines the use of these results for the calculation methods and approaches to assess the fire-resistance limits.

Considering the burning of wood of long-term natural aging and its fire properties as an important direction in the science of wood, it should be emphasized that many aspects of this problem remain unresolved. Thus, it is of interest to establish the effect on the fire hazard and fire resistance of wooden structures, as well as the effectiveness of fire protection not only the duration of their operation but also the conditions of operation. The conditions and factors influencing the occurrence of wood smoldering combustion and its transition to the flame regime are not clear. It remains to be understood to what extent the wood combustion process and the critical conditions of its existence will be affected by the relative content in the wood composition of not only the main components but also the components of these components and their chemical nature, taking into account possible transformations as a result of long-term natural aging of wood material. Insufficient attention has been paid to the influence of the nature of fire retardants on the indicators of fire danger and the process of charring, the study of physical and chemical processes that ensures the effectiveness of fire protection in the fire action on wood of varying degrees of natural aging.

The authors note an increasing trend in world practice to fundamental study of the burning process and fire hazard characteristics of wood, taking into account the influence of various factors. The successful solution of this problem is associated with a deep understanding of the regularities of smoldering and flame burning of wood, the development of fire, and the conditions of termination of this process. The development of not only more perfect models of ignition and combustion of wood but also experimental work on the kinetics and mechanism of wood decomposition depending on various factors will contribute to the progress in creating effective types and methods of fire protection.

Index

S. Andrey Borisovich et al., *Fire Hazard and Fire Resistance of Wooden Structures*,
https://doi.org/10.1007/978-3-031-24074-4

S

T

V

W